RESEARCH IN SURFACE FORCES

Edited by Academician B. V. Deryagin

Volume 2

Three-Dimensional Aspects of Surface Forces

A SPECIAL RESEARCH REPORT / TRANSLATED FROM RUSSIAN

Springer Science+Business
Media, LLC

RESEARCH IN SURFACE FORCES

ISSLEDOVANIYA V OBLASTI POVERKHNOSTNYKH SIL

ИССЛЕДОВАНИЯ В ОБЛАСТИ ПОВЕРХНОСТНЫХ СИЛ

RESEARCH IN SURFACE FORCES

Edited by Academician B. V. Deryagin
Director, Laboratory of Surface Phenomena
Institute of Physical Chemistry
Academy of Sciences of the USSR

Volume 2

Three-Dimensional Aspects of Surface Forces

Translated from Russian by
Paul Porter Sutton
Department of Chemistry
North Carolina State University

Springer Science+Business Media, LLC
1966

The Russian text, which comprises the Proceedings of the Second
Conference on Surface Forces sponsored by the Institute of Physical
Chemistry of the Academy of Sciences of the USSR, and which
originally was printed for the Institute by Nauka Press in Moscow in
1964, has been extensively corrected and updated by the editor for
this edition.

Library of Congress Catalog Card Number 62-15549

© 1966 Springer Science+Business Media New York
Originally published by Consultants Bureau in 1966.
Softcover reprint of the hardcover 1st edition 1966

ISBN 978-1-4899-5792-4   ISBN 978-1-4899-5790-0 (eBook)
DOI 10.1007/978-1-4899-5790-0

CONTENTS

SECTION I

THEORETICAL PROBLEMS IN SURFACE PHENOMENA

TWO- AND THREE-DIMENSIONAL ASPECTS
OF SURFACE PHENOMENA

B. V. Deryagin

Institute of Physical Chemistry, Acad. Sci., USSR,
Laboratory for Surface Phenomena

The fact that the development of the science of surface phenomena has largely tended to emphasize the two-dimensional aspects of these effects at the expense of the three-dimensional is understandable and has had its own advantage. The two-dimensional aspects of the science trace back to Gibbs' treatment of hetero-geneous equilibria in terms of total surface excess of the various thermodynamic functions with no allowance for possible functional variation perpendicular to the phase interface, and the Langmuir concepts of short-range surface interaction and predominant significance of the monomolecular layer.

The discovery of surface effects involving interactions over many molecular diameters has, however, made it clear that the Langmuir concepts are not only inadequate, but may even be entirely incorrect, at least in their extreme form. This does not, of course, prevent the continued use of these concepts to obtain good first approximations for the study of many problems. New data do show, however, that systems and states cannot be adequately described by the Gibbs' thermodynamics when they involve films whose depths are of the same order of magnitude as the range of surface interaction, and this despite the fact that the thermodynamics itself is well-grounded and exact. At the same time, it becomes more and more obvious that these are the systems and states which require study.

The necessity for such study goes back in the first instance to the fact that the particles of soils, pastes, foams, limiting emulsions, and other concentrated, highly dispersed systems are invariably separated by thin films. The specific properties of these systems cannot be understood without taking account of the equally spe-cific behavior of these films and the forces set up within them. The study of systems of particles separated by thin films has even greater significance for an understanding of mechanisms and kinetics in such highly important processes as wetting, heterogeneous new-phase formation, adhesion, peptization, coagulation, and heterocoagulation.

Taken alone, the two-dimensional aspects of surface phenomena prove to be entirely inadequate for a treatment of these various kinetic processes and a requirement is thus imposed which does not apply to the meth-ods usually employed in studying other heterogeneous equilibria. A. N. Frumkin has shown, for example, [1] that even incomplete wetting under adsorptional equilibrium cannot be understood without an analysis of thin polymolecular film behavior.

The Gibbs' thermodynamics also proves to be inadequate for a theory of colloidal particle adhesion; here, account must be taken of the variation of system energy with depth of separating film, an energy barrier analo-gous to the activation energy for processes at the molecular or atomic level being the controlling factor. The inhibiting action of such an "activating" potential barrier can be extensive enough to bring the process in ques-tion (coagulation, for example) almost to a standstill, thereby assuring stability in the system under study.

From this it does not follow, however, that two-dimensional factors (the states of monolayers of molecules or ions at the interface between a thin film and its neighboring phases, for example) can be ignored. The state of the surface monolayer is usually a factor of great significance in determining the possibility of the existence of polymolecular films, fixing the properties of these films and thereby influencing the process controlled by them.

This situation finds exact mathematical expression in the theories of semiconduction and hydrophobic colloidal stability where the boundary conditions correspond to the two-dimensional aspects and differential equations describing the variation of the electrical potential within the film to the three-dimensional aspects, of the problem. Here both aspects must be considered together in order to obtain a quantitative description of the interesting phenomena and processes. Various examples taken from the work of our own and other laboratories will serve to illustrate this point.

It has been shown in [2] that the establishment of good electrical contact between wires crossed in an electrolytic solution requires the application of a certain minimum force to overcome a repulsive or splitting force in the electrolyte film separating the wires. The measured value of this force barrier is closely dependent on the polarization-induced potential on the wires. Polarization is known to alter the surface charge by covering the surface with a rarefied monolayer of ions of the one type or the other. The monolayer charges on the two wires interact through ionic atmosphere overlap in the electrolyte films thus affecting the potential barrier for wire adhesion.

It must be noted that the surface potential of the wires no longer affects the potential barrier for adhesion when the concentration of the electrolyte solution is high, the barrier height increasing markedly at a certain definite concentration. This is clear indication that the three-dimensional aspects of the properties of the thin film can be independent of the two-dimensional, the latter being determined by van der Waals forces which are insensitive to the monolayer state in the polymolecular film. A. D. Sheludko in Sofia has shown [3] that repulsive forces with a considerable range of interaction are also set up in free foam films.

It should also be pointed out that the effect of two-dimensional factors on the stability of free films and films bounded by nonsolid phases will depend on whether film thinning is by liquid extrusion without surface stretching or by simultaneous stretching of film and surfaces. Although the effect of two-dimensional factors (the properties of the adsorbed monolayers, for example) on film stability is the same as for a film entrapped between solid phases in the first case, it is entirely different in the second; here the adsorbed monolayers impede film stretching as a whole (according to the Marangoni—Gibbs principle, for example) and local rupturing may result. An exact two-dimensional mechanism must apply to local rupture from thermal fluctuations in very thin films (the free film is best treated as two adsorbed monolayers [4]), all the particles moving in a plane parallel to the film surface and the film depth remaining constant. This two-dimensional rupture mechanism has been theoretically treated by Yu. V. Gutop and the present author [4]; A. D. Sheludko has shown [5] that it passes over to a three-dimensional mechanism in thick films. However, local rupturing in the adsorption layer with predominance of two-dimensional factors can be assumed even in the latter case. Thus the interplay of two- and three-dimensional factors is rather complex in the free film.

A second example can be found in the experiments of Yu. M. Popovskii on the effect of a liquid medium in glass surface adhesion. It has been found that a polymolecular film capable of withstanding the force of compression cannot be formed in benzene unless a monolayer of nitrobenzene has first been laid down on the glass. Such a monolayer effect cannot be accounted for by van der Waals forces, as has already been pointed out, and the marked alteration in benzene properties must therefore be assumed to arise from the fact that a structure is induced into the adhering benzene layers by the nitrobenzene monolayer. Other examples can also be adduced to show that the state of the glass affects the properties of liquid films adhering to it. Thus optical measurements carried out in this laboratory have shown that the depth of a polymolecular film in equilibrium with its saturated vapor can be markedly altered by subjecting the glass to glow-discharge or by modifying its surface with chemical reagents. The alteration of the adsorption isotherm resulting from modification of the adsorbing surface has been thoroughly studied by A. V. Kiselev and his coworkers.

The experiments of G. L. Mikhnevich and his students have shown [7] the probability of crystallization center formation in supercooled salol and other liquids to be closely dependent on the surface state of the glass. This effect can extend 10 μ, or more, into the supercooled liquid. It can be eliminated by treating the glass with a weak fluoric acid solution, but reappears under the action of an electrical discharge. There can be no doubt that these effects arise from an alteration in the number of surface hydroxyl groups. The action of hydrogen bonds and (possibly) dipole moments alters the molecular orientation in the adhering liquid phase layer; this, in turn, alters the orientation in the following layer, and so on.

The existence of such orientation is observed as a variation of structurally sensitive properties on moving through the liquid toward a solid wall. Thus the liquid viscosity measurements of V. V. Karasev [8] have shown that there is an abrupt breakdown of molecular orientation in surface films of simple liquids at certain definite distances from a wall, the effect being analogous to a phase change of Type I. On the other hand, there is a wide variation of the depth of the surface phases of characteristic structures and properties which are formed in the neighborhood of an interface. There are surely some cases in which orientation is limited to the first mono-layer and others in which structural phase alteration extends for distances of the order of several tenths of a micron. Instances of this last type are observed in the measurements of N. N. Zakhavaeva [9] on hexaethoxy-decane and tetraethoxydecane.

It is unfortunate that present theories are not in a position to give a quantitative prediction of surface phase depth in a liquid of known properties. Here one must start from the observation that the depth of oriented layer generally increases as the molecular weight and asymmetry of the liquid rise. Thus liquid crystal orienta-tion can extend out for an unlimited distance from the glass. The increase in depth of this "anomalous" layer which results from supercooling is naturally ascribed to a rise in molecular weight.

The outstanding experiments of N. N. Fedyakin [10] have convincingly indicated that polymolecular wall films can show special liquid structures which are homogeneous over distances of the order of 100 A. Here it was shown that water in glass capillaries less than 200 A in radius contracts linearly to 0℃ without passing through maximum density at 4℃. This fact can be explained only by assuming the water to have acquired a most dense packing which is maintained as the temperature is raised. V. V. Karasev has shown [11] a similar alteration in the thermal expansion of water entrapped in the pores of powdered silica gel particles.

It is undoubtedly true that boundary phases play a very significant role in the passage from adsorptional layer to bulk phase which results when the film depth is increased by raising the pressure. A thermodynamic treatment of this problem with L. M. Shcherbakov based on experimental data has shown that two surface states, a single surface phase of unique structure, and a bulk phase of normal structure underlain with a surface phase are possible over certain vapor pressure intervals. This type of polymorphism in a wetting adsorption layer gives rise to a hysteresis loop for condensation on, and vaporization from, the plane surface. An explanation can also be advanced here for the fact that liquid films of finite and infinite depth can both be in equilibrium with a single saturated vapor, even though the free energies are not the same in the two cases.*

Allowance for such polymorphism makes it possible to distinguish two types of incomplete wetting, thus extending the analysis begun by A. N. Frumkin. The un-wet portion of the surface is covered with a single oriented surface layer in one of these cases, and by two phases, the upper unoriented, in the other. There is also a difference in the wetting boundary in the two cases, with a discontinuity in the angle in the first case and a discontinuity in both the angle and film depth in the second.

These are effects of more than academic interest, since they lie at the basis of incomplete wetting and wetting hysteresis. These last are matters of practical significance for the thermotechnics of drop condensation (where the thermal exchange is considerably better than in film condensation) and for production of nonsweat-ing glass. Flotation, a problem of even greater economic significance, is directly related to the kinetics of film rupture and boundary angle formation.

The various studies of capillary condensation effects are all closely tied up with the exact determination of boundary angles and hystereses. The great significance of these effects has been clearly brought out in the remarkable work of N. N. Fedyakin [13] which has shown that the properties of columns of water, methyl alcohol, acetic acid, and acetone, produced by direct introduction of the respective liquids into glass capillaries 1-2 μ in radius are different from those of similar columns obtained by holding capillaries in atmospheres incompletely saturated with the vapors of the liquid in question.

In distinction to Type I columns, Type II columns do not show the same coefficients of thermal expansion as the bulk liquids, though their expansion is just as regular as that of water in capillaries of 200 A radius or greater. The viscosities of the anomalous columns are, on the other hand, several times higher (10 times, in

---

*See, also, the paper of Z. M. Zorin (this volume, p. 134).

the case of water) than the viscosities listed for the bulk liquids in tables of physical constants [14]. The very method of formation of these columns is such as to indicate vapor pressures some percent less than predicted by the Kelvin equation. An obvious explanation for this effect can be had by assuming the oriented structure of the surface film to be maintained as the polymolecular layer is built up to a depth considerably in excess of that corresponding to equilibrium and then converted into a column under the special conditions applying to this process. These conditions are imposed by the fact that the working volumes are small, other studies (on supercooled water, for example) having shown that metastable states can persist for long periods of time in such situations. The concepts presented here constitute a working hypothesis for interpreting experiments by K. V. Chmutov [15] and I. Shereshevskii [16] in which departures form the Kelvin Equation in the direction of anomalously low vapor pressures have been shown to accompany capillary condensation in the fissures and capillaries of glass and quartz.

The principle of crystal-chemical conformity advanced by P. D. Dankov indicates that the ability of a solid body to initiate a metastable structure in a phase growing on its surface will be at a maximum when phase and substrate are structurally identical. Here a growing phase incorporating long range order of the type associated with the crystalline or liquid crystal states can take on any depth whatsoever, even though the free energy of the phase is higher than the free energy of another, more stable, modification. This accounts for the possibility of unlimited growth in seed crystals (metastable modifications), a property which has found various important applications in our laboratory. This is obviously a limiting case of phase formation under the action of surface forces, such as has already been discussed. Here we have illustration of the fact that surface forces are fixed by the nature of the interface and vary widely in their effectiveness and range of interaction, as well as indication of the ease of passage from two-dimensional aspects to three.

The interplay of these aspects of surface phenomena at a gas phase interface has great practical significance and should be considered at least briefly. Here it is only in the close approach of extremely thin layers that pronounced potential barriers arise to impede the thinning of the gaseous film. The work of [17] has unexpectedly shown that van der Waals forces give rise to a repulsion which comes into play at large, but perfectly well defined, distances. The effect is weakly expressed, however, and has only theoretical interest at the present time; whether it offers an explanation for the failure of soap bubbles to coalesce when held in contact for long periods of time under equilibrium conditions (Dewar experiment) is open to question.

The situation changes, however, with the breakdown of equilibrium at the interfacial surface. S. S. Dukhin has shown, for example, [18] that droplet streaming leads to breakdown of equilibrated ion and dipole layers, thus eliminating electrical charge screening in the internal sphere to give rise to an external electric field. This nonequilibrium, external field can affect particle capture from the drop stream and the adhesion of drops with different fall ratio.

Colored indicators have been used [19] to show that coagulation in aerosol mixtures of aqueous solutions of different electrolytic compositions is predominantly to form heterogeneous drops, a result which has been confirmed by other methods [20]. It is obvious that this effect arises from the interaction of the nonequilibrium fields surrounding the falling droplets. Quantitative confirmation of this conclusion has not been obtained as yet. The diffusional processes accompany vaporization and condensational particle growth give another mechanism for the appearance of repulsive forces between particles suspended in a vapor, as was first shown in our own experiments [21]. These forces can be theoretically evaluated when the particle diameters are small in comparison with the distances of mutual separation [22]. Calculations show these forces to be largely determined by effects within the gas−vapor mixture, pure surface factors coming into play only in so far as accommodation coefficients depend on the nature and state of the particle (droplet, for example) surface.

The equilibrium forces of molecular attraction can, of course, exert a significant effect on particle capture from the gas or liquid stream, even in the absence of a potential barrier. This situation arises from the fact that only inertial forces are available for overcoming the thinning resistance of the gaseous film separating particles and streaming body when the van der Waals forces have an infinitesimally low range of action. The dissipative (viscous) resistance of the film cannot be overcome when the interaction radius is low and the proximity time limited. The effectiveness of capture is then determined by the surface forces alone, equilibrium forces of molecular attraction being included among the latter. *

Thus it is clear that the three-dimensional aspects of surface effects must be drawn on to obtain correct answers to the problems in this field. "Nonequilibrium" surface forces are significant at the interface between a liquid, another liquid, a gas, or a solid, but account must usually be taken of "equilibrium" surface force effects, as well. This applies, in the first instance, to the electrical fields which arise when an electrolyte flows past an impediment. The present author and S. S. Dukhin have shown [24] that thinning is predicted (Dorn effect) when the classical Smoluchowskii field theory is extended to allow for the concentration changes and ionic diffusion currents which arise in flow. These latter effects radically alter the portion of the theory applying to liquid interfaces (liquid—liquid and liquid—gas). A detailed treatment of this problem has been given by S. S. Dukhin in a paper appearing in this volume (see p. 54). The ion diffusion current alters the nature of the forces acting on the suspended particle, a new expression being obtained for these forces when the presence of this current is taken into account. These forces were previously evaluated from the assumed electric field by drawing on the theory of electrophoresis, but the electrolyte concentration gradient is now considered to be the primary factor and the electric field only secondary, so that the movement of the suspended particle becomes entirely different, sui generi. We have designated this effect as diffusion phoresis [25]; it comes into play not only in processes involving flow past a barrier, but also in the ionic deposition method for latex film formation which has been discussed by A. A. Korotkova (this volume, p. 51).

The effect of the diffusional electrical field on mineral particle capture by a floating bubble has been pointed out in [26] and can serve as an example of the practical significance of work in this field.

Literature

1. A. N. Frumkin, Zh. Fiz. Khim. 12:33 (1938).
2. T. N. Voropaeva, B. V. Deryagin, and B. N. Kabanov, Dokl. Akad. Nauk SSSR 128:981 (1959); Kolloidn. Zh. 24:396 (1962); Izv. Akad. Nauk SSSR Otd. Khim. Nauk, No. 2:257 (1963).
3. A. D. Sheludko, Dokl. Bolgar. Akad. Nauk 9:11 (1956); Proc. Koninkl. Ned. Akad. Wetenschap., Ser. B, No. 1:76 (1962).
4. B. V. Deryagin and Yu. V. Gutop. Kolloidn. Zh. 24:431 (1962).
5. A. D. Sheludko, Proc. Koninkl. Ned. Akad. Wetenschap., Ser. B., No. 1:76 (1962).
6. B. V. Deryagin, V. V. Karasev, and V. I. Gol'danskii, Dokl. Akad. Nauk SSSR 57:697 (1947).
    B. V. Deryagin and Z. M. Zorin, Dokl. Akad. Nauk SSSR 98:93 (1954); Zh. Fiz. Khim. 29:1755, 1910 (1955).
7. G. L. Mikhnevich and V. G. Zaremba, Kolloidn. Zh. 24:491 (1962).
8. B. V. Deryagin and V. V. Karasev, Dokl. Akad. Nauk SSSR, 62:761 (1948); Kolloidn. Zh. 15:365 (1953); Dokl. Akad. Nauk SSSR 101:289 (1955); Proc. of the Second Intern. Congress of Surf. activ. London (1957); Zh. Fiz. Khim. 33:100 (1959).
    B. V. Deryagin and E. F. Pichugin, Dokl. Akad. Nauk SSSR 63:53 (1948).
    Transactions, Second All-Union Conference on Friction and Wear in Machines, Vol. 3, Moscow—Leningrad, Izd. AN SSSR (1949), p. 101.
9. B. V. Deryagin, N. N. Zakhavaeva, S. V. Andreev, A. A. Milovidov, and A. M. Khomutov, the collection: Studies in Surface Forces, Moscow, Izd. Akad. Nauk.SSSR (1961), p. 139; Kolloidn. Zh. 24:289 (1962).
10. N. N. Fedyakin, Dokl. Akad. Nauk SSSR 138:1389 (1961).
    V. V. Karasev, B. V. Deryagin, and E. N. Efremova, Kolloidn. Zh. 24:471 (1962).
11. B. V. Deryagin, V. V. Karasev, and E. N. Efremova, Kolloidn. Zh. 24:471 (1962).
12. B. V. Deryagin and L. M. Shcherbakov, Kolloidn. Zh. 23:40 (1961).
13. N. N. Fedyakin, Kolloidn. Zh. 24:497 (1962).
14. B. V. Deryagin and N. N. Fedyakin, Dokl. Akad. Nauk SSSR, No. 2:147 (1962).

---

* Molecular surface forces are equally significant for particle deposition from a flowing liquid onto a barrier. This fact has been brought out by the detailed experimental and theoretical work of V. Makkrle [23] on suspension filtration through granular materials.

15. K. V. Chmutov, Zh. Fiz. Khim. 9:57 (1937); Kolloidn. Zh. 11:44 (1949).

16. I. L. Shereshevskii, J. Am. Chem. Soc. 50:2966 (1928); 101:1315 (1950).

17. B. V. Deryagin and G. A. Martynov, Dokl. Akad. Nauk SSSR 144:825 (1962).

18. S. S. Dukhin and B. V. Deryagin, Dokl. Akad. Nauk SSSR 121:503 (1958); Kolloidn. Zh. 21:37 (1959); 22:587 (1960).

19. V. A. Fedoseev, B. A. Manakin, and Z. M. Domentianova, Kolloidn. Zh. 14:470 (1952).

20. D. R. Benton and G. A. Elton, Discussion Faraday Soc., No. 30:68 (1960).

21. B. V. Deryagin and P. S. Prokhorov, Dokl. Akad. Nauk SSSR 54:511 (1946); 81:637 (1951).
   P. S. Prokhorov and V. N. Yashin, Kolloidn. Zh. 10:122 (1948).

22. B. V. Deryagin and S. S. Dukhin, Dokl. Akad. Nauk SSSR 112:407 (1957); 115:126 (1957).
   P. S. Prokhorov and L. F. Leonov, Discussions Faraday Soc., No. 3:124 (1960).

23. Vladimir Mackrle, L'Étude du phénomène d'adhérence colmatage dans le milieu poreux. Thèse présentée à la faculté des sciences de l'Université de Grenoble (1960).

24. B. V. Deryagin and S. S. Dukhin, Dokl. Akad. Nauk SSSR 129:1328 (1959); Kolloidn. Zh. 21:37 (1959).
   B. V. Deryagin, Kolloidn. Zh. 22:148 (1960).

25. B. V. Deryagin, S. S. Dukhin, and A. A. Korotkova, Kolloidn. Zh. 23:53 (1961).
   S. S. Dukhin, Kolloidn. Zh. 24:446 (1952).

26. B. V. Deryagin and V. D. Samygin, In collection: Of the Scientific Papers of the Gintsvetmet, Vol. 19, Moscow, Metallurgizdat (1962); p. 240.
   B. V. Deryagin and S. S. Dukhin, Bull. Inst. Mining Met., No. 651 (1961); Trans. Mining Met. 70, Part 5:221 (1960-1961).

THE THERMODYNAMICS AND STABILITY OF FREE FILMS

B. V. Deryagin,
Corresponding Member, Acad. Sci., USSR

G. A. Martynov and Yu. V. Gutop
Institute of Physical Chemistry, Acad. Sci., USSR,
Laboratory for Surface Phenomena

I. There are two fundamentally different approaches to the problem of the stability of those free symmetric films which are the structural elements of foams. The one of these considers that the slowness of the thinning process itself accounts for film stability. The Gibbs thermodynamics ascribes this slowness to the stretching resistance of the quasi-elastic adsorptional surface layers, and to the flow resistance of the viscous liquid enclosed between these layers. Gibbs' treatment [1] is also limited to the film which is bounded by two surfaces of discontinuity and so thick that its interior properties are those of the bulk phase.

A second approach was opened up by the work of one of the present authors with A. S. Titievskaya [2] where it was shown that the free film could come to thermodynamic equilibrium even when acted on by thinning forces. Such equilibrium can be maintained indefinitely, although it is metastable in the sense that the associated minimum in Gibbs free energy (thermodynamic potential) is higher than the free energy of the system after film rupture. The possibility of extended film existence can clearly be explained by assuming the state of metastable film equilibrium to be separated from absolutely unstable states following film rupture by a high energy barrier. Such metastable film equilibrium would clearly be impossible in the case considered by Gibbs since it requires that the free energy of a system of fixed surface area vary with the film depth. This, in turn, implies that the surface zones intersect in the interior of the film so that the interior properties are not longer those of the bulk phase (i.e., no longer characterized by the values of the intensive variables, concentration, and chemical potential).

Treatment in terms of a film delineated by two surfaces of discontinuity at the gas—film interfaces is certainly not applicable under intersection of the surface zones. Although the phases bounding the film are identical in this case, the general method of treating the arbitrarily structured layer on the surface of discontinuity between two different phases is still applicable, as Gibbs himself [3] noted. This point was not explored however, either by Gibbs or his successors.

The present article outlines a thermodynamics of free films based on this idea. The advantage of this approach is that it eliminates the concept of geometric film depth. This concept is lacking in exact meaning and becomes more and more inaccurate and arbitrary as the film depth approaches the value of the bimolecular palisade of the stabilizer molecules, where indefinitely stable films are often met. Elimination of the geometric depth makes the treatment applicable to films of any thickness. A complete physical theory of film stability naturally requires the introduction of molecular-kinetic concepts, but this can rationally be done after establishing the basis for a thermodynamic treatment of the problem.

II. The general case of the film containing an arbitrary number of components is quite involved, and discussion here is therefore limited to the case in which one component, 1, ("soap") is present in the film alone and the other, 2, ("water" or "gas") is common to the film and the surrounding phase.

We follow Gibbs and study a system consisting of a plane film surrounded on both sides by a bulk phase. The free energy, F, of such a layered system is a function of the absolute temperature, T, the system volume, V, the area of the separating interface, A, and the total number of particles, N_1, N_2, and N_3. N_1 will designate the number of particles of that component which is lacking in the bulk phase surrounding the film. One then has

$$dF = -SdT - PdV + \sigma dA + \sum_{i=1}^{i=3} \mu_i dN_i,$$

(1)

where S is the system entropy, P is the bulk phase pressure, σ is the film tension, and μ_i is the chemical potential of the i-th component.

The independent variables V, N_2, and N_3 will be replaced by the intensive variables, P, μ_2, and μ_3, and the bulk terms eliminated through the state function of the system

$$F^S = F + PV - \mu_2 N_2 - \mu_3 N_3$$

(2)

to obtain

$$dF^S = -SdT + VdP - N_2 d\mu_2 - N_3 d\mu_3 + \sigma dA + \mu_1 dN_1.$$

(3)

The pressure in the bulk portion of the system is a function of the temperature, T, and the chemical potentials μ_2 and μ_3. Applying the Gibbs—Duhem Equation to the bulk phase, one has

$$-s^V dT + dP - n_2^V d\mu_2 - n_3^V d\mu_3 = 0,$$

(4)

s^V, n_2^V, and n_3^V being the bulk entropy, and numbers of particles 2 and 3, per unit volume. One obtains from (3) and (4),

$$dF^S = -S^S dT - N_2^S d\mu_2 - N_3^S d\mu_3 + \sigma dA + \mu_1 dN_1,$$

(3')

S^S, N_2^S, and N_3^S being the surface excesses of the various quantities, defined with Gibbs as

$$N_{2,3}^S = N_{2,3} - n_{2,3}^V \cdot V, \qquad S^S = S - s^V \cdot V.$$

When the third component, the gas, is neither present in the film nor adsorbed on its surface, Eq. (3') can be usefully rearranged by introducing the pressure, P, of the gaseous phase, eliminating $d\mu_3$ with the aid of (4), and assuming that $N_3^S = -n_3^V \cdot V_S$, V_S being the volume of the film to obtain

$$dF^S = -S^f dT + \sigma dA + \mu_1 dN_1 - N_2^f d\mu_2 + V_S dP.$$

(3")

Here $S^f = S^S + s^V \cdot V_S$ (film entropy) and $N_2^f = N_2^S + n_2^V \cdot V_S$; or, passing to specific quantities:

$$dF_S = -S_f d\Gamma_1 + \mu_1 d\Gamma_1 - \Gamma_2^f d\mu_2 + h dP,$$

h being the film thickness.

This approach to film equilibrium is analogous to the treatment developed in [4].

The surface excesses are uniquely determined for the general case of the film bounded by identical phases, and their values are independent of the particular position assigned to the geometrical interfacial surface. If component 3 is neither present in the film nor adsorbed on its surface, $N_3^S = -n_3^V \cdot V_S$, V_S being the effective film volume. The value of V_S is somewhat arbitrary in the case of thin films. When (3') is integrated holding the intensive variables T, μ_2, μ_3, σ, μ_1 constant, one obtains

$$F^S = \sigma A + \mu_1 N_1.$$

(5)

When the film degenerates into a bulk phase, $A \to 0$, and $N_1 \neq 0$, and F^S takes the value $F_0^S = (\mu_1)_0 \cdot N_1$, where

$$(\mu_1)_0 = \varphi \ (T, \mu_2, \mu_3) \tag{6}$$

is independent of N_1. The experiments of A. D. Sheludko and his coworkers [5] have shown the possibility of the existence of a metastable film ($A \neq 0$) with $N_1 = 0$, which is to say, a film consisting of two volatile components.

The introduction of quantities referring to unit film surface area gives

$$F_S = \sigma + \mu_1 \Gamma_1, \tag{7}$$

where

$$F_S = \frac{F^S}{A} \qquad \text{and} \qquad \Gamma_1 = \frac{N_1}{A}$$

is the surface concentration of that component which is present in the film but lacking in the bulk phase. We will now derive the Gibbs–Duhem Equation for the film. If (5) is differentiated, the result obtained equated to the right-hand member of (3') and divided by A, one finds that

$$S_S \, dT + d\sigma + \Gamma_1 \, d\mu_1 + \Gamma_2 \, d\mu_2 + \Gamma_3 \, d\mu_3 = 0, \quad \Gamma_i = N_i / A. \tag{8}$$

III. Let the equation of state (fundamental equation, in Gibbs' terminology) of the film have the form.

$$\sigma = \sigma(T, \Gamma_1, \mu_2, \mu_3). \tag{9}$$

From (3), it follows that

$$\sigma = \left(\frac{\partial F^S}{\partial A} \right)_{T, N_1, \mu_2, \mu_3} \tag{10}$$

Integration of this relation gives

$$F^S = \int_0^A \sigma \, dA + F_0^S = \int_0^A \sigma dA + N_1 \varphi(T, \mu_2, \mu_3) \cdot \tag{11}$$

Using the film area per molecule, $a = 1/\Gamma_1 = A/N_1$, or $A = N_1 \cdot a$ and $dA = N_1 da$, rather than the area as the variable of integration, and dividing by A, one obtains

$$F_S = \Gamma_1 \int_0^a \sigma da + \Gamma_1 \varphi(T, \mu_2, \mu_3). \tag{12}$$

The chemical potential, μ_1, is determined from (3') by differentiating F^S with respect to the number of particles N_1, holding T, A, μ_2, and μ_3 constant, the result being

$$\mu_1 = \left(\frac{dF^S}{dN_1} \right)_{T, A, \mu_2, \mu_3} = \left(\frac{\partial F_S}{\partial \Gamma_1} \right)_{T, \mu_2, \mu_3} = \int_0^a \sigma da - a\sigma + \varphi, \tag{13}$$

when account is taken of the fact that $a = A/N_1$ in differentiating over the upper limit of the integral.

Integrating the right-hand member of Eq. (13) by parts, one finally obtains

$$\mu_1 = - \int_{\sigma_0}^\sigma a d\sigma + \varphi \ (T, \mu_2, \mu_3), \tag{14}$$

where σ_0 is the film tension for $a \to 0$ and $\Gamma_1 \to \infty$ or twice the surface tension, γ, of the bulk phase into which the film degenerates. This latter phase is a solution in which the particle number N_1 is fixed and the numbers of particles N_2 and N_3 determined by the given values of the chemical potentials, μ_2 and μ_3. The relation $a = a [\sigma, T, \mu_2, \mu_3]$ is then covered by the equation of state (9).

It is clear the φ is the chemical potential of the nonvolatile component 1 of a bulk degenerate phase which is in equilibrium with the bulk phase surrounding the film. This film differs from the bulk phase surrounding the film in that it contains particles of type 1; it may possibly, though not necessarily, be in a different state of aggregation, but has the same values of the chemical potentials, μ_2 and μ_3, as the latter. These conditions will, in general uniquely determine the mole fractions of the components, which is to say, the bulk film composition. These potentials vary according to Eq. (14) during film thinning, this being the basic equation for free film stability and equilibrium.

IV. The numbers of particles, N_2^S and N_3^S, are fixed by (3) as

$$N_2^S = -\left(\frac{\partial F^S}{\partial \mu_2}\right)_{T, A, N_1, \mu_3} \qquad \text{and} \qquad N_3^S = -\left(\frac{\partial F^S}{\partial \mu_3}\right)_{T, A, N_1, \mu_2} \tag{15}$$

Or, using the expressions of (5) and (7) for F^S and F_S,

$$\Gamma_2 = -\Gamma_1 \left(\frac{\partial}{\partial \mu_2}\right)[\int_0^a \sigma da + \varphi(T, \mu_2, \mu_3)]_{T, a, \mu_3}, \tag{16'}$$

$$\Gamma_3 = -\Gamma_1 \left(\frac{\partial}{\partial \mu_3}\right)[\int_0^a \sigma da + \varphi(T, \mu_2, \mu_3)]_{T, a, \mu_2} \tag{16''}$$

Equation (13) can now be drawn on to obtain

$$\Gamma_2 = -\left(\frac{\partial}{\partial \mu_2}\right)(\sigma + \mu_1 \Gamma_1)_{T, \Gamma_1, \mu_3}, \tag{17'}$$

$$\Gamma_3 = -\left(\frac{\partial}{\partial \mu_3}\right)(\sigma + \mu_1 \Gamma_1)_{T, \Gamma_1, \mu_2}. \tag{17''}$$

Equations (17') and (17") differ from the Gibbs Adsorption Theorem in that a term $\mu_1 \Gamma_1$ is added to σ. This distinction disappears when the film is formed from components 2 and 3 alone, since then $\Gamma_1 = 0$. The fact remains, however, that these equations give absolute values of the surface concentrations, regardless of the arbitrary location of the geometric surface of separation, as is characteristic for interfacial adsorption between two different phases. Equation (8) can also be used to obtain the relations

$$\Gamma_2 = -\left(\frac{\partial \sigma}{\partial \mu_2}\right)_{T, \mu_1, \mu_3}, \tag{17'''}$$

$$\Gamma_3 = -\left(\frac{\partial \sigma}{\partial \mu_3}\right)_{T, \mu_1, \mu_2}, \tag{17''''}$$

which are identical with the Gibbs Equation. Differentiation of σ with the condition $\mu_1 = \text{const}$ is here less useful than would be differentiation with $\Gamma_1 = \text{const}$.

V. We will now consider the conditions required for thermodynamic equilibrium in the free film. For an arbitrary irreversible process, the equality sign in the expression for the differential of the system free energy must be replaced by an inequality sign

$$dF < SdT - PdV + \sigma dA + \sum_{i=1}^{i=3} \mu_i dN_i \tag{18}$$

Considering the temperature, chemical potentials μ_2 and μ_3, total film surface area and amount of first component in the film to be constant, one obtains

$$d(F + PV - \mu_2 N_2 - \mu_3 N_3) = dF^S < 0 \cdot \tag{19}$$

Thus the function F^S tends to diminish at fixed T, P, μ_2, μ_3, A, and N_1, and equilibrium is established when the minimum is reached.

This is equivalent to the equilibrium condition

$$\Sigma \Delta F^S > 0, \tag{20}$$

where ΔF^S is the change in F^S resulting from departure from equilibrium, and summation is extended over all portions of the film which can be assumed homogeneous. The requirements of contancy of total film area, A, and total number of particles, N_1, lead to

$$\Sigma \Delta A = 0, \ \Sigma \Delta N_1 = 0 \cdot \tag{21}$$

In order that the summations of (20) and (21) each contain two terms, the entire film is now divided into two parts, the second much larger than the first. The expression applying to the second part, under variation of the area, A, and number of particles, N_1 can be written as

$$(\Delta F^S)_2 = \sigma(\Delta A)_2 + \mu_1 (\Delta N_1)_2 \cdot$$

The Gibbs—Duhem Equation (8) is applicable to the first part, since the ratios $\Delta A_2 / A$ and $(\Delta N_1)_2 / (N_1)_2$ are small in value. The relations of (21) can be drawn on, and this expression rewritten as

$$(\Delta F^S)_2 = -\sigma(\Delta A)_1 - \mu_1 (\Delta N_1)_1 \tag{22}$$

the equilibrium condition of (20) then becoming

$$\Delta F^S - \sigma \Delta A - \mu_1 \Delta N_1 > 0 \cdot \tag{23}$$

Here the parenthesis with subscript 1, $(\quad)_1$, has been omitted throughout. All the terms of Eq. (23) refer to the smaller section of the film.

By developing the expression of (23) as a power series and breaking this series off with the second member, one obtains

$$\frac{\partial^2 F^S}{\partial A^2} \ (\Delta A)^2 + 2 \left(\frac{\partial^2 F^S}{\partial A \cdot \partial N_1} \right) \ \Delta A \cdot \Delta N_1 + \frac{\partial^2 F^S}{\partial F_1^2} \ (\Delta N_1^2) > 0 \cdot$$

or, in view of (3'),

$$\left(\frac{\partial \sigma}{\partial A} \right)_{N_1} (\Delta A)^2 + 2 \left(\frac{\partial \sigma}{\partial N_1} \right)_A \Delta A \cdot N_1 + \left(\frac{\partial \mu_1}{\partial N_1} \right)_A (\Delta N_1)^2 > 0 \cdot \tag{24}$$

From (14), the expression for the chemical potential, it follows that

$$\left(\frac{\partial \mu_1}{\partial N_1} \right)_A = -a \left(\frac{\partial \sigma}{\partial N_1} \right)_A \cdot$$

13

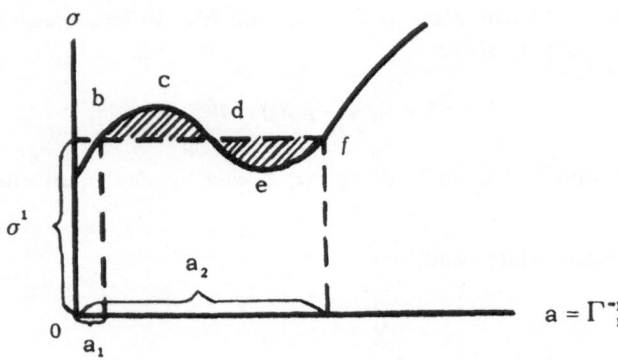

Fig. 1. The equilibrium coexistence of two strips of free film
of unequal thickness.

Since σ depends only on the ratio $a = 1/\Gamma_1 = A/N_1$ at fixed μ_2 and μ_3,

$$\left(\frac{\partial\sigma}{\theta A}\right)_{N_1} = \frac{1}{N_1}\frac{\partial\sigma}{\partial(A/N_1)} = \frac{1}{N_1}\frac{\partial\sigma}{\partial(1/\Gamma_1)} = -\frac{\Gamma_1^2}{N_1}\left(\frac{\partial\sigma}{\partial\Gamma_1}\right) = -\frac{\Gamma_1^2}{N_1}A\left(\frac{\partial\sigma}{\partial N_1}\right)_A = -\Gamma_1\left(\frac{\partial\sigma}{\partial N_1}\right)_A.$$

Thus

$$\left(\frac{\partial\sigma}{\partial A}\right)_{N_1}\left(\frac{\partial\mu_1}{\partial N_1}\right)_A = \left(\frac{\partial\sigma}{\partial N_1}\right)_A^2, \tag{25}$$

since $a\Gamma_1 = 1$. The discriminant of the square form (24) thus reduces to zero, and the latter takes the form

$$\left(\frac{\partial\sigma}{\partial A}\right)_{N_1}(\Delta A - a\Delta N_1)^2 > 0. \tag{24'}$$

Passing by the trivial case of $\Delta A/\Delta N_1 = A/N_1$ where there is, in general, no change, the criterion for film stability results as

$$\left(\frac{\partial\sigma}{\partial A}\right)_{N_1, T, \mu_2, \mu_3} > 0. \tag{26}$$

This relation can be written in two equivalent forms by taking account of the fact that $\Gamma_1 = N_2/A > 0$; namely

$$\left(\frac{\partial\sigma}{\partial a}\right)_{T, \mu_2, \mu_3} > 0, \tag{26'}$$

and

$$\left(\frac{\partial\sigma}{\partial\Gamma_1}\right)_{T, \mu_2, \mu_3} < 0. \tag{26''}$$

It follows that the tension should diminish with increasing concentration, Γ_1. It is important to note that the condition of (26) is satisfied at depths where the film possesses none of the properties of the bulk phase, the value of σ for the thick film becoming $\sigma_2 = 2\gamma$ (γ is the surface tension of the degenerate bulk film) and depending, therefore, on μ_2 and μ_3, but not on Γ_1. Condition (26'') is, then, fundamentally different from the Gibbs—Marangoni Equation. This latter shows $\partial\sigma/\partial\Gamma_1$ for the thick film to differ from zero, the time required for establishing solvent (water) equilibrium with the surrounding phase being assumed much larger than the excitation (stretching) time of the portion of the film under study. Thus the Gibbs—Marangoni Equation differs from (26'') in that it does not assure free film stability under continued action, and cannot explain the strict uniformity of the depth of the horizontal portions of soap films in extended equilibrium.

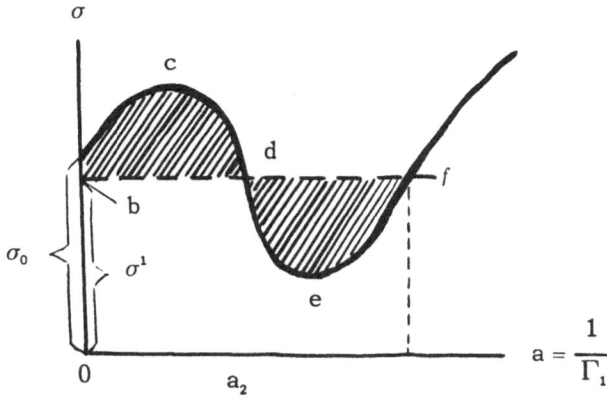

Fig. 2. Equilibrium between free film and bulk phase at finite contact angle.

VI. Only this approach offers an explanation of the coexistence of film sections whose depths are uniform but different from one another [6]. The necessary condition for such coexistence is equality of the chemical potentials, μ_1, μ_2, μ_3, and tensions, σ', of the two film portions, this being equivalent to a requirement of equal Γ_1 values, which is to say, equality of $a = a_1$ and $a = a_2$. These requirements can be satisfied only if σ is a non-monotonic of a function such as shown in Fig. 1. The expression of Eq. (14) for the chemical potential μ_1 is unsuitable for an $\sigma(a)$ relation of this kind, since the inverse function $a(\sigma)$ is then no longer single-valued, and Eq. (13) is therefore used in its place. Two relations for the determination of a_1 and a_2 can be obtained by equating the values of the chemical potential μ_1 for the two phases and considering the film tensions identical (and equal to σ', see Fig. 1):

$$\sigma(a_1) = \sigma(a_2) = \sigma',$$

(27)

$$\int_{a_1}^{a_2} \sigma \, da - \sigma'(a_2 - a_1) = 0 \, .$$

(28)

Geometrically, this implies equality of the two hatched areas, bcd and def, of Fig. 1, on which basis determination of the unknowns a_1 and a_2 can be made at fixed temperature, pressure, and chemical potentials, μ_2 and μ_3.

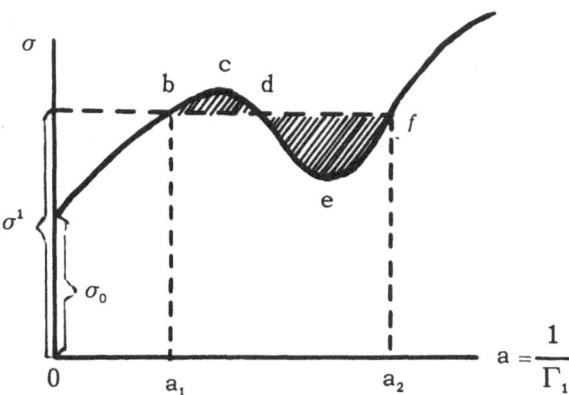

Fig. 3. Nonequilibrium coexistence of two strips of
free film.

Certain forms of the σ vs a relation permit a situation (Fig. 2) in which equality of the hatched areas implies $a_1 = 0$, the point b falling below the $\sigma(a)$ curve on the axis of ordinates. Reasoning in analogy with the well-known work of Frumkin [7], it can be shown that a film for which $a = a_2$ will be in equilibrium only with a degenerate bulk phase on which its contact angle, θ, satisfies the relation

$$\cos \theta = \sigma'/\sigma_0.$$

(29)

There will be a shift in the position of the separating boundary between two neighboring film sections which satisfy (27), the condition for mechanical equilibrium, but not (28). The case of Fig. 3 where $\mu_1(a_2) < \mu_1(a_1)$ is readily seen to be such that component 1 must pass into the thicker part of the film, the area of this portion of the film increasing while that of the thicker portion diminishes. This corresponds to the black spot expansion [8] which is frequently observed as a spontaneous process in soap films, and to the boundary displacement between black and gray film sectors studied in [6].

Literature

1. J. W. Gibbs, Collected Works, Vol. 1, "Thermodynamics," pp. 300-314, Longmans, Green and Co., New York (1931).
2. B. V. Deryagin and A. S. Titievskaya, Scientific Research Papers of the Chemical Institutes of the Academy of Science, USSR, for 1940, Moscow—Leningrad Akad. Nauk SSSR (1941), p. 135; Dokl. Akad. Nauk SSSR 89:1041 (1953); Kolloidn. Zh. 15:416 (1953); Discussions Faraday Soc., No. 18:24 (1954); Proceedings of the Second International Congress of Surface Activity, Vol. 1 (1957), p. 211; Kolloidn. Zh. 22:398 (1960).
3. J. W. Gibbs, loc. cit., p. 305, bottom.
4. G. A. Martynov and B. V. Deryagin, Kolloidn. Zh. 24:480 (1962).
5. A. D. Sheludko, Dokl. Bolgar. Akad. Nauk 9:11 (1956).
6. E. M. Duyvis and J. T. Overbeck, Proc. Koninkl. Ned. Akad. Metenschap., Ser. B. 65:26 (1961). J. A. Kitchener, Endeavour 22:118 (1963).
7. A. N. Frumkin, Zh. Fiz. Khim. 12:1 (1938).
8. B. V. Deryagin et al., J. Colloid Sci. 19:113 (1964).

THE SPLITTING PRESSURE AND EQUILIBRIUM OF FREE FILMS

B. V. Deryagin and Yu. V. Gutop

Institute of Physical Chemistry, Acad. Sci., USSR,
Laboratory for Surface Phenomena

It has been shown in [1] that the equation of state of the free film can be developed by studying the σ, T, Γ_1, μ_2, and μ_3 relation which determines the form of the fundamental equation. Practical (but not theoretical!) difficulties arise here in: 1) the determination of Γ_1, and, 2) the measurement of σ, especially when dealing with thick films where σ is close to σ_0, and a small change in σ corresponds to marked alteration in the film depth and value of Γ_1. It has been shown in [2] that the first of these difficulties can be overcome by introducing a radioactive indicator into the first component. The second difficulty can be partially overcome by determining, possibly through γ-ray adsorption, the variation of excess surface density ($m_1\Gamma_1 + m_2\Gamma_2 + m_3\Gamma_3$, m_i being the mass of the i-th molecule) with height. It is clear that $\sigma - g \int (m_1\Gamma_1 + m_2\Gamma_2 + m_3\Gamma_3) dz$ = const, and the value of σ can therefore be obtained. For experimental study of the equation of state of the thick film, however, it is preferable to investigate the equilibrium established in a free film in contact at each end with a bulk liquid, L, this, in turn, separated by two identical meniscii from the gaseous phase surrounding the film.

Equilibrium of this kind was first treated by one of us in [3]; it has considerable interest in its own right and the forces of attraction and repulsion acting within the film can be analyzed through its study. The principal advantage of this technique is in the case of establishing exact equilibrium between bulk phase and film, the dimensions of the latter being quite small (of the order of 1 mm^2, or less).

For the sake of definiteness we will consider a schematic experiment, somewhat different from that of [3, 4] but more convenient for a theoretical treatment (see Fig. 1). The hollow concave-convex capsule, C, has in its center circular "windows" between which a free film, F will form when the meniscus curvature has taken on the appropriate value. This curvature, the values of Γ_1, and the film depth, can all be varied by altering the liquid level in arm B of the U-tube by changing the pressure on the liquid in arm U. The level difference, h, between film plane and the liquid meniscus in arm B determines the splitting pressure, Π, of the film. The value of Γ_1 can be determined by the use of radioactive indicators, and the film depth calculated from the coefficient of optical reflection and the theory of thin film interference [3, 5]. The value of μ_2 is held constant by having a solution, L_0, of fixed concentration ("bulk degenerate film") in the chamber K. This will, in principle, also hold the value of μ_3 constant. This same result can be easily realized by bringing the contents of chamber K into contact with the surrounding atmosphere through a small opening in the cover T. At equilibrium, the thermodynamic potentials μ_2 and μ_3, must each be identical in phases A and L_0 and the film F while the potential μ_1 must be the same in the liquid phase L_0 and in the film.

The meniscus curvature gives rise to a pressure discontinuity at the interface between A and L_0. This pressure drop is related to the curvature $\varepsilon = 1/R_1 + 1/R_2$ through the Laplace equation:

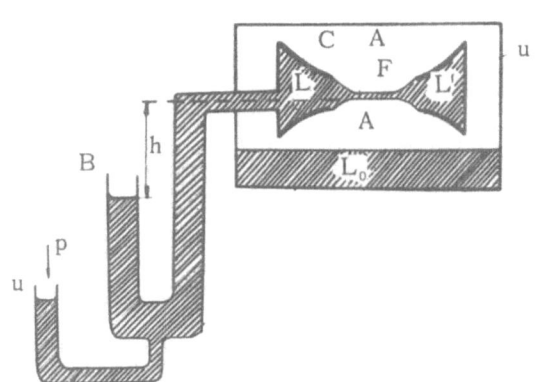

Fig. 1. Scheme for developing the $\pi(\Gamma_1)$ or $\pi(h)$ relation.

17

$$P_2 - P_1 = \gamma E, \tag{1}$$

where P_2 is the pressure in phase A, P_1 is the pressure in the surrounding liquid phase, L, γ is the surface tension at the interface between the two phases, and R_1 and R_2 are the principal radii of curvature of the surface. As a result of this pressure discontinuity, the internal pressure acting on the film surface at equilibrium exceeds by a certain amount, Π, the hydrostatic pressure on phase A of which it is a part and with which it communicates through the end sections. The ideas developed by one of us [6] indicate that Π is also an overall measure of the repulsive forces acting within the film and tending to thicken it. This excess is designated as the splitting pressure of the film. Π is a function of T, Γ_1, (or μ_1), μ_2, and μ_3, just like all other thermodynamic properties of the film. Two assumptions can be made concerning the formal localization of this excess pressure, Π, in the film [7]:

1. Π is the pressure drop between the film and the bulk phase A.

2. Π is the pressure drop at the interface between the film and the neighboring phases, this drop being in addition to that called for by the Laplace equation for the nonplanar film. The first approach is the one which will be adopted here. Alteration of the liquid level in arm B with P_2 held constant will change P_1 and thereby alter the pressure drop Π by the same amount.

The potentials μ_2 and μ_3 are maintained constant in the film and in the end portions of the surrounding liquid phase through contact with the neighboring bulk gas phase, A, and the layer, L_0, of solution of fixed composition. Alteration of P_1 must therefore change the value of the potential μ_1 in the film and in the neighborhood of its ends. The lateration of bulk liquid phase composition and meniscus surface tension resulting from variation in μ_1 and P_1 can be neglected.* In order to make the treatment exact, it can be assumed that these alterations in composition are realized by the exchange of volatile components with the phase, L, of fixed composition.

It follows from the Gibbs—Duhem equation and Eq. (14) of [1] that

$$dP_1 = d\Pi = \rho_1 d\mu_1 = a\rho_1 d\sigma, \tag{2}$$

ρ_1 being the molecular concentration of component 1 in the liquid phase. By integrating (2) with the conditions $\mu_2 = \text{const}$ and $\mu_3 = \text{const}$, and neglecting the change in ρ_1, one obtains

$$\Pi \cong \rho_1(\varphi - \mu_1)\rho_1 \int_{\sigma_0}^{\sigma} a d\sigma, \tag{3}$$

where φ (introduced in Eq. (6) of [1]) becomes equal to μ_1 when $a \to 0$, $P_1 \to P_2$, and $\Pi \to 0$. If a is now developed as a power series in $\sigma - \sigma_0$, and account is taken of the fact that $\sigma - \sigma_0 \to 0$, when $a \to 0$, one obtained from (3):

$$\Pi = \frac{1}{2}\left(\frac{\partial a}{\partial \sigma}\right)_{\sigma = \sigma_0} \cdot \rho_1(\sigma - \sigma_0)^2. \tag{4}$$

---

*This is quite similar to what was done in the experiments of A. D. Sheludko and D. Ekserova [4]; there, however, it was a matter of determining the equilibrium value of the film thickness rather than varying the pressure Π and observing the kinetics of film thinning and the maximum. The experiments of B. V. Deryagin and A. S. Titievskaya [3] differed from those under discussion here in so far as P_1 was held constant and P_2 varied, principally through alteration of the partial pressure and chemical potential of component 3. This lack of constancy in μ_3 should make for complications in the relation developed above. These changes were not large, however, and could be neglected since there was only slight variation in μ_3, N_2, and N_3 and dissolved gases have little effect on the properties of water. The lack of constancy of μ_3 will certainly have no effect on the derived relations if the third component is neither present in the film nor adsorbed on the film surface.

Conversely, (2) can be used to obtain an exact expression for σ in terms of Π:

$$\sigma - \sigma_0 \equiv \Delta\sigma = \int_0^\Pi \frac{d\Pi}{a\rho_1},$$

(5)

where μ_2 = const and μ_3 = const. Since $a = 1/\Gamma_1$ can be measured by the method of radioactive indicators, while Π is determined from the difference in levels and ρ_1 is almost constant, Eq. (5) can be used to express the very small difference $\sigma - \sigma_0$ as a function of Γ_1, μ_2, and μ_3, or μ_1, μ_2, and μ_3. This permits the fundamental equation to be developed quite accurately, even for the thick film where the value of $\Delta\sigma$ is low and direct measurements can be carried out only with difficulty. Knowing $\Delta\sigma$ as a function of Γ_1, μ_2, and μ_3, Eq. (14), (16'), and (16") of [1] can be drawn on to find μ_1, Γ_2, and Γ_3:

$$\mu_1 = \varphi(T_1,\mu_2,\mu_3) - \int_\sigma^{\sigma-\sigma_0} ad(\sigma - \sigma_0),$$

(6)

$$\Gamma_2 = -\Gamma_1 \frac{\partial\varphi}{\partial\mu_2} - \frac{\partial\sigma_0}{\partial\mu_2} - \int_0^1 \frac{\partial\Delta\sigma(\epsilon a)}{\partial\mu_2} d\epsilon,$$

(6')

$$\Gamma_3 = -\Gamma_1 \frac{\partial\varphi}{\partial\mu_3} - \frac{\partial\sigma_0}{\partial\mu_3} - \int_0^1 \frac{\partial\Delta\sigma(\epsilon a)}{\partial\mu_3}.$$

(6")

Contributions arising from the bulk properties of the liquid, are covered by the terms of these equations in the function $\varphi(T, \mu_2, \mu_3)$, contributions from the surface properties by terms in σ_0 and contributions from surface forces related to the splitting pressure, Π, through (5) by terms in $\sigma - \sigma_0 = \Delta\sigma$, these latter coming into play when the film is so thin that no part of it, not even that lying in the plane of symmetry and furthest removed from the surface, has the properties of the bulk phase. The fact that Eqs. (5), (6), (7'), and (7") can be used to express the unknown film parameters, μ_1, Γ_2, and Γ_3, through the known values Γ_1, μ_2, and μ_3, once the function $a(\Pi)$ has been developed, shows the fundamental role played by the splitting pressure in thin film thermodynamics.

The significance of the splitting pressure traces back to the fact that Eq. (26) of [1], the necessary condition for stability (more exactly, metastability) of films of any depth including Perron bimolecular films, can be expressed through the differential equation (2) as

$$\left(\frac{\partial\Pi}{\partial a}\right)_{T,\mu_2,\mu_3} > 0,$$

(8')

$$\left(\frac{\partial\Pi}{\partial\Gamma_1}\right)_{T,\mu_2,\mu_3} < 0.$$

(8")

These conditions are applicable to films of any thickness whatsoever since Π has been give a strict phenomenological, rather than intuitive, definition. The question of necessary conditions for free film metastability should be kept separate from the companion problems of degree of stability and mean life of the film, since these latter relate to the rupture mechanism. The authors of this article have solved these problems for a simple two-dimensional fluctuation rupture mechanism [8]. There is no doubt, however, that breakdown of the condition of (8), or its equivalent, Eq. (26) of [1], will render the film thermodynamically metastable and markedly reduce its life, regardless of the film rupture mechanism and the factors tending to delay rupture (e.g., enhanced viscosity). This is easily understood since breakdown of these conditions must have the result that any accidentical thinning of the film with attendant reduction in Γ_1 will lead to a local increase in σ with an accompanying stretching of the film and a progressive reduction in Γ_1, thus setting up a chain rupture reaction. The thermodynamic condition for film stability is therefore basic in one sense, although it does not fully determine the life time of the film. These remarks are certainly not limited to the special case of the three-component system which has been under discussion here, but are still more widely applicable. Thus the conclusion that

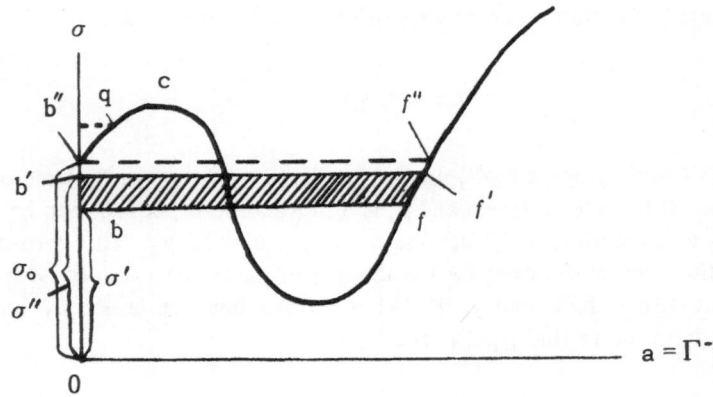

Fig. 2. The effect of μ_1 on the equilibrium of two coexistent strips
of free film of different thickness.

thermodynamic metastability is a necessary condition for extended life of a film between two identical liquid phases, is indicative of the role played by thermodynamic factors in determining the stability of concentrated emulsion foams.*

The principal significance of the experimental technique schematically illustrated in Fig. 1 is that it offers the possibility of studying the alteration in depth, splitting pressure, Π, value of Γ_1, etc., (that is to say, the state parameters of the film) arising from a change in μ_1.

Using the diagram of Fig. 2 we will now consider theoretically the special case of the alteration of the equilibrium between film and bulk liquid caused by a change in μ_1 (at μ_2, μ_3 = const), which is, itself, the result of an alteration in the meniscus curvature. Equation (14) of [1] shows an alteration of μ_1 at constant μ_2, μ_3, approximated by

$$\Delta\mu_1 \cong -\frac{\Delta\Pi}{\rho_1} \qquad (9)$$

displaces the representative point, f, for the film to a new position, f', such that

$$-\int_{f}^{f'} a\,d\sigma = \Delta\mu_1. \qquad (10)$$

The value of the integral of this expression is represented by the hatched area in Fig. 2. Displacement of the point f increases the specific area, a, of component 1 and thereby raises the film tension. This, in turn, increases $\cos\theta = \sigma''/\sigma_0$, a function of the meniscus curvature, and the angle θ at point f'' corresponding to the predetermined value of $\Delta\mu_1 = K$ reduces to zero. Further reduction in the value of μ_1 leads to continued

---

*The work of Sonntag [9] on rupture processes and the stability of films of hydrocarbon solutions of surface active substances between mercury meniscii, refers to Figs. 3 and 4 for the thinning kinetics and states that these indicate the possibility of film stability even when the splitting pressure is negative at all thicknesses. There is a clear misunderstanding here. The equation used for calculating the splitting pressure, Π, in this work, namely $d/dt\,(1/d^2) = a(P_0 - \Pi)$, indicates that $\pi = P_0$, so that the pressure must be positive for constancy of film depth.

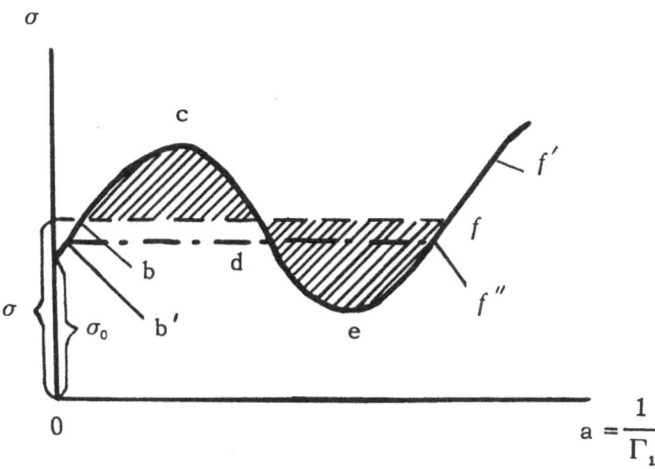

Fig. 3. The effect of μ_1 on the equilibrium between free film and bulk liquid phase.

increase in a and σ, with the contact angle remaining equal to zero. This does not involve breakdown of the mechanical equilibrium of the film and bulk phase at the film ends since it is readily shown that Eq.(2) contains

$$do = \frac{\Gamma_1}{\rho_1} d\Pi,$$

$$(11)$$

the condition for mechanical equilibrium in any small wedge-shaped portion of the film.* By the integration of (11) it can be shown that $\sigma > \sigma_0$ for a film in which $\Pi > 0$ for all values of Γ_1. Designating $\Gamma_1/\rho_1 \equiv H$ as the virtual film thickness, Eq. (11) can be brought to the form

$$do = H d\Pi,$$

$$(11')$$

which is the same relation as was derived earlier in [10] for the single component film. The concept of geometrical thickness is indefinite in the case of the very thin film, but the ratio Γ_1/ρ_1 is always strictly defined and can, in principle, be measured with any desired degree of accuracy. It is clear that Γ_2 will diminish with μ_1, and the film thickness fall off accordingly.

A film represented by a point on the part of the $\sigma(a)$ curve lying to the left of the maximum can also be in equilibrium with the bulk liquid when $\theta = 0$. The value of σ for this film will obviously be higher than for a film represented by a point f lying further to the right and corresponding to the same value of μ_1, so that coexistence of these films would imply that mechanical equilibrium could not be established. Since high values of σ correspond to high values of Γ_1, the film with low Γ_1 stretches under the action of the adjacent thick film sections and rapidly thins, while the thick section becomes still thicker by contracting its surface area. The thermodynamic potential of the thicker section is thereby increased, so that reduction in the surface area favors the outward flow of component 1 with reduction of chemical potential. This must inevitably lead to the disappearance of the thicker of the two sections. Moreover, the expression for the film free energy can be drawn on to show that there is only low probability of fluctuation effects on the thicker section of the film.

It can also be shown that there is still less likelihood of the appearance of a thick section on a film represented by a point falling to the left of f, and in equilibrium with the bulk phase convex meniscii.

---

*Detailed proof of this point will be given in another paper which will consider equilibrium in a film of varying thickness, and behavior in an external field.

Films corresponding to the interval b"q in Fig. 2 can exist in metastable equilibrium for extended periods of time; they are obtained by the gradual thinning of a thick solution layer with accompanying reduction in the value of μ_1. An increase in the value of a also entails an increase in the probability of spontaneous formation of a thinner and more nearly stable film section such as would be associated with the branch of the curve falling to the left of the point f. Such section would gradually spread and finally occupy the entire film surface. This spreading could be accompanied by the establishment of a nonzero contact angle provided the alteration in the value of μ_1 was not too great. Similar effects involving the formation and growth of dark patches on soap films have been reported in numerous papers [11] on the basis of experimental procedures quite like those indicated in Fig. 1. The case of Fig. 3 is that in which the contact angle of the film in equilibrium with concave meniscii ($\Pi > 0$) is equal to zero, this film corresponding to a point, f', lying to the left of f so that the areas bcd and def are equal. The state and properties of this film alter with μ_1 in exactly the same qualitative manner as in the case just considered, and there is also the possibility of the formation and growth of dark patches. Films represented by points f'' lying still further to the left of f also show the possibility of discontinuous formation (point b') of a thicker "island," which can expand to cover the entire surface of the original thinner film. It is, however, necessary that the ordinate of f'' be greater than F_0, i.e., that the contact angle be zero.

Literature

1. B. V. Deryagin, G. A. Martynov, and Yu. V. Gutop, Kolloidn. Zh. (in press).
2. J. M. Corkil et al., Trans. Faraday Soc. 57, Pt. 5:821 (1961); Proc. Roy. Soc., A 273:84 (1963).
3. B. V. Deryagin and A. S. Titievskaya, Dokl. Akad. Nauk SSSR 89:1041 (1953); Discussions Faraday Soc., No. 18:85 (1954); Proceedings of the Second International Congress on Surface Activities Vol. 1 (1957), p. 211; Kolloidn. J. 22:398 (1960); J. Colloid Sci. 19:113 (1964).
4. See A. D. Sheludko et al., Kolloidn. Zh. 155:39 (1957); 165:148 (1959); 168:24 (1960); 175:150 (1961); Dokl. Akad. Nauk SSSR 123:1074 (1958).
5. See refs. 2, 3, and 4.
    E. M. Duyvis and J. T. Overbeck, Proc. Knonikl. Ned. Akad. Wetenschap., Ser. B. 65:26 (1962).
6. B. V. Deryagin, Kolloidn. Zh. 17:207 (1955); see also ref. 3.
7. B. V. Deryagin, Acta Phys. Khim. 12:181 (1940). See p. 191.
    S. B Nerpin, doctoral dissertation, LIIVT, Leningrad (1957).
    S. B. Nerpin and B. V. Deryagin, Dokl. Akad. Nauk SSSR 99:1029 (1954); 100:17 (1955).
8. B. V. Deryagin and Yu. V. Gutop, Kolloidn. Zh. 24:431 (1962).
9. H. Sonntag, Z. Physik. Chem. 221:373 (1962).
10. B. V. Deryagin, Acta Physicochim., URSS 12:181 (1940); Zh. Fiz. Khim. 14:137 (1940).
11. B. V. Deryagin et al., J. Colloid Sci. 19:113 (1964).
    E. M. Duyvis and J. T. Overbeek, Proc. Koninkl. Ned. Akad. Wetenschap., Ser. B. 65:26 (1962).
    J. A. Kitchener, Endeavour 22:118 (1963).

THE PRESSURE DISTRIBUTION IN THIN FILMS

B. V. Deryagin

Institute of Physical Chemistry, Acad. Sci., USSR,
Laboratory for Surface Phenomena

The two preceding papers have characterized the thin film in terms of the external parameters alone. Interest attaches not only to a study of the conditions for overall equilibrium, but also to analysis of the pressure distribution within the film, when the film is no longer extremely thin. Something of this kind has been attempted by the Dutch physicists, e.g., Bakker [1], for the transition layer at the liquid—gas interface. This work led to the equation

$$\gamma = \int_0^\infty [P_N - P_T(z)]\, dz,$$

(1)

in which γ is the surface tension at the liquid—gas interface, p_N is the external pressure directed normally to the interface, this pressure being transmitted unaltered to any depth, z, if the elementary liquid layers are free of long-range forces, and $p_T(z)$ is the tangential pressure directed normally to a surface located at depth z below the interface and perpendicular to it. Applicability of this equation assumes, however, that certain conditions are fulfilled.

1. Such "continuous" macroscopic treatment requires that each monolayer make a relatively small contribution to the calculated value of γ, a condition which is certainly not fulfilled in the outermost monolayer where there is a rapid alteration of forces over distances of the order of a molecular diameter.

2. All effects from molecular forces, including those in the transition layer, have until recently been considered to result from long-range action over distances of the order of the layer depth. Bakker, in fact, assumed the p_N of Eq. (1) to be constant and independent of z, even within the transition layer. This is equivalent to including the molecular attraction forces in a component of the normal pressure acting on each surface. Only if the linear dimensions of the surface are much larger than the interaction radius of the molecular forces is this procedure consistent with the requirement that the pressure be uniquely determined. Under these conditions, however, a surface oriented at right angles to the interface would cover the entire transition layer, and could therefore not be used to define the local pressure, $p_T(z)$, within the layer. Bakker's treatment thus suffers from internal contradictions, as well as from a formalism which appears when it is a matter of evaluating a transition layer depth of the order of several angstroms [1].

All of these defects in the equation and formulation of Bakker are eliminated by replacing the classical long-range molecular forces by the Lifshits concept [2] of van der Waals forces arising from electromagnetic field fluctuation. This makes it possible to replace the microscopic description in terms of distance-dependent molecular interaction forces by a macroscopic description in terms of an electric field distribution, constant, on the average, over not-too-short time intervals, and effective in material bodies as well as vacuum [3]. A special instance of such treatment is found in the calculation of the potentials arising from electrostatic interaction in double ionic layers.

The current form of macroscopic treatment does not, however, give an acceptable interpretion of short-range forces, the principal contribution here being from the interaction of nearest neighbor molecules in condensed bodies. This approach will not, then, make it possible to use Eq. (1) for calculating the interfacial tension between condensed media, this tension depending in the first instance on surface monolayer interaction.

There is, however, no reason why such treatment could not be applied to effects arising from molecular and electrostatic interactions in polymolecular films (and cracks), action being affected here over distances of the order of many molecular diameters. Thus the expression for σ, the tension of a thin film (or crack) whose depth, h, is much larger than the molecular diameter, can be set up on the basis of reasoning similar to that employed by Bakker in his derivation of Eq. (1), and a correction term, σ_0, introduced to allow for those short-range molecular interaction effects which cannot be handled macroscopically:

$$\sigma = \sigma_0 + \int_{-\infty}^{+\infty} [P_N - P_T(z)]\, dz.$$

(2)

Here σ_0 is to be understood as the contribution to the thin polymolecular interphase layer tension which arises from short-range interaction forces.

The terms p_N and p_T of Eq. (2) represent the forces on the respective surfaces arising from the electrostatic field of the ion atmospheres and from the mean stationary electromagnetic field which the theory of Dzyaloshinskii, Lifshits, and Pitaevskii [3] associates with the attraction of microscopic objects. The isotropic osmotic pressure makes no contribution to (2) and has not been included for this reason.

The final result is that Eq. (2) can be replaced by

$$\sigma = \sigma_0 + \int_0^h \frac{\epsilon}{4\pi} E^2 dz + \int_{-\infty}^{+\infty} (\sigma_{yy} - \sigma_{zz})\, dz.$$

(3)

The first integral of this expression is to extend over the depth of the electrolyte layer, since the layer surface has been assumed to have fixed potential, and σ_{zz} and σ_{yy} are the diagonal components of the field strength tensor for the stationary but fluctuating electromagnetic field, inside and outside the layer. With the approach of h to infinity, the integrals tend to the respective values $\mathcal{J}_e$ and $\mathcal{J}_m$, and the left-hand member approaches 2γ, so that one has

$$2\gamma = \sigma_0 + \mathcal{J}_e + \mathcal{J}_m.$$

(4)

$\mathcal{J}_m$ requires special treatment since σ_{yy} was not evaluated in [2, 3].

A good test of Eq. (3) is found in the fact that it can serve [4] as a basis for the derivation of the thermodynamic relation

$$\frac{d\sigma}{h} = d\Pi,$$

(5)

Π being the layer splitting pressure. When use is made of the expressions for E and Π which have been obtained for surfaces separated by a layer of electrolyte and charged to low potentials [5], the electrostatic component, σ, calculated from (3) will exactly satisfy (5). The relation is also applicable to the elementary case of surfaces immersed in a dielectric liquid and maintained at a fixed potential difference. No simple check can be given for the $\mathcal{J}_m$ component of the film tension in Eq. (5).

It should be noted that (5) will be replaced by the incorrect relation σ/h = Π if Eq. (2) is used for σ and the Shcherbakov expression [6] drawn on for Π. From this it follows that this expression for Π is in error, despite the fact that the fundamental idea involved in its derivation is closely related to that developed in the present work.

Literature

1. G. Bakker, "Kapillarität und Oberflächenspannung," Handbuch d. Experimentalphysik, Vol. 6, Leipzig (1928).
2. E. M. Lifshits, e.g., Usp. Fiz. Nauk 64:493 (1958).
3. I. E Dzyaloshinskii, E. M. Lifshits, and L. P. Pitaevskii, Zh. Eksperim. i Teor. Fiz. 37:229 (1959).

4.   Acta Physicochim. URSS 12:181 (1940).
      See also the paper by B. V. Deryagin, G. A. Martynov, and Yu. A. Gutop in this collection, p. 9.
5.   B. V. Deryagin, Izv. Akad. Nauk SSSR, No. 5:1153 (1937); URSS 10:333 (1939).
6.   L. M. Shcherbakov, In collection: Research in Surface Forces, Vol. II, Izd. Akad. Nauk SSSR, Moscow
      (1964), p. 11.

EVALUATION OF THE EXCESS FREE ENERGY OF SMALL OBJECTS

L. M. Shcherbakov

Tula Polytechnic Institute

A more or less rigorous thermodynamic description [2, 3] of microheterogeneous systems has been obtained from the general concept of Type II capillary effects advanced in earlier papers [1, 2]. The point of view adopted here is that a single term ($\sigma_\infty S$) proportional to the interfacial area is inadequate to account for deviations from additivity in the free energy distribution (and other potential distributions) of a reasonably small object ($r \leq 10^{-5}$ cm), other terms depending on the dimensions of the object in question being required for this purpose. The "excess" free energy is here to be associated with the object as a whole rather than the interface. This opens the way for thermodynamic treatment at dimensions so low ($r \leq 10^{-6}$) that the Gibbs' thermodynamics breaks down because of the impossibility of distinguishing between "bulk" and "surface."

Practical application of the thermodynamic apparatus* developed here presupposes the existence of various equations for calculating the excess free energies of small objects. The methods developed by N. N. Bogolyubov [4] would probably yield the most exact relations that could be obtained here. The lack of data on correlation functions for the transition zone separating the liquid and vapor phases makes it impossible to carry out actual calculations of this type at the present time [5]. Recourse must therefore be had to cruder approximation methods.

It is to be noted that the calculation of liquid surface tensions requires information on the alteration of density (and other parameters) within the transition zone, and thus continues to be a complex problem even after simplifying assumptions have been introduced. A discussion of various pertinent methods and the difficulties associated with them has been given in [5, 6]. The problem of developing a relation between surface tension and drop dimensions is still more complex, as can be seen from the lack of agreement in the results obtained by various authors.

From this point of view, it would seem that a somewhat simplified formulation of the problem could be obtained from the conceptions which we have developed. The theory of Type II capillary effects treats the entire excess free energy (or other potential) associated with the group of droplet-forming molecules, this excess being calculated with respect to the free energy of a similar number of molecules taken from within an infinitely extended liquid phase. It is clear that this approach emphasizes droplet dimensions, the variation of density in the transition zone being much the same for drops of different dimensions and therefore a factor of secondary importance. The thermodynamic theory of perturbations developed by L. Landau and E. Lifshits [7] can be used for evaluating the excess free energy of the small droplet.

The finiteness of the droplet dimensions† is primarily reflected in the intermolecular interaction energy. Here the hamiltonian of the g particles composing the small drop can be written as

$$E = E_0 + \Delta U,$$

---

*It is not the individual small object — medium system which is to be considered when the number of particles in the object is small, but an assemblage of such systems.

†Discussion centers around the small drop for the sake of definiteness, but the same methods can be applied to other small objects as well.

E_0 being the "unperturbed" hamiltonion of g particles taken out of the bulk liquid, and ΔU the "excess" potential energy of the drop particles as compared with a similar number of particles taken from the infinitely extended liquid phase.

The state integral for the system under discussion can therefore be written as

$$Z = \frac{1}{g!} \int \exp\left(-\frac{E_0 + \Delta U}{kT}\right) dX,$$

dX being an element of phase space.

Expressing the integrand as a power series in $\Delta U/kT$, and breaking this series off with the second member, gives

$$Z = Z_0 + Z' = Z_0(1 + Z'/Z_0),$$

Z_0 being the "unperturbed" value of the state integral corresponding to zero "perturbation," ΔU, and Z' the addition to the state integral

$$Z' = -\frac{1}{g!} \int \left(-\frac{\Delta U}{kT}\right) e^{-E_0/kT} \, dX$$

arising from the perturbation. The assumption is that $Z' \ll Z_0$ yields an expression for the free energy of the small droplet, namely:

$$F = -kT \ln Z = -kT \ln Z_0 - kT \ln(1 + Z'/Z_0)$$

or

$$F \cong F_0 - kT\frac{Z'}{Z_0} = F_0 + \frac{\int \Delta U e^{-E_0/kT} \, dX}{\int e^{-E_0/kT} \, dX}.$$

The last term of this expression is simply the mean value of ΔU calculated for the "unperturbed" Gibbs distribution. Thus

$$\psi = F - F_0 = \overline{\Delta U}, \tag{1}$$

which is to say that the excess free energy of the small object is the mean excess potential energy calculated for the "unperturbed" distribution. *

The evaluation of $\overline{\Delta U}$ is most easily carried out by using a radial distribution function, $\rho(r)$, to express the probability of finding one particle at a distance between r and r + dr from another. It is clear that $\rho(r) \to 1$ as $r \to \infty$. The existence of so-called close order in the liquid particle distribution will require that $\rho(r)$ also pass through various maxima and minima in the neighborhood of the reference particle. The pair-wise particle interaction will be described in terms of the potential model of [8]:

$$u(r) = \left\{ \begin{array}{ll} \infty \quad . \ . \ . \ . \ . \ . \ . \ . \ . \ . \ . \ .(r < a) \\ \dfrac{A}{r^s} - \dfrac{B}{r^q} \quad . \ . \ . \ . \ . \ . \ .(r > a) \end{array} \right\}. \tag{2}$$

---

*The application of perturbation methods requires that the perturbation energy per particle be small in comparison with kT. The orders of magnitude in the present case are such that $\psi = \sigma_\infty \cdot 4\pi R^2$ and the number of particles $g = (4/3)\pi R^3/v_\infty = (4\pi/3)(R^3\delta/M)$.

Here σ_∞ is the surface tension, δ the density, and M the molecular weight of the liquid; R is the drop radius. Estimates based on $\sigma_\infty = 20$ erg/cm^2, M = 50 g/mole, $\delta = 1$ g/cm^3, and T = 300°K, R = 10^{-5} cm, show that $(\psi/g)/kT < 0.01$.

TABLE 1

Liquid	Working temp., °K	a, Å	$\dfrac{B}{4a^6 k}$, °K	λ_∞, joule/g [from Eq. (8)]	$\dfrac{\lambda^*_{exp}}{\lambda_\infty}$
Carbon tetra-chloride	273	5.881	327	96	2.24
n-Octane	398	7.451	320	142	2.24
Benzene	273	5.270	440	199	2.24
n-Hexane	273	5.909	413	164	2.27
n-Butane	253	4.997	410	191	2.08
Cyclohexane	333	6.093	324	159	2.26

*Using λ_{exp} values for temperatures as close as possible to the cited.

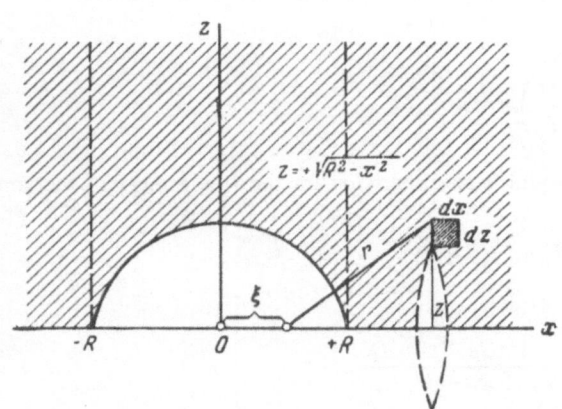

For the calculation of the excess energy of a small droplet.

Here, a is an effective diameter, A and B are positive constants, and s and q are whole numbers such that s > q, the values chosen for these numbers being q = 6; s = 9; 10; 12; 15 and q = 7; s = 28. A rather crude first approximation to the radial distribution function is then given by the expressions:

$$\rho(r) = \left\{ \begin{array}{ll} 0 & \ldots \ldots (r < a) \\ 1 & \ldots \ldots (r > a) \end{array} \right\}. \quad (3)$$

We first use this distribution function to calculate the specific heat of vaporization of the bulk liquid. The statistical theory of liquids [9] gives the mean potential energy of the particle of massive liquid as

$$\bar{u} = \frac{1}{2} n_1 \int_0^\infty u(r) \rho(r) \cdot 4\pi r^2 \, dr, \quad (4)$$

n_1 being the particle concentration.

The approximations adopted for u(r) and $\rho(r)$ lead to

$$\bar{u} = \frac{n_1}{2} \int_a^\infty \left(\frac{A}{r^s} - \frac{B}{r^q} \right) 4\pi r^2 \, dr = 2\pi n_1 \left(\frac{A/a^{s-3}}{s-3} - \frac{B/a^{q-3}}{q-3} \right). \quad (5)$$

An equilibrium value of the parameter a is obtained by minimizing this expression. The fact that a and the particle concentration are so related that $n_1 \approx 1/a^3$, leads to the equation

$$\frac{A/a^{s-3}}{s-3} = \frac{B/a^{q-3}}{q-3}. \quad (6)$$

An expression for the mean potential energy of a molecule in the interior of the bulk liquid can be obtained by combining Eq. (6) with (5):

$$\bar{u} = -\frac{2\pi n_1}{q-3} \left(1 - \frac{q}{s} \right) B/a^{q-3}. \quad (7)$$

A similar expression can be developed for the mean potential energy of a vapor molecule.

28

TABLE 2

Liquid	Working temp., °K	a, Å	$B/4a^6 R$, °K	M, g/mole	δ, g/cm³	σ, erg/cm² [from Eq. (16)]	$\dfrac{\sigma_{exp}}{\sigma_\infty}$
Carbon tetra-chloride	273	5.881	327	153.8	1.63	24.6	0.93
n-Octane	313	7.451	320	114.0	0.62	16.3	0.82
Benzene	273	5.270	440	78.0	0.90	25.2	0.80
n-Hexane	273	5.909	413	86.0	0.68	17.6	0.86
n-Butane	253	4.997	410	58.0	0.60	15.3	0.88
Cyclohexane	333	6.093	324	84.1	0.73	18.9	0.88

The latent heat of vaporization, λ_∞, per molecule of massive liquid can be calculated from the change in mean potential energy accompanying isothermal passage of a molecule from bulk liquid to vapor. Drawing on (7) and taking account of the fact that $n_1 \gg n_2$ when working far from the critical point, one finds that

$$\lambda_\infty \cong \theta_1 \pi n_1 B / a^{q-3},$$

(8)

where

$$\theta_1 = \frac{2}{q-3} \left(1 - \frac{q}{s} \right).$$

(8')

It is unfortunate that the lack of information concerning the a and B values of condensed systems makes it impossible to directly check this equation against the experimental data. The equation can, however, be drawn on in the inverse problem of determining one of these two parameters from the heat of vaporization. It is interesting that our calculations using gas phase a and B values not only give λ_∞ values of the correct order of magnitude for six nonpolar liquids (Table 1) but show the $\lambda_{exp}/\lambda_\infty$ ratio to be constant, at least within the limits of experimental error. This confirms the suggestion [8] that transition from liquid to gaseous phase profoundly alters the energy parameter B without essentially affecting the effective diameter a, thus permitting Eq. (8) to be used for finding reduced B values.

We now pass to a determination of the "excess" free energy of the small droplet. Here, thermodynamic analysis simply shows [10] the relation between excess energy and number of drop particles, g, to be covered by a monotonically increasing curve which is convex to the g axis. More detailed information can be obtained only by drawing on experimental data (which is not yet available) or by resorting to statistical methods. It follows from what has already been said that the excess free energy of the small object is equal to the mean excess potential energy. The averaging called for here can be carried out in the following manner.

Let us consider a small droplet of radius R. The excess potential energy of an interior molecule located at a distance ξ from the drop center (see figure) is given by

$$\Delta = n_1 \int_{V_k} u_{11}(r)\, \rho(r)\, d\tau + n_2 \int_{V'} u_{12}(r)\, \rho(r)\, d\tau - n_1 \int_{V_K + V'} u_{11}(r)\, \rho(r)\, d\tau$$

or

$$\Delta = - n_1 \int_{V'} u_{11}(r)\, \rho(r)\, d\tau + n_2 \int_{V'} u_{12}(r)\, \rho(r)\, d\tau.$$

(9)

Here, u_{11} is the energy of pair-wise interaction of the liquid molecules, u_{12} is the corresponding interaction energy for molecules of liquid and vapor, $d\tau$ is a volume element distant r from the molecule in question, $\rho(r)$ is the radial distribution function, and V' is the result of diminishing the total volume by the volume of the droplet. The second member of the equation can be neglected and the assumption made that

$$\Delta \cong - n_1 \int_{V'} u_{11}(r)\, \rho(r)\, d\tau.$$

(10)

29

conditions far from the critical point being such that $n_1 \gg n_2$. Assuming $u_{11}(r)$, and the radial function, $\rho(r)$, for the model potential to be approximated by (2) and (3), and taking the infinitely thin ring resulting from rotation of the dxdz surface around the X-axis (see figure) as the volume element, one has

$$\Delta = -\pi n_1 A \int\limits_Q \frac{2z\,dx\,dz}{[(x-\xi)^2 + z^2]^{s/2}} + \pi n_1 B \int\limits_Q \frac{2z\,dx\,dz}{[(x-\xi)^2 + z^2]^{q/2}} \;.$$

(11)

The plane of integration, Q, can be further broken down into three parts, namely: 1) the rectangular region satisfying the conditions $-\infty \leq x \leq -R$, $0 \leq z \leq \infty$; 2) the rectangular region satisfying the conditions $R \leq x \leq \infty$, $0 \leq z \leq \infty$; and 3) the region included between the lines $x = -R$ and $x = +R$, the semicircle $z = +\sqrt{R^2 - x^2}$, and the straight line at infinity, $z = \infty$ (see figure). Equation (3) can then be drawn on to obtain the following expression for the mean excess potential energy of the entire droplet:*

$$\overline{\Delta U} = \frac{1}{2}\,n_1 \int\limits_0^{R-a} \Delta \cdot 4\pi\xi^2\,d\xi.$$

(12)

The integrations called for in (11) and (12) can be carried out and Eqs. (1) and (6) used to obtain the following expression for the excess free energy of a liquid having a model potential in $(6-s)$:

$$\psi = 4\pi R^2 \cdot \frac{\pi n_1^2 B}{24a^2} \left\{ (1-\theta_2)\left(1-\frac{a}{R}\right) + \left[\left(1-\frac{a}{R}\right)\left(\frac{a}{2R-a}\right)^2 - \right.\right.$$

$$\left. - \frac{\theta_2}{s-6}\frac{a^2}{R^2} - \frac{a^2}{R^2}\ln\frac{2R-a}{a}\right] - \theta_2\left(1-\frac{a}{R}\right)\left(\frac{a}{2R-a}\right)^{s-4} +$$

$$\left. + \frac{\theta_2}{s-6}\frac{a^2}{R^2}\left(\frac{a}{2R-a}\right)^{s-6} \right\},$$

(13)

with

$$\theta_2 = \frac{q(q-2)(q-4)}{s(s-2)(s-4)}\,.$$

(13')

Powers of a/R higher than the third can be neglected in Eq. (13) since $R \geq R_0 > a$; furthermore, calculation shows that the terms in square brackets are practically insignificant. Equation (13) therefore simplifies to

$$\psi = 4\pi R^2 \frac{(1-\theta_2)\,\pi n_1^2 B}{24a^2}\left(1-\frac{a}{R}\right).$$

(14)

This result is consistent with the thermodynamically derived conclusions concerning the $\psi = \psi(g)$ curve which were referred to above. The lack of experimental data precludes the possibility of a more detailed comparison. We have therefore checked the method through calculations on the splitting pressures of thin films, the results obtained being in agreement with those of other investigators [11-13].

Indirect confirmation of the validity of the derived relation can be reached somewhat differently.

The excess free energy of the drop is given by $\psi = 4\pi R^2\sigma_\infty$ when $R \gg a$ and the expression for the specific free surface energy of the bulk liquid then becomes

$$\sigma_\infty = (1-\theta_2)\,\pi n_1^2 \cdot B/24a^2\,,$$

(15)

---

*The coefficient $\frac{1}{2}$ arises from the fact that each particle is counted twice in the complete integral.

30

TABLE 3

Liquid	Working temp., °K	σ_∞, erg/cm²	λ_∞, J/g	δ, g/cm³	a, A [from Eq. (19)]	a, A (for the gas)	a/a_g
Carbon tetra- chloride	293	27.0	213	1.59	4.942	5.881	0.84
n-Octane	313	19.8	297	0.62	6.694	7.451	0.90
Benzene	273	31.6	448	0.90	4.879	5.430	0.90
n-Hexane	273	20.5	373	0.68	5.033	5.909	0.85
n-Butane	253	17.3	398	0.60	4.509	4.997	0.90
Cyclohexane	333	21.3	358	0.73	5.075	6.093	0.83
Carbon disulfide	273	35.3	374	1.29	4.555	4.438	1.02
Argon	87	12.0	158	1.37	3.451	3.465	0.99

expressing the particle concentration through Avogadro's number, N, and the molar volume, M, of the liquid, one has

$$\sigma_\infty = \frac{(1-\theta_2)\,\pi}{24}\left(\frac{N\delta}{M}\right)B/a^2.$$

(16)

This last expression has been used in calculating specific free surface energies for the six nonpolar liquids of Table 1. These calculations were based on the a and B values of [6], the latter being increased by a factor of 2.25; other values were taken from handbook data [14]. It is seen from Table 2 that the calculated results are in good agreement with the experimental data. This agreement could be improved still further by making use of reduced a values.

These results can be used and the free excess energy of a droplet of nonpolar liquid with the model potential of (2) approximated through the expression

$$\psi \cong 4\pi R^2 \cdot \sigma_\infty \left(1 - \frac{a}{R}\right).$$

(17)

The derived σ vs R relation,

$$\sigma = \sigma_\infty \left(1 - \frac{a}{R}\right),$$

(18)

is equivalent to the well-known Tolman Equation and identical with that established earlier in [15, 16].

It has been pointed out above that liquid phase a and B values are somewhat different from the accepted vapor phase values. Equations (8) and (16) can be used to obtain reduced values of these parameters. This two-equation system leads to the following expression for the reduced value of a:

$$a = \frac{24\theta_1}{1-\theta_2}\frac{\sigma_\infty}{\lambda_\infty n_1};$$

passage to molar quantities gives

$$a = \frac{24\theta_2}{1-\theta_2}\frac{\sigma_\infty M}{\lambda_\infty^{mol}\delta},$$

(19)

λ_∞^{mol} being the molar heat of vaporization of the bulk liquid.

Values of the parameter a obtained through Eq. (19) are given in Table 3 to illustrate this point. The calculations in question here were based on the model potential of (6)-(9) since this gives the best concordance between reduced and tabulated a values. Values of a for the gaseous state are given for comparison [10].

Conclusions

1. The Landau—Lifshits thermodynamic perturbation theory has been used to reduce the determination of the excess free energy of a small object to a calculation of the excess potential energy from the bulk liquid distribution function.

2. Derivation has been given of an equation relating the heat of vaporization of the bulk liquid and the model potential constants for liquid particle interaction.

3. Derivation has also been given of an approximation formula

$$\psi = 4\pi R^2 \cdot \sigma_\infty \left(1 - \frac{a}{R} \right)$$

which can be used to determine the excess free energy of small droplets of nonpolar liquid.

4. A method has been suggested for evaluating the parameter a from the heat of vaporization and surface tension of the bulk liquid, two directly measurable quantities.

The author would like to conclude by expressing his thanks to B. V. Deryagin, Corresponding Member, Academy of Science, USSR, for discussions and valuable advice.

Literature

1. L. M. Shcherbakov. Trudy Tul'sk. Mekhan. Inst. 7:117 (1955).
2. L. M. Shcherbakov. In collection: Research in Surface Forces, Moscow, Izd. Akad. Nauk SSSR (1961), p. 28. [English translation: Consultants Bureau, New York (1963)].
3. L. M. Shcherbakov. Kolloidn. Zh. 23:215 (1961).
4. N. N. Bogolyubov. Problems of Dynamic Theory in Statistical Physics, Moscow—Leningrad, Gostekhizdat (1946).
5. I. Z. Fisher. The Statistical Theory of Liquids, Moscow, Gosfizmatizdat (1961).
6. G. Hirschfelder, C. Curtis, and R. Byrd. The Molecular Theory of Gases and Liquids, Moscow, IL (1961).
7. L. Landau and E. Lifshits. Statistical Physics, Moscow—Leningrad, Gostekhizdat (1951), p. 112.
8. E. A. Melwyn-Hughes. Physical Chemistry, Vol. 1, Moscow, IL (1962), p. 269.
9. T. Hill. Statistical Mechanics. Moscow, IL (1960), p. 216.
10. L. M. Shcherbakov, P. P. Ryazantsev, and N. P. Filippova. Kolloidn. Zh. 23:338 (1961).
11. B. V. Deryagin and S. V. Nerpin. Dokl. Akad. Nauk SSSR 99:1029 (1954).
12. S. V. Nerpin and B. V. Deryagin. Dokl. Akad. Nauk SSSR, 100:17 (1955).
13. I. E. Dzyaloshinskii, E. M. Lifshits, and L. P. Pitaevskii. Zh. Eksperim. i Teor. Fiz. 37:229 (1959).
14. L. K. Martens (editor). Technical Encyclopedia, A Handbook of Physical, Chemical, and Technological Data, Vols. 1, 2, 5, 7, 10, Moscow (1929-1930).
15. L. M. Shcherbakov. Kolloidn. Zh. 14:379 (1952).
16. L. M. Shcherbakov and A. S. Bolotin. Uch. Zap. Kishinevsk. Gas. Univ. 11:153 (1954).

THE EFFECT OF THE ENERGY OF THE WETTING PERIMETER ON THE BOUNDARY CONDITIONS

L. M. Shcherbakov and P. P. Ryazantsev

Tula Polytechnic Institute

Crystal thermodynamics [1, 2] considers not only the surface energy but also a linear edge energy. This concept comes into play, for example, in the theory of nucleation and crystal growth. It would be natural then to suppose that an excess free energy would also be associated with a wetting perimeter. Interest thus attaches to a study of the effect of the linear energy of wetting on the equilibrium value of the contact angle.

It has been shown earlier [3] that incomplete wetting is most pronounced when the surface of the wetted body is covered by a surface phase in the form of an oriented, single phase, polymolecular layer. Experimental determination of liquid viscosities by the blowing method [4, 5] indicate that an oriented surface phase of this type persists even after contact with the bulk liquid has been established. Our study of the drop — surface phase equilibrium therefore assumes that the drop rests directly on a surface phase.

Theoretical treatment of the boundary conditions is generally based on what is essentially the Gibbs—Curie principle of minimizing the free surface energy at constant drop volume. Application of this principle assumes the drop to be isolated, thus excluding the possibility of mass interchange with the surrounding medium or the oriented layer covering the solid surface. It is scarcely likely, however, that such conditions are actually realized in practice. Derivation of the boundary conditions is therefore based on the general principles of thermodynamic equilibrium, e.g., on a minimization of the thermodynamic potential.

Let us now consider a drop (subscript 1) — oriented surface layer on nondeformable solid body — vapor (subscript 2) system. The various parts of this system will be supposed delineated by equimolecular separating surfaces so chosen as to eliminate the possibility of autoadsorption. It will also be assumed that the drop is large enough to justify neglect of Type II capillary effects.* The variable portion of the thermodynamic potential of this system can be represented by

$$\Phi = \mu_1 g + \mu_1 \Gamma A + \mu_2 (N - g - \Gamma A) + \sigma_\infty S + \omega' Q + \omega (A - Q) + \varkappa L. \qquad (1)$$

Here μ_i is the chemical potential (per particle) of the i-th bulk phase,† N is the total number of particles in the system, g is the number of particles in the drop, σ_∞ is the specific free surface energy, S is the free surface area of the drop, Γ is the "depth" of the oriented polymolecular layer, expressed as the number of particles per unit surface area, ω is the specific excess free energy, $A - Q$ is the surface area of the free portion of the oriented layer, ω' and Q are the same, but evaluated for the portion of the surface phase in contact with the drop, $\varkappa$ is the linear free energy density, and, L is the length of the wetting perimeter.

By varying Φ with p,T, N,A = const, setting $\delta\Phi$ equal to zero, and carrying out certain simple rearrangements, one obtains

$$\left. \begin{array}{c} \dfrac{\partial\omega}{\partial\Gamma} = \dfrac{\partial\omega'}{\partial\Gamma} , \\[2mm] \delta\left[(\mu_1 - \mu_2)g + \sigma_\infty S + (\omega' - \omega)Q + \varkappa L\right] = 0. \end{array} \right\} \qquad (2)$$

---

* These effects have been treated in [3, 6].

† It is to be noted that $\mu_{drop} = \mu_{layer} = \mu_1$ [3].

Taken together, these equations and supplementary conditions fix the equilibrium state of the system. The first of these equations expresses the physically obvious fact that the splitting pressure must be constant throughout the equilibrated base layer (the same for the free portion, as for the portion in contact with the drop).

The second equation of System (2) is an instance of the rather complex Bolza problem from the calculus of variations [7]. The present problem can be considerably simplified by drawing on the obvious symmetry of the drop with respect to the vertical axis. With this in mind, we now introduce a system of cylindrical coordinates. Let $z = z(r)$ be the equation of the generating curve for the free surface of the drop. The area of the surface and base of the drop, its volume, and wetting perimeter can then be written as

$$S = 2\pi \int_0^{r_0} r \sqrt{1 + p^2}, \quad Q = 2\pi \int_0^{r_0} r\, dr, \quad V_1 = 2\pi \int_0^{r_0} zr\, dr, \quad L = 2\pi \int_0^{r_0} dr,$$

p being the partial derivative $\partial z / \partial r$, and r_0 the radius of the wetting perimeter.

Introduction of these expressions carries the second equation of System (2) over to

$$\delta \int_0^{r_0} \left[\frac{\mu_1 - \mu_2}{v_1} zr + \sigma_\infty r \sqrt{1 + p^2} + (\omega' - \omega) r + \varkappa \right] dr = 0,$$

(3)

v_1 being the specific volume per liquid particle.

The resulting variation problem (known as "the variable ends problem") can be reduced to solution of the Ostrogradskii—Euler Equation:

$$\frac{\partial \psi}{\partial z} - \frac{d}{dr} \frac{\partial \psi}{\partial p} = 0,$$

(4)

ψ being the integrand of Eq. (3).

The solution required here must satisfy a condition of transversality

$$\psi_\Pi - \left(\frac{\partial \psi}{\partial p} \right)_\Pi \cdot p_\Pi = 0$$

(5)

at the point $z_\Pi = 0$ on the wetting perimeter. Substitution of the expression for ψ into the Ostrogradskii—Euler Equation, followed by certain rearrangements, carries (5) over to the familiar equation for the free surface area of the drop. This indicates that the linear energy of the wetting perimeter has no effect on the form of the free surface of the drop.

Substitution of the expression for ψ into the condition for transversality, leads to

$$\sigma_\infty r_0 \sqrt{1 + p_\Pi^2} + (\omega' - \omega) r_0 + \varkappa - \sigma_\infty r_0 \frac{p_\Pi^2}{\sqrt{1 + p_\Pi^2}} = 0$$

or

$$\frac{1}{\sqrt{1 + p_\Pi^2}} = \frac{\omega - \omega'}{\sigma_\infty} - \frac{\varkappa/\sigma_\infty}{r_0}.$$

(6)

$1/\sqrt{1 + p_\Pi^2}$ is, however, the cosine of the angle between the Z axis and the normal to the free drop surface. This angle becomes equal to the contact angle, θ, (i.e., the angle between the meniscus surface and the surface of the oriented surface phase) at points on the wetting perimeter. From this it follows that

$$\cos \theta = \frac{\omega - \omega'}{\sigma_\infty} - \frac{\varkappa/\sigma_\infty}{r_0}.$$

(7)

The last member of this equation approaches zero when $r_0 \to \infty$, and

$$\cos \theta_\infty = \frac{\omega - \omega'}{\sigma_\infty}.$$

It is then possible to rewrite Eq. (7) as

$$\cos \theta = \cos \theta_\infty - \frac{\varkappa/\sigma_\infty}{r}.$$

(8)

The results obtained here indicate that introduction of the linear free energy of the wetting perimeter will diminish $\cos \theta$ and thereby increase the contact angle itself. The fact that $\varkappa \ll \sigma_\infty$ suggests that such effect would be observable only in very small drops, although it would probably be masked by Type II capillary effects.

Literature

1. V. K. Semenchenko. Surface Effects in Metals and Alloys, Moscow, Gostekhizdat (1957).
2. O. M. Poltorak, Zh. Fiz. Khim. 34:3 (1960).
3. B. V. Deryagin and L. M. Shcherbakov. Kolloidn. Zh. 23:40 (1961).
4. B. V. Deryagin and V. V. Karasev. Dokl. Akad. Nauk SSSR 101:289 (1955).
5. V. V. Karasev and B. V. Deryagin. Zh. Fiz. Khim. 33:100 (1959).
6. L. M. Shcherbakov and P. P. Ryazantsev, Zh. Fiz. Khim. 34:2120 (1960).
7. G. A Bliss. Lectures on Calculus of Variations, Moscow, Gostekhizdat (1950), p. 222.

THE THEORY OF THE FLOTATIONAL RUPTURE OF WETTING FILMS AND ITS APPLICATION TO THE KINETICS OF FLOTATIONAL ADHESION

V. V. Deryagin and Yu. V. Gutop

Institute of Physical Chemistry, Acad. Sci., USSR
Laboratory for Surface Phenomena

The adhesion of mineral particles and gas bubbles in flotational ore enrichment involves rupture of the water films separating particles and bubbles with establishment of a nonzero contact angle. We propose to develop the theory of this process. Our treatment will be simplified by assuming the liquid film to rest on an infinitely extended plane base. We will also assume that the film behaves like a two-dimensional system in rupture, the original film depth being maintained all around the "hole." Behavior of this kind would be most nearly reproduced in monomolecular films.

The work of D. M. Tolstoi [1] will be drawn on in treating the liquid flow associated with the widening of the hole in the ruptured film. Here it was shown that the flow of a viscous liquid over a solid base could be accompanied by liquid slippage at the solid interface, this movement being impeded by a frictional force proportional to the slip rate,

$$f = kv \tag{1}$$

(v is the liquid velocity and k the coefficient of friction for the base).

Determination of the coefficient of friction, k, of Eq. (1) is through the expression

$$k = \frac{\eta_0}{a} \cdot \frac{1}{e^{\frac{1}{\theta} S_m \sigma_{12}(1-\cos\varphi)/\theta} - 1} \cdot \tag{2}$$

Here η_0 is the bulk velocity of the liquid, a is the mean separation of the molecules, S_m is the surface area of the sphere bounding the microvoid, calculated per molecule ($S_m = \sim \pi a^2$), σ_{12} is the surface tension at a vacuum interface, φ is the contact angle established on the base, and θ is the product of the absolute temperature and the Boltzmann constant.

The essential distinction between a liquid film on a solid base and an analogous free film is in the special meaning attaching to the film tension, p_0, which figures in the expression $p_0 S$ for the work of forming a hole of area S (we take no account of the work required for forming the hole perimeter). The work of hole formation can be expressed through the surface tensions at the phase interfaces. Thus the work of hole formation in the monomolecular film is given by $(\sigma_{32} - \sigma'_{32})S$, σ_{32} being the surface tension of the clean solid base, and σ'_{32} the surface tension of this same base carrying the monomolecular film.

From this it follows that

$$p_0 = \sigma'_{32} - \sigma_{32}. \tag{3}$$

A first approximation to the work of hole formation in a polymolecular film of depth h which shows two-dimensional behavior in rupture can be obtained in terms of the surface tensions, σ_{31} and σ_{12}, for the upper and lower film faces, respectively. This work is

$$\left[(\sigma_{32} - \sigma_{31} - \sigma_{12}) - \int hd\Pi(h) \right] \cdot S,$$

$\Pi(h)$, the film splitting pressure, being a function of the film depth

$$p_0 = \sigma_{31} + \sigma_{12} - \sigma_{32} + \int hd\Pi(h).$$

The right-hand member of this equation can be expressed in terms of the contact angle, φ, and an equation for p_0 obtained:

$$p_0 = \sigma_{12}(1 - \cos\varphi) + \int_h^\infty hd\Pi(h). \tag{3'}$$

The integral of the splitting pressure can be neglected in a first approximation, so that

$$p_0 = \sigma_{12}(1 - \cos\varphi). \tag{3''}$$

The value of σ_{12} for water will be assumed to be ≈ 70 dyne/cm in subsequent calculations. When so formulated, this problem can be treated by those methods which were used in [2] to obtain an expression for the mean life of the free thin film in fluctuational rupture. The calculation of life of the film on the base is completely analogous to the free film calculations and will not be reproduced here for this reason. We can at once write down an expression for the film on the base which is entirely analogous to the corresponding expression for the free film [2], namely:

$$\alpha = \frac{\sqrt{\theta}e^{\frac{\pi/\gamma^2}{p_0\theta}}}{2Dc_0\sqrt{p_0}}\left[1 + A\left(\gamma\sqrt{\frac{\pi}{p_0\theta}}\right)\right]. \tag{4}$$

Here, p_0 is the film tension at a point far removed from the rupture, this calculated from Eq. (3), α is the mean life time of the film, γ is the linear tension of the hole perimeter, c_0 is the concentration of holes of molecular radius, per unit radius interval and 1 cm^2, A is the error integral, and D is a coefficient formally analogous to a diffusion coefficient (see [2] for additional details).

Because of the frictional force at the base, this coefficient cannot be evaluated from the expression applying to the free film ($D_0 = \theta/4\pi\eta$, with η representing the two-dimensional film shear viscosity). The D coefficient for the film on the base will be calculated below.

The equation for two-dimensional liquid flow with a frictional force on the base has the form (sonic approximation)

$$\frac{\partial p}{\partial t} + \rho_0 c^2\left(\frac{\partial v}{\partial r} + \frac{v}{r}\right) = 0, \tag{5'}$$

$$-\frac{\partial p}{\partial r} + (\zeta + 4/3\eta)\left(\frac{\partial^2 v}{\partial r^2} + \frac{1}{r}\frac{\partial v}{\partial r} - \frac{v}{r^2}\right) - kv = 0. \tag{5''}$$

It is assumed, as in [2], that the velocity v depends only on the distance, r, from the center of the hole; ζ designates the second two-dimensional viscosity of the liquid, a quantity which comes into play in the uniform expansion of the film in its own plane; ρ_0 is the liquid density, c is the velocity of sound ($c^2 = \partial p/\partial\rho$), and, p is the two-dimensional film pressure. The treatment of the liquid movement will once again be a quasi-stationary approximation with $\partial v/\partial t = 0$ (a solution can be obtained with this member retained, but the problem is much more complex).

TABLE 1

β	δ	$y=10^{-3}$	$y=10^{-2}$	$y=10^{-1}$	$y=1$	$y=10$
1	3/2	1.00	0.98	0.98	0.58	0.170
1	1	1.00	0.97	0.85	0.49	0.130
1	1/2	0.99	0.95	0.75	0.34	0.070
1	1/10	0.97	0.80	0.38	0.10	0.015
0.1	3/2	1.00	0.98	0.84	0.48	0.090
0.1	1	0.99	0.97	0.79	0.38	0.055
0.1	1/2	0.99	0.94	0.68	0.23	0.034
0.1	0.1	0.96	0.78	0.34	0.05	0.005
0.01	1.5	0.99	0.96	0.79	0.32	0.050
0.01	1	0.99	0.95	0.75	0.24	0.040
0.01	0.5	0.99	0.92	0.59	0.14	0.0098
0.01	0.1	0.95	0.83	0.24	0.04	0.003
0.001	1.5	0.99	0.96	0.74	0.25	0.03
0.001	1.0	0.99	0.95	0.66	0.18	0.02
0.001	0.5	0.97	0.90	0.51	0.11	0.01
0.001	0.1	0.94	0.67	0.18	0.02	0.0029

The requirement of equality of forces on both sides of the perimeter of the expanding hole is expressed by the boundary condition

$$\left[p - 2\eta \frac{\partial v}{\partial r} + \frac{\tau}{R} + (2/3\eta - \zeta)\left(\frac{\partial v}{\partial r} + \frac{v}{r}\right) \right]_{r=R} = 0,$$

(6)

R being the hole radius.

A solution to Eq. (5) in the form

$$v = V \frac{K_1(\varkappa r)}{K_1(\varkappa R)}$$

(7)

will now be sought. Here V is the rate of expansion of the newly formed hole, $\varkappa$ is an unknown flow parameter having the dimensions of a reciprocal length, and K_1 is a Bessel function of several variables which diminishes exponentially at large distances (Macdonald function).

Substituting from (7) into the continuity condition of (5'), differentiating, and then drawing on the recurrence formula for the Bessel function,* gives

$$\frac{\partial p}{\partial t} = \rho_0 c^2 \varkappa \cdot \frac{V}{K_1(\varkappa R)} K_0(\varkappa r).$$

Integration of this expression with the condition that $p = -p_0$ when $r \to \infty$ gives

$$p = -p_0 + \rho_0 c^2 \cdot \varkappa \cdot K_0(\varkappa r) \int_{R_c}^{R} \frac{dR}{K_1(\varkappa R)},$$

(8)

---

* $\dfrac{1}{r}\dfrac{d}{dr}[rK_1(\varkappa r)] = -\varkappa K_0(\varkappa r)$.

TABLE 2

$\varphi°$	P_0, dyne/cm	$\gamma=2\cdot10^{-5}$	$\gamma=10^{-5}$	$\gamma=0.7\cdot10^{-5}$	$\gamma=0.5\cdot10^{-5}$	$\gamma=0.3\cdot10^{-5}$	$\gamma=0.1\cdot10^{-5}$
0	0	∞	∞	∞	∞	∞	∞
55	30	$4.2\cdot10^{-11}$	$1.05\cdot10^{-11}$	$0.5\cdot10^{-11}$	$2.5\cdot10^{-12}$	$0.9\cdot10^{-12}$	$0.1\cdot10^{-12}$
82	60	$2.1\cdot10^{-11}$	$0.5\cdot10^{-11}$	$2.5\cdot10^{-12}$	$1.2\cdot10^{-12}$	$0.5\cdot10^{-12}$	$0.5\cdot10^{-13}$
106	90	$1.4\cdot10^{-11}$	$3.5\cdot10^{-12}$	$1.7\cdot10^{-12}$	$0.8\cdot10^{-12}$	$0.3\cdot10^{-12}$	$0.3\cdot10^{-13}$
135	120	10^{-11}	$2.5\cdot10^{-12}$	$1.2\cdot10^{-12}$	$0.6\cdot10^{-12}$	$0.2\cdot10^{-12}$	$0.2\cdot10^{-13}$

R_C being the critical hole radius [1]. The expressions for v and p are now substituted into the Navier—Stokes Equation, (5"). Factoring out the common spatial part of this expression, one has

$$\left[\rho_0 c \varkappa^2 \int_{R_c}^{R} \frac{dR}{K_1(\varkappa R)} + \frac{\varkappa^2 (\zeta + \frac{4}{3}\eta) V}{K_1(\varkappa R)} - \frac{kV}{K_1(\varkappa R)} \right] K_1(\varkappa r) = 0.$$

The rate of expansion of the newly formed hole can be obtained by setting the polynomical in square brackets equal to zero. The region of interest for the V(R) function is that in which R is close to R_C. Strictly speaking a value of the coefficient D is obtained only for the point $R = R_C$. The value of D at R_C can be used in place of D(R) without any considerable error, however, the exponential factor in Eq. (10) of [2] rapidly falling off to zero at distances larger than R_C. V can therefore be expressed as a power series in $R - R_C$, and this series broken off with the first member, to obtain

$$V = \frac{\rho_0 c^2 \varkappa^2}{k - \varkappa^2 (\zeta + {}^4/_3\eta)} (R - R_c).$$
(9)

A second relation of this type can be obtained from the boundary condition (6)

$$V = \frac{p_0 - \rho_0 c^2 \cdot \varkappa R_c \dfrac{K_0(\varkappa R_c)}{K_1(\varkappa R_c)}}{2\eta + \varkappa R_c (\zeta + {}^4/_3\eta) \dfrac{K_0(\varkappa R_c)}{K_1(\varkappa R_c)}} R - R_c.$$
(10)

The equating of right-hand members of (9) and (10), gives an expression $x = \varkappa R_c$:

$$\frac{x^2}{y - x^2} \cdot \frac{\beta - x\dfrac{K_0}{K_1}}{\delta + x\dfrac{K_0}{K_1}}.$$
(11)

The following changes of variable have been carried out here:

$$y = \frac{kR_c^2}{\zeta + {}^4/_3\eta}, \quad \beta = \frac{p_0}{\rho_0 c^2}, \quad \delta = \frac{2\eta}{\zeta + {}^4/_3\eta}.$$

All quantities entering (11) are dimensionless.

A somewhat more useful form of (11) can be obtained by simple transformations:

$$\beta - \frac{\beta + \delta}{y} x^2 = x\frac{K_0}{K_1}.$$
(11')

39

A calculation similar to that of [2] now leads from the velocity V, and the Einstein Equation to the "diffusion coefficient" D:

$$D = \frac{\theta}{4\pi\eta} \cdot \frac{\delta}{\beta} \cdot \frac{\beta - x\dfrac{K_0}{K_1}}{\delta + x\dfrac{K_0}{K_1}}.$$

(12)

The value of x in question here is that satisfying Eq. (11').

The effect of the frictional force at the base is shown by the factor multiplying $\theta/4\pi\eta$ in the expression for D:

$$M = \frac{\delta}{\beta} \frac{\beta - x\dfrac{K_0}{K_1}}{\delta + x\dfrac{K_0}{K_1}}.$$

(13)

Equation (11') is solved numerically. The results of such solution are presented in Table 1.

Table 1 shows, just as was to be expected, that the mean life of the film increases with the frictional force at the base (it is to be recalled that the mean life is inversely proportional to M). It can, however, be seen from Eq. (2) that there are various possibilities for an increase in the frictional force. In particular, the frictional force will increase as the contact angle, φ, approaches zero. Here the increase in mean life is also due in part to an increase in the exponent in Eq. (4), the film tension, p_0, being related to the contact angle through Eq. (3").

M differs but little from 1 when the frictional force is low, being a function of all of the other parameters which enter (11'). The figures of our Table 1 make it clear that the life of the film at fixed y and β increases with the second viscosity. This is easily understood, since the presence of the second viscosity leads, speaking crudely, to a force which opposes the stretching forces, p_0, and thereby impedes expansion of the newly formed hole.

The kinetics of the adhesion of mineral particles to gas bubbles has been experimentally studied in [3]. An exponential relation between α and a quantity inversely proportional to the temperature was obtained for many minerals and various liquid media. These results are in good agreement with Eq. (4), since the quantities standing before the exponential in this equation are temperature functions which vary only slowly in comparison with the exponential. Table 2 gives values of the exponent of Eq. (4), each multiplied by θ, for various values of the other parameters in the equation.

The values of p_0 shown in this table were calculated from Eq. (3") with $\sigma_{12} \approx 70$ dyne/cm. The figures shown for $\gamma = 0.3 \cdot 10^{-5}$-$10^{-5}$ agree well with the apparent activation obtained in [3]; with water films, the alteration for various minerals is of the order of several units times 10^{-12}.

The author would like to conclude by expressing his thanks to G. A. Martynov for his useful discussions.

Conclusions

1. An expression has been obtained for the mean life of the thin film, allowance being made for the existence of liquid slippage at the interface and a frictional force on the solid base.

2. The mean life of the film increases with the frictional force, all other conditions being held constant.

Literature

1. D. M Tolstoi, Doctoral dissertation, Moscow, Inst. Fiz. Khim., Akad. Nauk SSSR (1953).
2. B. V. Deryagin and Yu. V. Gutop, Kolloidn. Zh. 24(4) (1962).
3. M. A. Eigeles, Dokl. Akad. Nauk SSSR 129(1):177 (1959).

SECTION II

ELECTROSURFACE FORCES

THE IONIC – ELECTROSTATIC COMPONENT
OF THE SPLITTING PRESSURE

S. V. Nerpin and N. F. Bondarenko

Leningrad Agrophysics Institute

The Ionic – Electrostatic Field at an Interface in an Electrolytic Solution

The theory of the diffuse double layer has been developed by Gouy [1] and Chapman [2].

This theory draws on: 1) the Poisson Equation, to relate field potential and charge density,

$$\Delta \psi = -\frac{4\pi}{D}\,\rho,$$

(1)

(ψ is the field potential, D is the dielectric susceptibility, and ρ is the charge density), and 2) the Boltzmann Equation, to describe the diffuse distribution of ions under the action of the electric field at the interface.

Earlier work on the ion distribution in solution [1–4] assumed a distribution law of the same form as that applying to gas molecules in an external field, no account being taken of effects arising from the medium itself. It would seem worthwhile to adopt a more general approach to the theory of the diffuse double layer in which both the energy of the solvent molecules in the external field, and the effect of the molecular force field on the charged particle distribution would be taken into account.

In the presence of an external field, the chemical potential, μ', of solute molecules (or ions) can be written as [5]

$$\mu' = kT \ln C + f(p, T) + u_u(x, y, z) = \text{const.}$$

(2)

Here, $C = n_u/n_v$, n_u and n_v being the numbers of molecules of solute and solvent, respectively; f is a function depending only on the pressure (p) and temperature (T); u_u is the potential energy, a quantity essentially fixed by external fields; and, k is the Boltzmann constant.

By differentiating the expression of (2) with respect to x, the normal to the interfacial surface, drawing on the fact that thermodynamic equilibrium requires constancy of temperature, T, and assuming that

$$\frac{\partial u_u}{\partial z} = \frac{\partial u_u}{\partial y} = 0,$$

one finds

$$\frac{kT}{C}\frac{dC}{dx} + \frac{\partial f}{\partial p}\frac{dp}{dx} + \frac{du_u}{dx} = 0.$$

(3)

Since the thermodynamic potential of the solution can be expressed as [5]

$$\Phi_s = n_c\mu_0(p, T) + nkT \ln \frac{n_u}{n_v} + n_u f(p, T)$$

(4)

(μ_0 is the chemical potential of the pure solvent), and

$$\frac{\partial \Phi_s}{\partial p} = V_s,$$

(V_s is the volume of the solution), one also finds that

$$V_s = n_v \frac{\partial \mu_0}{\partial p} + n_u \frac{\partial f}{\partial p}.$$

This last expression indicates that $\partial f/\partial p$ is the volume per solute molecule, a quantity which will be designated by V_{solute}. The molecular volume, $\partial \mu_0/\partial p$, will be designated by V_C.

The condition for equilibrium of the solvent molecules, i.e., $\mu_C = \text{const}$, will now be used to developed the function $p(x)$.

It is a well-known fact that the chemical potential of the solvent of a dilute solution is expressed by [5]:

$$\mu_C = \mu_0 - kTC = \text{const}, \tag{5}$$

so that

$$\frac{d\mu_C}{dx} = \frac{d\mu_0}{dx} - kT\frac{dC}{dx} = 0. \tag{6}$$

With an external field, the differential of the chemical potential, μ_0, can be written as [5]

$$d\mu_0 = SdT + V_C dp + du_C, \tag{7}$$

or, with $T = \text{const}$,

$$\frac{d\mu_0}{dx} = V_C \frac{dp}{dx} + \frac{du_C}{dx}. \tag{8}$$

Here, u_C is the energy of the solvent molecule in the external field.

Substitution of the expression of (8) into (6) gives

$$V_C \frac{dp}{dx} + \frac{du_C}{dx} - kT\frac{dC}{dx} = 0. \tag{9}$$

Using (9), and substituting the expression for dp/dx into (3), leads to

$$\left(\frac{kT}{C} + \frac{V_u}{V_C}kT\right)\frac{dC}{dx} - \frac{du_C}{dx}\frac{V_u}{V_C} + \frac{du_u}{dx} = 0. \tag{10}$$

A $\rho(\psi)$ relation obtained from Eq. (10) can be combined with the Poisson Equation, (1), to give the system for the ionic—electrostatic field. We limit ourselves here to deriving the general form of (10) and indicating the assumptions required for passing to the $\rho(\psi)$ relation usually employed in the theory of the diffuse double layer. * The second term in the parentheses of Eq. (10) can be neglected when $C \ll (V_C/V_u)$ as is the case in dilute solutions where $C \ll 1$.

In order to obtain the usual relations, it is necessary to make a second assumption to the effect that du_C/dx is equal to zero, which is to say, that the external field has no effect on the energy of the solvent molecules, even if these are polar. Under such conditions, Eq. (10) takes the form:

---

* The authors propose to discuss removal of the limitations generally imposed on (10) in a special communication.

$$\frac{kT}{C}\frac{dC}{dx} + \frac{du_\mathrm{u}}{dx} = 0.\tag{10'}$$

The third assumption required for the development of the familiar theory of the diffuse double layer is to the effect that the molecular force field has no effect on the solute molecules or ions.

The energy of the ion in the electrical field can then be written as

$$u_\mathrm{u} = \pm\, ez_i\psi,\tag{11}$$

e being the charge on the electron and the z_i the electrovalences of the ions.

Drawing on (11), and taking account of the fact that the potential field, ψ, is equal to zero, and the ion concentration equal to the bulk concentration, C_{i0}, when $x = \infty$, one passes directly from (10') to

$$\ln\frac{C_i}{C_{i0}} = \pm\frac{\psi ez}{kT} \quad\text{or}\quad C_i = C_{i0}\exp\left(\pm\frac{\psi ez}{kT}\right).\tag{12}$$

The density, ρ_i, of ion charges of like sign at any point in the solution is fixed by the expression

$$\rho_i = \pm\, n_i z_i e,\tag{13}$$

n_i being the number of ions of given charge per unit volume of solution.

Since $n_i/n_C = C_i$, it follows that (13) can be replaced by

$$\rho_i = \pm\, C_i n_c z_i e.$$

The total ionic charge density in a binary electrolyte is given by

$$\rho = \rho_1 + \rho_2 = n_c e\,(C_2 z_2 - C_1 z_1).$$

Equation (12) can be drawn on to obtain

$$\rho = n_c e\left[z_2 C_{02}\exp\left(-\frac{\psi ez_2}{kT}\right) - z_1 C_{01}\exp\left(\frac{\psi ez_1}{kT}\right)\right].\tag{14}$$

By taking account of the fact that $C_{01} = n_{01}/n_C$ and $C_{02} = n_{02}/n_C$, the charge density can be written as

$$\rho = e\left[z_2 n_{02}\exp\left(-\frac{\psi ez_2}{kT}\right) - z_1 n_{01}\exp\left(\frac{\psi ez_1}{kT}\right)\right].\tag{15}$$

Electrical neutrality of the solute molecules requires that $z_1 n_{01} = z_2 n_{02}$, and (15) can, therefore, be rewritten as

$$\rho = ez_1 n_{01}\left[\exp\left(-\frac{\psi ez_2}{kT}\right) - \exp\left(\frac{\psi ez_1}{kT}\right)\right] = ez_2 n_{02}\left[\exp\left(-\frac{\psi ez_2}{kT}\right) - \left(\frac{\psi ez_1}{kT}\right)\right].\tag{16}$$

The numbers of ions, n_{01} and n_{02}, will be expressed in terms of the solution concentration, γ (in gram-moles), and the decomposition factors, k_{p_1} and k_{p_2}.

The number of molecules, n_m, per 1 cm^3 of solution is given by

$$n_\mathrm{m} = \gamma N,\tag{17}$$

N being the Avogadro number.

The number of molecules involved in breakdown with the formation of n_{01} and n_{02} ions per 1 cm^3 of solution can be represented by

$$n_{\mathrm{m}} = \frac{n_{01}}{k_{p_1}} = \frac{n_{02}}{k_{p_2}}.$$

(18)

Equations (17) and (18) can be combined to give

$$n_{01} = \gamma k_{p_1} N, \quad n_{02} = \gamma k_{p_2} N.$$

(18')

Substitution of the expression for ρ from (16) into (1), gives *

$$\Delta \psi = \frac{a}{2} [\exp(\psi b z_1) - \exp(-\psi b z_2)].$$

(19)

Here

$$a = \frac{8\pi}{D} e k_{p_1} z_1 \gamma N = \frac{8\pi}{D} e k_{p_2} z_2 \gamma N,$$

$$b = \frac{e}{kT}.$$

(20)

For a one-dimensional electrostatic field in the neighborhood of a plane surface, Eq. (19) is replaced by

$$\frac{d^2 \psi}{d x^2} = \frac{a}{2} [\exp(z_1 b \psi) - \exp(-z_2 b \psi)].$$

(21)

This is the electric potential equation usually used for calculating the interaction of the bounding charged surfaces of the electrolyte film [3, 4, 6].

The assumptions involved in the derivation of Eq. (21) are clearly brought out in the remarks concerning passage from (10) to (10').

The Interaction of Similarly Charged Particles

The ionic—electrostatic theory of the interaction of two similarly charged surfaces was developed by B. V. Deryagin in his studies on lyophobic colloid stability [7], and then extended in collaboration with L. D. Landau [3]. The same equation has also been used to calculate the interaction of oppositely charged surfaces [4, 6].

Other studies of ionic—electrostatic interaction are reported in the papers of A. N. Frumkin [8], E. Verwey and I. Overbeck [9], and E. Childs [10]. The condition of equality of ionic chemical potentials, in the bulk and between the films, was not rigorously imposed in any of this work.

The weak point here was that the chemical potentials of the solute molecules (or ions) were calculated for the space between the charged surfaces without taking account of the fact that equilibrium can be established only if an external force is applied to the surfaces of these particles. It will now be shown that rigid imposition of the conditions for thermodynamic equilibrium must introduce a correction term in the familiar expression for the ionic—electrostatic component of the splitting pressure.

The equilibrium conditions for a system consisting of two charged particles in an electrolytic solution can be written as

$$\mu_c = \mu_\infty,$$

(22)

---

* The values of the dielectric susceptibility is assumed constant throughout the body of the solution, and orientation of the polar molecules in the electric field is not considered in the equation for this reason.

$$\mu'_C = \mu'_\infty, \tag{23}$$

where μ_∞ and μ'_∞ are the respective chemical potentials of solvent and solute in the bulk, and μ_C and μ'_C are the corresponding potentials in the layer, the latter being corrected for the fact that an external pressure must be applied to establish equality of chemical potentials of solute and solvent in layer and in bulk.

Equation (22) can be drawn on in the general case, and the force of interaction, R_i, per unit surface area of the two films written as

$$R_i = (f_{n\,ex} - f_{n\,u}) + (p_{os.\,in} - p_{os.\,ex}), \tag{24}$$

f_n being the Maxwell stress, and p the osmotic pressure.

The subscript "in" refers to the space between the films and the subscript "ex" to the external body of electrolyte.

The symmetrical case is treated by carrying the origin of coordinates into the plane of symmetry; then $\partial\psi/dx = 0$, and the Maxwellian stresses vanish on the planes at $x = 0$ and $x = \infty$ so that Eq. (24) can be replaced by

$$R_i = p_{os.\,in.} - p_{os.\,ex} = \delta p_{os}\,. \tag{25}$$

It is a well-known fact that

$$\delta p_{os} = kT\,(n_0 - n_\infty), \tag{26}$$

n_0 being the total number of ions of both signs per unit of solution at $x = 0$ and n_∞ the same, but for $x = \infty$.

The n_0 of Eq. (26) is fixed by the equality of (23). The bulk chemical potential of the solute can, according to Eq. (2), be written as

$$\mu'_\infty = kT \ln C_\infty + f\,(p_\infty,\ T). \tag{27}$$

In the interspace where the solute is acted upon by the electrical field of the charged surfaces, the chemical potential, μ'_0, $z_1 = z_2 = z$ for the case can be written as

$$\mu'_0 = kT \ln C_0 + f\,(p_0 T) \pm z\psi_0 e. \tag{28}$$

Here, ψ_0 is the double electric layer potential at the middle of the interspace, and e is the charge on the electron.

If the pressure difference, $p_0 - p_\infty = \delta_{os}$, is assumed to be minute, and $f(p_0,\ T)$ expanded as a power series in δp_{os} with only the first two terms in this expansion retained, one has

$$f\,(p_0,\ T) = f\,(p_\infty,\ T) + \delta p_{os}\,\frac{\partial f}{\partial p}. \tag{29}$$

Here, as before,

$$\frac{\partial f}{\partial p} = V_u\,. \tag{30}$$

Equations (27-30) can now be drawn on, and (23) rewritten as

$$kT \ln C_\infty + f\,(p_\infty,\ T) = kT \ln C_0 + f\,(p_\infty,\ T) + \delta p_{os}\,V_u \mp z\psi_0 e, \tag{31}$$

from which it follows that

$$\ln \frac{C_0}{C_\infty} = \frac{-\delta p_{os}\,V_u \mp z\psi_0 e}{kT}\,. \tag{32}$$

47

For solutions of low concentration, $n_{C,o}/n_{C,\infty} \cong 1$, and (32) can be replaced by

$$n_0 = n_\infty \exp\left(\pm \frac{z\psi_0 e}{kT} \right) \exp\left(- \frac{\delta p_{os} V_u}{kT} \right).$$

(33)

Equation (33) can now be drawn on and (26) rewritten as

$$\delta p_{os} = kTn_\infty \left[\exp\left(\pm \frac{z\psi_0 e}{kT} \right) \exp\left(- \frac{\delta p_{os} V_u}{kT} \right) - 1 \right].$$

(34)

The value of $\exp(z\psi_0 e/kT)$ in (34) is determined by solving Eq. (21), which, for the case of the symmetric electrolyte, takes the form

$$\frac{d^2\psi}{dx^2} = \frac{a}{2} \left[\exp(zb\psi) - \exp(-zb\psi) \right],$$

(35)

where

$$b = \frac{e}{kT}, \quad a = \frac{8\pi}{D} en_0 z.$$

We now set

$$zb\,\psi = \psi',$$

(36)

and introduce the new variable

$$\xi = \frac{x}{\varkappa},$$

(37)

with the reciprocal depth of the ionic atmosphere

$$\varkappa = \frac{1}{d} = \sqrt{\frac{8\pi e^2 n_0 z^2}{DkT}}.$$

(37')

In view of (36) and (37), Eq. (35) can be rewritten as

$$2 \frac{d^2\psi'}{d\xi^2} = \exp(\psi') - \exp(-\psi').$$

(38)

Only those terms which depend on the counter ion need be retained (i.e., $\exp(-\psi')$ can be neglected) when $\psi' \gg 1$, as is the case when the value of the layer depth, H, is low. Equation (38) can then be replaced by

$$2 \frac{d^2\psi'}{d\xi^2} = \exp(\psi').$$

(38')

By multiplying both sides of (38') by $2(d\psi'/d\xi)d\xi$ and integrating, one obtains

$$\left(\frac{d\psi'}{d\xi} \right)^2 = \exp(\psi') + C_1.$$

(39)

The constant of integration, C_1, is evaluated from the conditions

$$x = 0, \quad \frac{d\psi'}{d\xi} = 0, \quad \psi' = \psi_0', \quad C_1 = -\exp(\psi_0').$$

(40)

Finally, (39) can be written as

$$\frac{d\psi'}{d\xi} = \sqrt{\exp(\psi') - \exp(\psi_0')}$$

(41)

and the variables separated to obtain

$$\frac{d\psi'}{\sqrt{\exp(\psi') - \exp(\psi_0')}} = d\xi.$$

(41')

Integration of (41') gives

$$\xi = 2\exp\left(-\frac{\psi_0'}{2}\right) \operatorname{arctg} [\exp(\psi' - \psi_0') - 1]^{1/2} + C_2.$$

(42)

For x = H/2, $\psi' = \psi_1'$, $\xi = \xi_1$

$$C_2 = \xi_1 - 2\exp\left(-\frac{\psi_0'}{2}\right) \operatorname{arctg} [\exp(\psi_1' - \psi_0') - 1]^{1/2}.$$

(43)

Let us set x = 0. Then $\psi' = \psi_0'$. For the case $\psi_1' \gg \psi_0' \gg 1$

$$\operatorname{arctg} [\exp(\psi_1' - \psi_0') - 1]^{1/2} \to \frac{\pi}{2}.$$

After considering the assumptions involved here, and allowing for (43), one finds that (42) can be replaced by

$$\frac{H}{2\varkappa} = \exp\left(-\frac{\psi_0'}{2}\right) \pi.$$

(44)

Replacing $\varkappa$ by the expression from (37'),

$$\frac{1}{\varkappa^2} = \frac{DkT}{8\pi e^2 n_0 z^2},$$

and raising both members of (44) to the second power,

$$\frac{\pi^2}{\exp(\psi_0')} = \frac{H^2 8\pi e^2 n_0 z^2}{4DkT}$$

(45)

gives

$$\exp(\psi_0') = \exp\left(\frac{z\psi_0 e}{kT}\right) = \frac{\pi^2 DkT}{2H^2 e^2 n_0 z^2}.$$

(46)

The unit of Eq. (34) can be neglected in the present case where $\psi_1' \gg \psi_0' \gg 1$. Then substitution of the expression for $\exp(z\psi_0 e/kT)$ in (34) leads to

$$\delta p_{OS} = \frac{\pi}{2} D \left(\frac{kT}{zHe}\right)^2 \exp\left(-\frac{\delta p_{OS} V_u}{kT}\right).$$

(47)

The equation of I. Langmuir [11],

$$\delta p_{OS} = R(H) = \frac{\pi}{2} D \left(\frac{kT}{He}\right)^2,$$

(48)

is the special case of (47) applying when z = 1 and the value of $\delta p_{OS} V_u/kT$ is low.

Literature

1. G. Gouy. J. Phys. 9:457 (1910); Ann. Phys. 7:129 (1917).
2. D. L. Chapman. Phil. Mag. 25:475 (1913).
3. B. V. Deryagin and L. Landau. Zh. Eksperm. i Teor. Fiz. 2:802 (1942); 15:662 (1945).
4. B. V. Deryagin and V. G. Levich. Dokl. Akad. Nauk SSSR 98:985 (1954).
5. L. Landau and E. Lifshits. Statistical Physics, Moscow, Gostekhteoretizdat (1951).
6. B. V. Deryagin. Kolloidn. Zh. 16:6 (1954).
7. B. V. Deryagin, Izd. Akad. Nauk SSSR, Ser. Khim. No. 5:1153 (1937).
8. A. N. Frumkin and A. Gorodetskaya. Zh. Fiz. Khim. 12(5-6) (1938).
9. E. Verwey and I. Overbeck. Theory of the Stability of Liophobic Colloids, New York (1948).
10. E. C. Childs. Trans. Faraday Soc. 50:1356 (1954).
11. I. Langmuir. J. Chem. Phys. 6:873 (1938).

THE ROLE OF DIFFUSIONAL PHORESIS IN FILM FORMATION BY IONIC DEPOSITION FROM RUBBER LATEXES

A. A. Korotkova

S. V. Lebedev Scientific-Research Institute for

Synthetic Rubber, Leningrad

The theory of diffusional phoresis developed by B. V. Deryagin and S. S. Dukhin [1] explains film formation by ionic deposition in the following terms. Immersion of a plate covered with a layer of electrolyte into a latex sets up a diffusional field with a linear profile of ion concentrations directed from the electrolyte surface into the latex body. The electrical field and field of chemical potentials which arise from this electrolyte ion concentration gradient induce diffusional phoresis in an entering particle of the rubber latex, this movement usually being such as to carry the particle toward the surface of the electrolyte. The theory of diffusional phoresis thus accounts for an accumulation of latex particles at the deposition surface with an increase in the concentration of latex gel as compared with the concentration in the bulk.

Clear-cut instances of diffusional phoresis mechanisms of film formation by ionic deposition are met in cases where the deposition of latex particles on the surface is not accompanied by gel formation. Thus, a deposit obtained from synthetic SKS-ZOU latex is dispersed and carried back into the latex by introducing an emulsifier into the solution.

Ionic deposition is widely employed in preparing rubber goods from latexes. Here, it is the practice to work under such conditions that deposition is accompanied by gel formation, the latex particles irreversibly adhering to a film of predetermined physicomechanical properties. Until recently, ionic deposition has been considered as a slow latex coagulation resulting from the action of electrolyte ions which have diffused from the deposition surface into the latex body [2-4]. This view is not consistent with the actual mechanism of the process.

The theory of diffusional phoresis gives the following expression for the amount of deposition, Γ, when the conditions are such that: 1) the latex concentration is low so that there is no possibility of particle interaction; 2) the deposition surface is an infinitely extended plane so that the depth of the film is small in comparison with its linear dimensions; and 3) the time of deposition is limited so that the diffusional field is quasi-homogeneous:

$$\Gamma = \frac{2\rho_0 RT}{\eta}\left(\frac{z^+\xi^-}{D^-}+\frac{z^-\xi^+}{D^+}\right)(C_0-C_1)\frac{(D'_{\text{eff}}\tau)^{1/2}}{\pi^{1/2}\left[1+\left(\frac{D'_{\text{eff}}}{D_{\text{eff}}}\right)^{1/2}\right]}.$$

Here ρ_0 is latex concentration, g/cm^3; η, viscosity of the dispersing medium, cp; C_0, concentration of the coagulating salt, mole/cm^3; D_{eff}, effective diffusion coefficient of the coagulating salt in the latex, where

$$D_{\text{eff}} = \frac{D^+D^-(z^+ + z^-)}{z^+D^+ + z^-D^-},$$

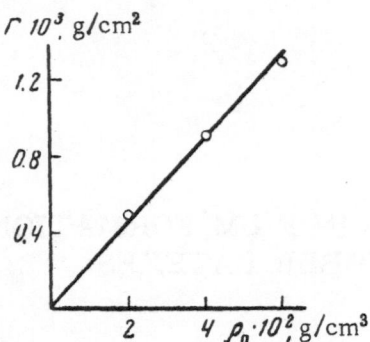

Fig. 1. The ρ_0 vs Γ relation for latex L-7. $C = 10^{-4}$ mole/cm^3 CaCl$_2$; t = 25 sec.

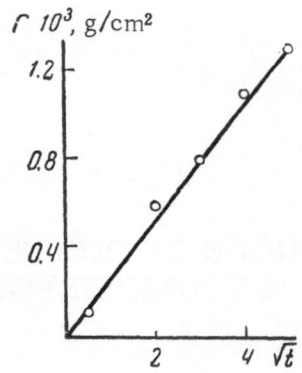

Fig. 2. The $\sqrt{t}$ vs Γ relation for latex L-7. $\rho_0 = 6 \cdot 10$ g/cm^3; $C_0 = 10^{-4}$ mole/cm^3 CaCl$_2$; $\bar{t}$ in sec.

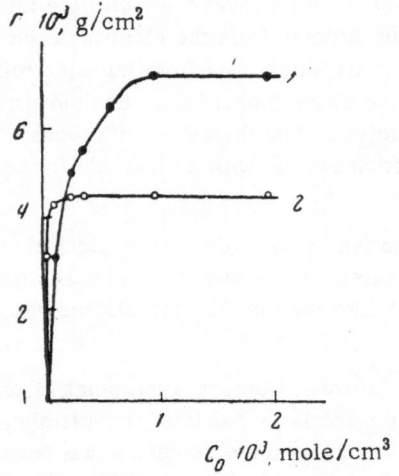

Fig. 3. The C_0 vs Γ relation for latex L-7. $\rho_0 = 10^{-1}$ g/cm^3; t = 60 sec; 1) experimental; 2) theroetical.

D'_{eff}, the same, but in the fixer; D^+ and D^-, diffusion coefficients for the electrolyte ions; z^+ and z^-, charges on the electrolyte ions; τ, time; and $\xi^{\pm} = \frac{1}{C_0} \int\limits_0^{\infty} \gamma^{\pm} h\, dh$, γ being the excess concentration of cations and anions at distance h in the double layer, calculated with respect to concentrations in the bulk.

Experiments carried out at conditions approximating those assumed in the development of the theory led to results consistent with the theoretical predictions.

Ionic deposition was carried out on 2 × 2.5 cm filter papers, the coagulating salts being taken in aqueous solution. Each paper was attached to a steel wire and weighed on a damped analytic balance, immersed, first in the solution of coagulating salt and then in the latex, and once more weighed. The raw gel was dried to constant weight at 100°C. The reported Γ values are mean results from three experiments.

The experimental results of Figs. 1 and 2 show Γ to be linearly dependent on the latex concentration and the square root of the time of deposition, τ.

The Γ values obtained here were, however, some 5-7 times greater than the predicted. These deviations may have arisen from failure to observe in the experiment one of the conditions assumed in the theory, deposition being carried out on a vertical, rather than a horizontal, surface. High Γ values may also have resulted from the fact that latex was picked up as the films were removed.

A diffusional phoresis theory for ionic deposition on a vertical surface has been developed by S. S. Dukhin [5] and improved by determining the amount of gel at the electrolyte—latex interface. This work led to the following expressions for Γ_h (horizontal surface) and Γ_v (vertical surface):

$$\Gamma_h = \frac{2\rho_0 B}{1-\varepsilon} \left(\frac{D_{eff}\, \tau}{\pi} \right)^{1/2},$$

$$\Gamma_v = \rho_0 \left[\frac{2BD_{eff}\, \tau}{(1-\varepsilon)\,\varepsilon} \right]^{1/2}.$$

Here

$$B = \frac{RT}{\eta}\left(\frac{z^{-}\xi^{+}}{D^{-}} + \frac{z^{+}\xi^{-}}{D^{+}}\right)(C_0 - C_1).$$

Calculation shows the rate of deposition on a vertical surface to be approximately three times as large as the rate of deposition on a horizontal surface. The value of Γ is increased by a factor of 2-3 when account is taken of the depth of gel which is formed.

Prime interest attaches to an experimental check of the relation between the deposition, Γ, and the electrolyte concentration, C_0. The theory shows the Γ value tending to a limit with increasing electrolyte concentration.

Changes in the experimental technique eliminated the possibility of latex adherence in removing the films.

Figure 3 shows the experimental and calculated values of Γ_V for vertical surface deposition and their dependence on the electrolyte concentration. The figure indicates, that the experimental and theoretical Γ_V vs C_0 relations are of the same type. The quantitative difference between Γ experimental and theoretical values was no more than 40%.

The indication is that the agreement between theory and experiment is good under these conditions.

Literature

1. B. V. Deryagin, S. S. Dukhin, and A. A. Korotkova. Kolloidn. Zh. 23:53 (1961).
2. H. B. Morris and R. F. Hinderer. Rubber Age (New York) 63:745 (1948).
3. D. M. Sandomirskii and V. V. Chernaya. Tr. N-Issl. Inst. Rezin. Prom.. 20(1) (1954).
4. H. Thiele. Discussions Faraday Soc. 18:294 (1954).
5. S. S. Dukhin. Kolloidn. Zh. 24:446 (1962).

AN ELECTRICAL DIFFUSION THEORY OF THE DORN EFFECT AND A DISCUSSION OF THE POSSIBILITY OF MEASURING ZETA POTENTIALS

S. S. Dukhin

Institute for General and Inorganic Chemistry,
Acad. Sci., USSR

Introduction

The theory of electrokinetic phenomena considers the difference in mean tangential velocities of the ions forming the double electric layer to be the factor chiefly responsible for the appearance of an electrical field in particle movement. The convective surface current which is set up in this layer requires an electric field extending beyond the boundaries of the layer for its maintenance. The mathematical description of these processes leads to the familiar equation [1] showing a fall potential proportional to the ζ-potential of the particle:

$$\varphi = \frac{3}{2} \cdot \frac{ua\varepsilon\zeta}{4\pi\varkappa} \cdot \frac{\cos\theta}{r^2}.$$

(1)

Here, u is the rate of fall of the spherical particle of radius a; $\varkappa$ is the specific electrical conductivity; and ε is the dielectric constant of the electrolyte.

It must be emphasized that Eq. (1) calls for the disappearance of the electric field of the solid particle when the ζ-potential is equal to zero.

The studies of A. N. Frumkin and V. G. Levich [2] on the fall potentials of mercury droplets in electrolytes showed good agreement between experimental data and theoretical predictions, thus bringing out a fundamental quantitative distinction in the case of the liquid interface. Here the electric field was shown to be related to the particle charge, q, rather than the ζ-potential, the spatial charge distribution in the double electric layer having no significance for a liquid interface which is moving as a whole. Thus the movement of the adsorbed layer as a whole proved more significant for an exact description of the electrokinetic phenomena associated with the liquid interface than the relatively small difference in mean rates of ionic movement in the double electric layer.

It must, however, be emphasized that the results of these special experiments on mercury droplets in electrolytes cannot be mechanically carried over to other systems. A reworking of the particle movement potential theory for more general systems, including floating bubbles and liquid droplets in nonconducting media [3], has shown the possibility of quantitatively new effects, establishing the previously unsuspected fact that diffusion can be a significant process in electrokinetic phenomena.

The only adsorption-desportion processes which needs consideration in the case of the mercury droplet are those in the single type of ion which forms the outer surface of the double electric layer. Here ionic diffusion can be forgotten and attention fixed on ion migration in the electric field. The situation is entirely different when the moving droplet surface can adsorb two types of ions from the electrolyte. Then maintenance of approximate electrical neutrality in the continually renewed double layer requires approximate equality in the numbers of positive and negative charges entering the layer from the body of the solution. To fulfill this condition in the absence of an electric field, the diffusion coefficients of the two ions must be identical. Inequality of these coefficients entails establishment of an electric field of such magnitude as to equalize the rates of movement of positive and negative ions to the surface. This effect is analogous to a diffusion potential.

Such an electric field is clearly independent of the ζ-potential, since it arises from a difference in diffusion coefficients rather than a difference in mean tangential velocities of positive and negative charges and is maintained even at zero ζ-potential, disappearing only when the diffusion coefficients become equal.

The discovery that diffusion processes can be significant for the external field of a particle carrying a liquid interface suggested a reexamination of the classical theory of the Dorn effect in light of both electrical migration and diffusion of ions [4]. This work led to an equation reducing to the classical Eq. (1) in the special case of equality of diffusion coefficients. Thus ion diffusion must be considered in any treatment of the solid particle and essentially modifies the basic formula relating the fall potential and the ζ-potential.

Derivation of the classical formula for the fall potential of the solid particle [1] (considered by many as applicable to effects associated with the liquid interface, as well) assumes the particle diameter to be such that the Reynold number (Re = au/ν ≪ 1, ν being the kinematic viscosity of the liquid) is small in comparison with unity, this condition being imposed to assure applicability of Stoke's velocity distribution.

An important consequence of the diffusion theory of the Dorn effect is that the potential distribution, and the entire reasoning for reasonably small particles with Pe ≪ 1 (Pe is the dimensionless Peclet number characterizing the relative effects of molecular and convective diffusion: Pe = au/D, with D representing the diffusion coefficient) is radically different from that applying to coarse particles with Pe ≫ 1.

The papers of B. V. Deryagin and the present author [5] have treated the electric fields of bubbles, but with consideration limited to the possibility of the participation of these fields in the elementary act of flotation. For this reason, no attention was given to the electric field of the bubble for which Pe ≪ 1; furthermore the theory for Pe ≫ 1 was limited to a first approximation in which both surface and solution were considered strictly neutral electrically, the field being localized in the boundary diffusion layer. This work led to conclusions of interest to both the theory and the practice of flotation, but these were difficult to check experimentally since measurement of the localized electric field at a point within the thin diffusion layer of the bubble was required.

The aims of the present work are somewhat different. Here the attempt will be made to determine the long-range electric field of the bubble, an easily measured quantity since the fields compound in a collection of bubbles to give an observable potential difference in the column of liquid. Interest attaches to the electric fields of small bubbles, since a boundary diffusion layer is not formed when Pe ≪ 1 and the long-range electric field is different from zero, even in a first approximation. This problem of determining the ζ-potential for a liquid−gas interface from potential measurements on floating bubbles has general theoretical interest and is, at the same time, basic to the study of mechanisms of flotational adhesion of mineral particles and bubbles [6], where the sign and magnitude of the bubble ζ-potential can be determining factors.

It should be pointed out that measurements of the ζ-potentials of liquid interfaces could not have been carried out earlier since the Dorn effect and bubble electrophoresis are masked by diffusional electric processes [4]. Such measurements have therefore had to await development of a theory embracing both the ζ-potential and the electrical diffusion effects.

The theory of the long-range electric field of the bubble proves to be most complex when Pe ≫ 1; a first approximation treatment being inadequate, and a second approximation quite involved.

Fundamental Equation and Boundary Conditions for a Theory of the Dorn Effect

The electric field of the bubble originates directly in localized deviations from electrical neutrality in the bubble surface, in departures from the equilibrium ion distribution in the double electric layer, and, (at Pe ≫ 1) in breakdown of electrical neutrality in the layer of solution adhering to the surface, these latter effects arising because of differences in the diffusion coefficients and mean convective velocities of the ions forming the double electric layer carried by the bubble surface. The excess surface charge need not be studied quantitatively, however, when it is a matter of explaining preferential adsorption of the ion of higher diffusion coefficient by the surface of the moving bubble, or calculating the electric field. These problems can, on the contrary, be treated simply by considering departures from electrical neutrality to be small in both the body of the system and in the adsorption layer, as must be the case since high electric fields would otherwise be set up and electrical neutrality reestablished. These facts are mathematically reflected in the equations

$$z^+C^+ - z^-C^- \neq 0, \quad z^+\Gamma^+ - z^-\Gamma^- \neq 0, \tag{2}$$

$$z^+C_1^+ - z^-C_1^- = 0, \tag{3}$$

$$z^+\Gamma_1^+ - z^-\Gamma_1^- = 0. \tag{4}$$

Here, z^+ and z^-, C^+ and C^-, and Γ^+ and Γ^- are the valences, concentrations, and adsorptions of the positive and negative ions of the binary electrolyte, and C_1^+ and C_1^-, and Γ_1^+ and Γ_1^- are the ion concentrations and adsorptions calculated in a first approximation assuming electrical neutrality.

It is convenient to represent the adsorptions of positive and negative ions by

$$\Gamma^+ = \int_0^\infty \gamma^\pm(x)\,dx, \tag{5}$$

x being the distance in the double electric layer measured from the layer surface, and $\gamma^\pm(x)$ the concentration excesses of ions in the adsorption layer as compared with the bulk, quantities which are to be combined to obtain the adsorptional condensation.

In order to avoid misunderstanding, it should be noted that the term electrical neutrality has somewhat different meanings when applied to the bulk and to the double electric layer. While bulk electrical neutrality implies equality of positive and negative ions in any infinitesimally small physical volume, the same is not true of neutrality of the double electric layer

$$z^+\gamma_1^+(x) - z^-\gamma_1^-(x) \neq 0.$$

Rather,

$$\int_0^\infty [z^+\gamma_1^+(x) - z^-\gamma_1^-(x)]\,dx = z^+\Gamma_1^+ - z^-\Gamma_1^- = 0.$$

It will also be assumed that the rate of liquid flow and the ion currents are both so low that one can operate with equilibrium values of $\gamma^+(x)$ and $\gamma^-(x)$, the double electric layer structure being unperturbed by these processes. An electric field can then arise even in the absence of an equilibrated double electric layer, i.e., even if $z^+\gamma^+(x) - z\gamma^-(x) \equiv 0$, and we will in what follows therefore frequently speak in terms of the more general concept of the adsorption layer.

We now turn to the formulation of expressions for the tangential ion currents. Convective transfer of the positive and negative ions of the adsorption layer gives rise to surface ion currents with densities

$$I_V^\pm = \int_0^\infty \gamma^\pm(x)\,V_\theta(\theta,\ x)\left(1 + \frac{x}{a}\right)dx, \tag{6}$$

$V_\theta(\theta, x)$ being the tangential velocity component distribution.

The general case involves not only convective surface currents, but also ion transfer along the surface through surface conduction and diffusion; allowing for all three components, the surface current densities can be expressed by

$$I^{\pm} = I_V^{\pm} - D_S^{\pm} \operatorname{grad} \Gamma^{\pm} \pm \frac{F}{RT} D_S^{\pm} z^{\pm} \Gamma^{\pm} E_\theta(a, \theta), \tag{7}$$

the $D_S^{\pm}$ being surface diffusion coefficients and E_θ (a, θ) representing the distribution of the electric field tangential components along the bubble surface.

In order that the surface density of ions remain constant under stationary conditions, the difference between ionic gain and loss as the result of flow must be compensated by the associated normal component of the bulk ion currents over each sector of the surface j_n^+ or j_n^-:

$$\operatorname{div}_S I^{\pm}(\theta) = -j_n^{\pm}, \tag{8}$$

div_S designated the surface divergence. Both diffusion and migration in the electric field contribute to the normal component of the ion current and the right hand members of the equations of (8) can therefore be written out as

$$\operatorname{div}_S I^{\pm}(\theta) = \left(D^{\pm} \frac{\partial C^{\pm}}{\partial_r} \mp \frac{F}{RT} D^{\pm} z^{\pm} C^{\pm} E_r \right)_a. \tag{9}$$

Here, E_r is the radial component of the electric field, F is the Faraday number, T is the absolute temperature, and R is the universal gas constant.

The surface electric current can be written as

$$I = F(z^+ I^+ - z^- I^-). \tag{10}$$

The boundary condition for charge continuity is

$$\operatorname{div}_S I(\theta) = i_n(\theta) = F(z^+ j_n^+ - z^- j_n^-), \tag{11}$$

i_n being the normal component of the total bulk current density, at the surface.

The present treatment is unique [1] (M. Smoluchowski mistakenly considers only the equation for charge continuity) in so far as it sets up separate continuity equations for each type of ion, thus making Eq. (11) a consequence of (9) rather than an independent condition.

Separate continuity conditions are also set up for each of the two types of ions in the body of the solution [7]

$$D^{\pm} \Delta C^{\pm} + \frac{F}{RT} D^{\pm} z^{\pm} \operatorname{div}(C^{\pm}\vec{E}) = \vec{V} \operatorname{grad} C^{\pm}, \tag{12}$$

$\vec{E}$ being the intensity vector for the electric field, and $\vec{V}$ the velocity distribution in the flux around the bubble.

One also obtains an expression for the electric current density

$$\vec{i} = F(z^+ C^+ - z^- C^-)\vec{V} - F(D^+ z^+ \operatorname{grad} C^+ - D^- z^- \operatorname{grad} C^-) + \frac{F^2}{RT}\left[D^+(z^+)^2 C^+ + D^-(z^-)^2 C^- \right]\vec{E}. \tag{13}$$

Account must finally be taken of the equation for charge continuity which follows from (12)

$$\operatorname{div} \vec{i} = 0. \tag{14}$$

It should be pointed out that classical electrokinetic theory replaces the complex system of (12) by the Laplace equation $\Delta \varphi = 0$, which is to say, assumes complete electrical neutrality in the double electric layer. It will become clear in the sequence that this approach is quite incorrect, especially when Pe ≫ 1.

The problem of the potential associated with the movement of a droplet of nonelectrolyte through an electrolyte can be formulated in exactly the same way. A similar formulation applies when the droplet is composed of an electrolyte and the medium is nonelectrolytic, but this is an internal, rather than an external, problem. Finally, the formulation can be readily generalized to apply to cases in which droplet and medium are both nonelectrolytic.

It will be considered in what follows that there is no possibility of ionic discharge in the double electric layer, which is to say that the potentials will be assumed to be so low that the polarization is ideal.

The First Approximation Theory of the Dorn Effect

The first approximation theory (more exactly, the complete electrical neutrality approximation) is based on the conditions of (3) and (4). C_1^+ and C_1^- can therefore be expressed through C (mole/cm^3) alone, and Γ_1^+ and Γ_1^- through the function Γ (mole/cm^2):

$$C_1^+ = Cz^-, \quad C_1^- = Cz^+;$$
(15)

$$\Gamma_1^+ = \Gamma z^-, \quad \Gamma_1^- = \Gamma z^+.$$
(16)

It will also be assumed that equilibrium is established between surface and bulk concentrations at the surface itself, which is to say that the relation between $C(a, \theta)$ and $\Gamma(\theta)$ is the same as that holding between C_0 and Γ_0. Γ_0 being the surface concentration in an adsorption layer which has come into equilibrium with a solution of bulk concentration C. Having already limited the discussion to the case of low concentrations, it can now be assumed that Γ_0 and C_0 are linearly related, so that

$$\Gamma(\theta)/C(a, \theta) = \Gamma_0/C_0 = \alpha.$$
(17)

We also limit discussion to the case of relatively small alterations in $C(a, \theta)$, a condition explicitly expressed by

$$[C(a, \theta) - C_0]/C_0 \ll 1,$$
(18)

where C_0 is the concentration at infinity. Equations (17) and (18) can be combined to give

$$[\Gamma(\theta) - \Gamma_0]/\Gamma_0 \ll 1.$$
(19)

Equation (15) can now be drawn on, and (9) rewritten as

$$\text{div}_s I = \text{div}_s [I_V^\pm - D_S^\pm z^\mp \text{ grad } \Gamma \mp \frac{F}{RT} D_S^\pm z^+ z^- \Gamma \frac{\partial \varphi}{\partial \theta}(a, \theta)] = -j_n^\pm = \left(D^\pm z^\mp \frac{\partial C}{\partial r} \pm \frac{F}{RT} D^\pm z^+ z^- C_0 \frac{\partial \varphi}{\partial r} \right)_{r=a}.$$
(20)

Equation (15) can be applied to (12) to obtain a much simpler expression for the distribution function $C(r, \theta)$

$$\vec{V} \text{ grad } C = D_e \Delta C,$$
(21)

(here, $D_e^{\cdot} = (z^+ + z^-) D^+ D^- /(D^+ z^+ + D^- z^-)$ is the effective diffusion coefficient), and to introduce the function

$$\psi(r, \theta) = Fz^+ z^- \left[\frac{F}{RT} (z^+ D^+ + z^- D^-) \varphi(r, \theta) + (D^+ - D^-) \frac{C(r, \theta) - C_0}{C_0} \right].$$
(22)

Further, Eq. (22) can be combined with (13) to obtain an expression for the current density

$$\vec{i}(r, \theta) = C(r, \theta) \text{ grad } \psi(r, \theta).$$
(23)

This permits (14), the condition for current continuity, to be brought to the form

$$\text{div}\,[C\,(r,\,\theta)\,\text{grad}\,\psi\,(r,\,\theta)] = 0. \tag{24}$$

The boundary condition of (11) and Eq. (24) can be considerably simplified with the aid of (18):

$$\frac{\partial \psi}{\partial r}\,(a,\,\theta) = \text{div}_S\,\left(\frac{I}{C_0}\right), \tag{25}$$

$$\Delta \psi = 0. \tag{26}$$

Thus the potential distribution, $\varphi_1(r,\,\theta)$, can be determined as soon as $C(r,\,\theta)$ is known. This last can be obtained by solving (21) with the boundary condition resulting from the elimination of $\partial \varphi / \partial r$ and $\psi(r,\,\theta)$ from (20); which in turn necessitates solution of (26) with the boundary condition of (25). Equation (22) can then be drawn on, and $\varphi_1(r,\,\theta)$ readily expressed in terms of the functions already obtained:

$$\varphi_1\,(r,\,\theta) = -\,\frac{RT}{F^2 z^+ z^-}\,\frac{\psi\,(r,\,\theta)}{z^+ D^+ + z^- D^-} + \frac{RT}{F}\,\frac{D^+ - D^-}{z^+ D^+ + z^- D^-}\,\frac{\Delta C\,(r,\,\theta)}{C_0}. \tag{27}$$

We will now implement this plan for Pe $\ll$ 1 and Pe $\gg$ 1.

The Theory of the Dorn Effect for Pe << 1

Since Pe $\ll$ 1, convective diffusion can be neglected in comparison with molecular diffusion in Eq. (21). This permits the replacement of (21) by the Laplace Equation

$$\Delta C\,(r,\,\theta) = 0. \tag{28}$$

A comparison of Eqs. (27) and (25), and (22), the defining equation for the function ψ, leads to the conclusion that a potential distribution consistent with Pe $\ll$ 1 and the condition of (18), must satisfy the Laplace Equation

$$\Delta \varphi\,(r,\,\theta) = 0. \tag{29}$$

We will now set up expressions for the $I_V^{\pm}$, making use of the velocity field equation.

A. N. Frumkin and V. G. Levich have shown [8] the movement of the bubble surface to be slowed by the presence of even infinitesimally small amounts of surface active substances, and have proposed a quantitative theory to cover this effect. Their equations apply however, only to the case in which the bubble and adsorbed layer are floating with velocity u while the bubble surface is moving with velocity v_0 at the equator

$$u = \frac{2}{3}\,\frac{\Delta \rho g a^2}{\mu}\,\frac{\mu + \mu' + \gamma}{2\mu + 3\mu' + 3\gamma}, \tag{30}$$

$$v_0 = \frac{1}{3}\,\frac{\Delta \rho g a^2}{2\mu + 3\mu' + 3\gamma}. \tag{31}$$

Here $\Delta \rho = \rho - \rho'$, ρ and ρ', μ and μ' are the respective densities and viscosities of liquid and gas, g is the acceleration of gravity, and, γ is the so-called retardation coefficient. Now it proves necessary to once more carry out the work of [8] to obtain equation for the distribution of tangential velocity components at the bubble surface

$$v_\theta = u\left[1 - \frac{2\mu + 3\mu' + 3\gamma}{4\,(\mu + \mu' + \gamma)}\cdot\frac{a}{r} - \frac{\mu' + \gamma}{4\,(\mu + \mu' + \gamma)}\cdot\frac{a^3}{r^3}\right]\sin \theta, \tag{32}$$

where u is determined from Eq. (30).

This result can be essentially simplified when $x = r - a \ll a$:

$$v_\theta(\theta, x) = v_0 \left[1 + \frac{x}{a} \left(1 + 3 \frac{\mu' + \gamma}{\mu} \right) \right] \sin \theta,$$

(33)

v_0 being found from Eq. (31).

This expression for the velocity distribution is substituted into the equation for the ion convective surface current to obtain

$$I_v^\pm = \int\limits_0^\infty \gamma^\pm(x) v_\theta(\theta, x) \frac{2\pi(a+x)\sin\theta}{2\pi a \sin\theta} \, d\theta = \int\limits_0^\infty \gamma^\pm(x) v_\theta(\theta, x) \left(1 + \frac{x}{a} \right) dx.$$

(34)

Equations (5), (33), and (34) taken together give

$$\mathrm{div}_S I_V^\pm = \frac{2v_0 \cos\theta}{a} \left[\Gamma^\pm + \frac{2 + 3 \frac{\mu' + \gamma}{\mu}}{a} C^\pm \xi^\pm \right],$$

(35)

$$\xi^\pm = \frac{1}{C^\pm} \int\limits_0^\infty \gamma^\pm(x) \, x \, dx,$$

(36)

the latter being related to the ζ-potential by [4]

$$\xi^+ - \xi^- = \varepsilon(\zeta^+ - \zeta^-)(4\pi F C z^+ z^-)^{-1} = -\varepsilon\zeta(4\pi F C z^+ z^-)^{-1}.$$

(37)

It should be noted that the factor $[1 + (x/a)]$ in the integrand of (34) replaces the $[1 + 3(\mu' + \gamma/\mu)]$ of (35) with $[2 + 3(\mu' + \gamma/\mu)]$.

Thus this refined equation must be used when $(\mu' + \gamma) < \mu$. The limiting case of $(\mu' + \gamma)/\mu \to \infty$ corresponds to the solid particle. Here it is not necessary to use the more exact equation since the coefficient $(1 + x/a)$ can be neglected in treating these particles.

The boundary conditions of (20) are quite complex, and the treatment will therefore be limited at first to cases in which surface diffusion and electrical conduction can both be neglected (the conditions under which such simplifications are justified will be established below).

The boundary condition for $\varphi(r, \theta)$ is then obtained by substituting the expression of (35) for $\mathrm{div}_S I_V^\pm$ into (20), casting out the terms arising from surface diffusion and the electrical conduction, and eliminating the member containing $\partial C / \partial r$:

$$\frac{\partial \varphi}{\partial r}(a, \theta) = -\frac{2\Delta\rho g a \cos\theta}{3\mu} \left[\frac{F C z^+ z^-}{a D_e \varkappa}(D^+ \xi^- - D^- \xi^+) + \frac{1}{2 + 3 \frac{\mu' + \gamma}{\mu}} \times \frac{RT}{F D_e} \cdot \frac{D^+ - D^-}{z^+ D^+ + z^- D^-} \frac{\Gamma_0}{C_0} \right].$$

(38)

Here, $\varkappa = F^2(RT)^{-1} z^+ z^-(z^+ D^+ + z^- D^-) C_0$ is the specific conductivity. The boundary condition for $C(r, \theta)$ can be obtained in a similar manner, eliminating the member containing $\partial\varphi / \partial r$:

$$\frac{\partial C}{\partial r}(a, \theta) = -\frac{\Gamma_0(D^+ z^+ + D^- z^-) + C_0 a^{-1} \left(2 + 3 \frac{\mu' + \gamma}{\mu} \right)(D^- z^- \xi^+ + D^+ z^+ \xi^-)}{D^+ D^-(z^+ + z^-) a} 2v_0 \cos\theta.$$

(39)

60

Equation (29) can be readily solved with the boundary condition (38), and (28) with the boundary condition (39), to give

$$\varphi(r,\theta) = \frac{\Delta\rho g a^2}{3\mu}\left[\frac{FCz^+z^-}{aD_e\varkappa}(D^+\xi^- - D^-\xi^+) + \frac{1}{2+3\frac{\mu'+\gamma}{\mu}}\cdot\frac{RT}{FD_e}\cdot\frac{D^+ - D^-}{z^+D^+ + z^-D^-} \times \frac{\Gamma_0}{C_0}\right]\frac{a^2}{r^2}\cos\theta,$$

(40)

$$C(r,\theta) = \frac{\Gamma_0(D^+z^+ + D^-z^-) + C_0 a^{-1}\left(2+3\frac{\mu'+\gamma}{\mu}\right)(D^-z^-\xi^+ + D^+z^+\xi^-)}{D^+D^-(z^+ + z^-)}v_0\frac{a^2}{r^2}\cos\theta.$$

(41)

The method of V. G. Levich [7, §95] can then be followed to obtain the formula for the retardation coefficient from the concentration distribution (41) and the velocity field equation (33):

$$\gamma = \frac{1}{3}RT\cdot\frac{\Gamma_0}{C_0}\cdot\frac{\Gamma_0(D^+z^+ + D^-z^-) + C_0 a^{-1}\left(2+3\frac{\mu'}{\mu}\right)(D^-z^-\xi^+ + D^+z^+\xi^-)}{D^+D^-(z^+ + z^-) - (3\mu a)^{-1}RT\Gamma_0(D^-z^-\xi^+ + D^+z^+\xi^-)}.$$

(42)

By drawing on (33), it can be readily shown that Eq. (41) reproduces the earlier expression for the fall potential of the solid particle [4] when $(\mu'+\gamma)/\mu \to \infty$. The first term of (40) represents the contribution to the potential of the moving bubble arising from the ζ-potential of the double electric layer, ξ^+ and ξ^- being expressed in terms of this parameter, while the second term represents the contribution arising from the diffusional electric effect, or adsorption.

The possibilities of neglecting surface diffusion and electrical conduction can be readily analyzed on the basis of Eqs. (40) and (41):

$$D_S\,\mathrm{grad}_\theta\Gamma \ll \Gamma v_\theta,$$

(43)

$$\frac{RT}{F}D_S\Gamma E_\theta \ll \Gamma v_\theta.$$

(44)

This is in order when

$$\frac{\Gamma_0}{C_0} \ll a\,(D_e/D_S) \approx a.$$

(45)

It is clear that surface diffusion and electrical conduction will be of interest when the condition fulfilled is exactly the opposite of (45), namely

$$\frac{\Gamma_0}{C_0} \gg a.$$

(46)

It will be shown below that $j_n^\pm$ can, in a first approximation be neglected in comparison with each term of the left-hand member of the boundary condition of (20) when the condition of (46) is satisfied, which is to say that this latter implies

$$j_n^\pm \ll \mathrm{div}_S\,(\Gamma^\pm v_\theta).$$

(47)

This conclusion will be confirmed later. From this it follows that the distribution of the adsorption and electric potential along the bubble surface can be determined through the equation

$$I^\pm = 0.$$

(48)

An easily solved equation for the adsorption distribution can be obtained by eliminating $\partial \varphi / \partial \theta$ from these expressions, the result being

$$\Gamma_0 - \Gamma(\theta) = \Delta\Gamma \cos\theta, \tag{49}$$

$$C_0 - C(a,\theta) = \frac{C_0}{\Gamma_0} \cdot \Delta\Gamma \cos\theta. \tag{50}$$

where

$$\Delta\Gamma = V_0 a\left[\frac{\Gamma_0}{D_{Sl}} + \frac{2+3(\mu'+\gamma)\mu}{a(z^++z^-)}\left(\frac{z^-\,\xi^+}{D_S^+} + \frac{z^+\xi^-}{D_S^-}\right)C_0\right]. \tag{51}$$

A simple boundary condition for the potential distribution results from the substitution of this expression for $\Gamma(\theta)$ into any one of the equations of (48). Solution of Eq. (29) with this condition yields

$$\varphi(r,\theta) = \frac{\Delta\rho g a^2}{3\mu(z^++z^-)}\left[\frac{FC z^+z^-}{a\varkappa D_{Sl}}(D_S^+\xi^- - D_S^-\xi^+) + \frac{1}{2+3\frac{\mu'+\gamma}{\mu}}\cdot\frac{RT}{FD_{Sl}}\frac{D_S^+ - D_S^-}{z^+D_S^+ + z^-D_S^-}\frac{\Gamma_0}{C_0}\right]\frac{a^2}{r^2}\cos\theta, \tag{52}$$

where

$$D_{Sl} = (z^+ + z^-)\,D_S^+D_S^-/(D_S^+z^+ + D_S^-z^-).$$

The only difference between the adsorption distribution of (49) and that found by V. G. Levich [7, §75] for the insoluble adsorption layer is in the value of the constant coefficient. This opens the possibility of determining the retardation coefficient, γ, by the method of [7, §75]:

$$\gamma = \frac{aRT}{3}\frac{\Gamma_0(D_S^+z^+ + D_S^-z^-) + C_0a^{-1}\left(2+3\frac{\mu'}{\mu}\right)(D^-z^-\xi^+ + D^+z^+\xi^-)}{D_S^+D_S^-(z^++z^-) - RTC_0\cdot 3(\mu a)^{-1}(D^-z^-\,\xi^+ + D^+z^+\xi^-)}. \tag{53}$$

A concentration distribution satisfying (28) with the boundary condition of (50) has the form

$$C(r,\theta) = \frac{C_0}{\Gamma_0}\,\Delta\Gamma\,\frac{a^2}{r^2}\cos\theta. \tag{54}$$

Substitution of the expressions for $\varphi(r,\theta)$ and $C(r,\theta)$ from Eqs. (52) and (54) into Eq. (20) for $j_n^{\pm}$, readily shows the surface ion currents resulting from migration and diffusion in the electric field to be of the same order of magnitude. It will be proved at the end of this paper that only the second term of (52) and the first term of (51) need be considered in this evaluation. Thus, substitution of (52) and (54) into the expression for $j_n^{\pm}$ at once shows the inequality of (47) to be applicable when the condition of (46) is satisfied.

By drawing on Eq. (51) in which the first member predominates, it can be readily proven that the condition of (18) does not limit the generality of treatment when $\Gamma_0/C_0 \gg a$:

$$\Delta\Gamma/\Gamma_0 \sim aV_0/D_{Sl} \lesssim \mathrm{Pe} \ll 1. \tag{55}$$

Equation (41) can be drawn on to obtain an analogous proof that condition (18) is also fulfilled when $\Gamma_0/C_0 \ll a$:

$$\frac{\Delta\Gamma}{\Gamma_0} \ll \mathrm{Pe}\,\frac{v_0}{u}\frac{D_e}{D_S}. \tag{56}$$

Thus the condition of (18) imposes no limitation on the generality of treatment when $\mathrm{Pe} \ll 1$.

A First Approximation Theory of the Dorn Effect at Low Values of the Reynolds
Number (Re < 1) and High Values of the Peclet Number (Pe ≫ 1)

A study of the entire second approximation theory of the Dorn effect shows that a treatment applying to high values of the Peclet number (Pe ≫ 1) will be complicated by the appearance of various secondary effects (a secondary double electric layer, for example). We will at first limit discussion to the first approximation, assuming complete local electrical neutrality in the electrolyte, so that the potential distribution, $\varphi_1(r, \theta)$ can be expressed through the two-member equation, (27).

Determination of the function $\psi(r, \theta)$ requires solution of Eq. (26) and (25). The right-hand member of the boundary condition (25) can be expressed in terms of the ζ-potential when surface diffusion and conduction are neglected, as can be readily seen by drawing on Eqs. (35) and (37):

$$\psi(r, \theta) = Fv_0 \left(2 + 3 \frac{\mu' + \gamma}{\mu} \right) z^+ z^- (\xi^+ - \xi^-) \frac{a^2}{r^2} \cos \theta, \tag{57}$$

$$\varphi(r, \theta) = \frac{RT}{F} \cdot \frac{g \Delta \rho a}{3\mu} \cdot \frac{\varepsilon \zeta}{4\pi F (z^+ D^+ + z^- D^-) C_0 z^+ z^-} \cdot \frac{a^2}{r^2} \cos \theta + \frac{RT}{F} \cdot \frac{D^+ - D^-}{z^+ D^+ + z^- D^-} \cdot \frac{\Delta C(r, \theta)}{C_0}, \tag{58}$$

where $\Delta C = C_0 - C(r, \theta)$.

It must be emphasized that the function $\Delta C(r, \theta)$ is localized in the thin diffusion boundary layer of the bubble, the depth of this layer being such that $\delta \ll a$. The electric field arising from this field of diffusion has been considered earlier in [10]; it, also, is localized within the boundary layer. It should be noted that it would be more natural to treat this as the diffusional electric component of the surface forces, and consider that it has no direct bearing on those long-range bubble fields which can sum over the floating aggregate to produce a potential difference at various elevations.

Since the second member of Eq. (58) actually makes no contribution to the experimentally measured potential differences, it would seem that the interpretation of these measurements at Pe ≫ 1 would be simplified, the entire effect being ascribed to the ζ-potential. It will, however, be shown below that account must be taken of secondary effects arising from departures from strict electrical neutrality in the surface and diffusion boundary layer of the bubble, and from surface diffusion and conduction.

A Second Approximation Theory of the Convective Diffusion Potential

The first approximation electric field developed from the assumption of complete local electrical neutrality [3, 4] is such that the rate of addition (or removal) of the slowly diffusing ions is reduced to the point where the electric current and associated charge $F(z^+ \Gamma_1^+ - z^- \Gamma_1^-)$ are both equal to zero over every part of the surface. This field, $\vec{E_1}$, arises, however, because of certain departures from neutrality in the surface and the body of the solution, the ion concentrations in the bulk and surface being somewhat different from those required for exact satisfaction of the condition of electrical neutrality.

The relation between charge distribution and electric field reflected by the electrostatic equations can be made the basis for a second approximation in which the adsorption, the electric field, and the concentration distributions are specified more exactly and C_2^+, C_2^-, Γ_2^+, Γ_2^-, and $\vec{E_2}$ evaluated. It is understood that this procedure will be acceptable only if the correction terms are small, so that $C_2^+ \ll C_1^+$, $C_2^- \ll C_1^-$, $\Gamma_2^+ \ll \Gamma_1^+$, $\Gamma_2^- \ll \Gamma_1^-$, and $|\vec{E_2}| \ll |\vec{E_1}|$. It is also to be noted that these corrections are of no interest in themselves since they are small and localized within the diffusion boundary layer. It will be seen, however, that such refinements are required in the description of processes localized in the diffusion boundary when it is a matter of developing a rigorous method for calculating the long-range electric field which extends beyond the boundary limits, and this is what is aimed for here.

The Poisson theorem and the well-known boundary condition for the normal components of the electric field at the interface between two dielectrics can be drawn on to develop a relation between the first approximation electric field and the second approximation correction terms for the partial concentrations and adsorptions

$$z^+C_2^+ - z^-C_2^- = (4\pi F)^{-1}\varepsilon \operatorname{div}\vec{E}_1,$$

(59)

$$z^+\Gamma_2^+ - z^-\Gamma_2^- = (4\pi F)^{-1}(\varepsilon E_{1r} - \varepsilon' E'_{1r}).$$

(60)

Here ε' and E'_{1r} are, respectively the dielectric constant of air and the radial component of the electric field from the inside of the bubble surface.

Corrections to the individual partial concentrations and adsorptions cannot themselves be obtained from Eqs. (59) and (60), but only their combinations; certain additional physical relations can be drawn on, however, to obtain the signs and magnitudes of these corrections as well as the signs and magnitudes of the electric field correction, $\vec{E}_2$.

It will be assumed that $D^+ > D^-$. The electric field, $\vec{E}_1$, must then so retard the positive ions that $z^+\Gamma_1^+ = z^-\Gamma_1^-$ over that part of the bubble surface on which adsorption is taking place. But a certain excess positive charge is required ($z^+\Gamma_2^+ > 0$) to produce this retardation. The field correction, $\vec{E}_2$, must be such as to assure the presence of this excess, which is to say that $\vec{E}_2$ must be directed toward the surface and be antiparallel to $\vec{E}_1$. If $\vec{E}_2$ increases the rate of positive ion feed, it must also decrease the rate of negative ion feed so that $z^-\Gamma_2^- < 0$. Similar arguments can be used for fixing the magnitude of $\vec{E}_2$. The magnitude of $\vec{E}_1$ is determined by the fact that this field regulates the feed of the adsorbing ions so as to maintain the equality $z^+\Gamma_1^+ = z^-\Gamma_1^-$. The field $\vec{E}_2$ alters the adsorption by $z^+\Gamma_2^+$, so that $|\vec{E}_2/\vec{E}_1| \approx z^+\Gamma_2^+/z^+\Gamma_1^+$. Since the partial adsorption corrections, Γ_2^+ and Γ_2^-, have been shown above to be of opposite sign, $|z^+\Gamma_2^+| < |z^+\Gamma_2^+ - z^-\Gamma_2^-|$; the present evaluation can thus be brought to a more convenient form and written as

$$|\vec{E}_2/\vec{E}_1| \leqslant \frac{|z^+\Gamma_2^+ - z^-\Gamma_2^-|}{z^+\Gamma_1^+}.$$

Extended study of the conditions for ion current continuity at the bubble surface leads to the conclusion that each of the ratios

$$\Gamma_2^+/\Gamma_1^+, \quad \Gamma_2^-/\Gamma_1^-, \quad C_2^+/C_1^+, \quad C_2^-/C_1^-, \quad E_2/E_1$$

will be of the same order of magnitude.

A more exact treatment of this problem involves solution of the second approximation equation for convective ion diffusion, with allowance for complex boundary conditions at the bubble surface. Familiarity with the details indicates that it would be quite difficult, or even impossible, to solve such problem by analytical methods.

Calculation of the Long-Range Electric Field of the Bubble for Pe >> 1

Complex numerical calculations are clearly required for development of the second approximation for the convective diffusion electric field of the bubble within the diffusion boundary layer. It should be kept in view, however, that the second approximation is of greatest interest in regions which are external to this boundary layer. On the other hand, solution here is considerably simplified by the fact that the region is free of diffusional currents and volume charges. The absence of bulk charges means that the potential distribution outside the secondary double electric layer, $\varphi^{(g)}$, must satisfy the Laplace equation

$$\Delta\varphi^{(g)} = 0.$$

(61)

The fact that the electrolyte lying outside the secondary electric layer is electrically neutral and free of diffusional currents simplifies the equation relating current and field strength to

$$\vec{i} = \varkappa_0\vec{E}^{(g)} = -\varkappa_0 \operatorname{grad}\varphi^{(g)},$$

(62)

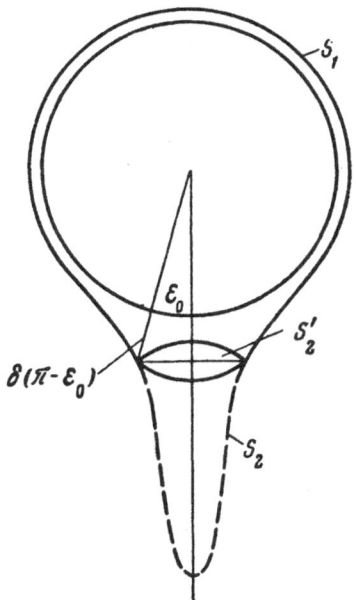

Schematic representation of the boundary
of the secondary double electric layer
(surface area S).

where

$$\varkappa_0 = F^2 (RT)^{-1} z^+ z^- (z^+ D^+ + z^- D^-) C_0.$$

This simple relation between current and potential permits the formulation of a suitable boundary condition for the solution of (61), namely: the normal component of the potential gradient at the boundary of the secondary double electric layer must be proportional to the normal component of the electric field at the same point, this latter value being determined by the processes occurring in the secondary layer itself:

$$\left. \frac{\partial \varphi^{(g)}}{\partial n} \right|_S = \varkappa_0^{-1} i_n \Big|_S .$$

(63)

Here, S is the bounding surface of the secondary double electric layer.

The potential distribution, $\varphi^{(g)}$, is completely determined by Eq. (61), the boundary condition of (63), and the boundary condition at infinity

$$\varphi^{(g)} \big|_{r \to \infty} = 0.$$

(64)

The problem of developing a function which would satisfy the Laplace equation and have a predetermined derivative at a surface is known in potential theory as the external Neiman problem [11]; its general solution has the form of the integral over surface S of the product of the given derivative and the so-called Green function for the external problem. The general solution of the present problem is therefore

$$\varphi^{(g)}(r, \theta) = (4\pi\varkappa_0)^{-1} \iint_S E_{2n} [a + \delta(\theta'), \theta'] G_1 [\theta', \theta, a + \delta(\theta'), r] \, dS(\theta'),$$

(65)

$G[\theta, \theta, a + \delta(\theta'), r]$ being the Green function and $dS(\theta')$ an element of area.

The fact that S, the external face of the secondary double electric layer, has the highly complex form shown schematically in the figure, is the principal source of difficulty in evaluating this integral.

We will divide this face into two zones, S_1 and S_2, by a circle passing through points equidistant from the bubble surface and the axis of symmetry $\theta = \pi$. The point of intersection of this circle with the plane of the paper, and the corresponding angle, $\pi - \varepsilon_0$, are shown in the figure. Reasonable accuracy in the calculations requires that Pe be several orders greater than unity, so that $\varepsilon_0 \ll 1$ and the diffusion boundary layer be thin at $\theta = \pi - \varepsilon_0$, $\delta(\pi - \varepsilon_0) \ll a$. The concept of diffusion layer depth, $\delta(\theta)$, loses all meaning in the lower zone where δ increases rapidly and without limit as the value of θ rises; here the characteristic parameter is the distance, h, separating a point on the surface S and the symmetry axis $\theta = \pi$. It would be more correct to refer to this as the laminar diffusion trail of the bubble, reserving the term diffusion boundary layer for the upper zone. The bulk electric charge is different from zero in both zones, so that the Laplace equation is applicable only to the boundary of S_2. The charge distribution in the lower zone is different from that in the upper zone; in fact, the earlier concept of the secondary double electric layer applies to the upper zone alone. The processes occurring in the laminar diffusion trail generally require special treatment. Thus the calculation of the long-range electric field of the bubble through Eq. (65) is best carried out by splitting the surface of integration into the two parts, S_1 and S_2, corresponding to the angular intervals 0, $\pi - \varepsilon_0$ and $\pi - \varepsilon_0 = \pi$:

$$\varphi = \int_{S_1} \frac{\partial \varphi}{\partial n} G_1 dS + \int_{S_2} \frac{\partial \varphi}{\partial n} G_1 dS.$$

(66)

It is almost impossible to evaluate the integral over S_2, the electric field distribution over this surface being unknown.

On the other hand the total electric current, or, what is the same thing, the total flux of electric field strength through the surface $S_1 + S_2$, is known to be

$$\int_{\pi-\varepsilon_0}^{\pi} i_r \left[a + \delta(\theta'), \theta' \right] \sin \theta' d\theta' = - \int_0^{\pi-\varepsilon_0} i_r \left[a + \delta(\theta'), \theta' \right] \sin \theta' d\theta',$$

$$\int_{S_2} \frac{\partial \varphi}{\partial n} dS = - \int_{S_1} \frac{\partial \varphi}{\partial n} dS. \tag{67}$$

It proves useful to introduce S_2', the spherical surface area of radius $a + \delta(\pi - \varepsilon_0)$ included within the angular interval from $\pi - \varepsilon_0$ to π, which closes S_1.

It is possible in principle to obtain the desired potential distribution outside $S_1 + S_2$ not only by integrating (66) but also by integrating in a similar manner over $S_1 + S_2'$. With this in view, we now consider an analogous problem with the following conditions. Let it once more be supposed that the boundary condition is specified over the surfaces S_1 and S_2, but that the desired function is defined in both directions from these surfaces and satisfies the Laplace equation over the entire space. It is clear that the solution of this problem will be identical with the desired solution over the inner region. The electric field strength will be assumed to alter continuously in passing through the surfaces S_1 and S_2. Solution of the external Neiman problem will then lead to the electric field strength and the potential distribution for the region inside of these surfaces. Thus study of this auxiliary problem shows the possibility of using the Green formula for the external Neiman problem to join the electric field distributions φ on S_2 and φ^* on S', the special superscript being attached to φ to distinguish the value in question from the true potential distribution in this region. Actually the information concerning the $\vec{E}^*$ distribution on S_2' is just as extensive as that concerning the $\vec{E}$ distribution on S_2', the flux of electric field strength, $\vec{E}^*$, through S_2 being equal to the flux of the electric field strength through S_2 as given by Eq. (67), since the flux through the closed surface composed of S_1 and S_2 is zero:

$$\int_{S_2'} \frac{\partial \varphi^*}{\partial n} dS = \int_{S_2} \frac{\partial \varphi}{\partial n} dS = - \int_{S_1} \frac{\partial \varphi}{\partial n} dS. \tag{68}$$

We now consider the Neiman problem for the region outside $S_1 + S_2'$, supposing the distribution of the electric field $\vec{E}$ over S_1 and $\vec{E}^*$ over S_2' to be known. From the Green function for this problem and the boundary conditions, a function φ^* of the form

$$\varphi^* = \int_{S_1} \frac{\partial \varphi}{\partial n} G_2 dS + \int_{S_2'} \frac{\partial \varphi^*}{\partial n} G_2 dS \tag{69}$$

can be developed.

It has been pointed out above, however, that φ^* must be identical with the desired potential distribution function, φ, over the region outside $S_1 + S_2$. Thus

$$\varphi = \varphi^* = \int_{S_1} \frac{\partial \varphi}{\partial n} G_2 dS + \int_{S_2'} \frac{\partial \varphi^*}{\partial n} G_2 dS. \tag{70}$$

The surface $S_1 + S_2'$ will depart only insignificantly from spherical form since consideration has been limited to cases in which the Peclet number is rather small and $\delta(\pi - \varepsilon_0) \ll a$ (so that $\delta(\theta') < \delta(\pi - \varepsilon_0)$ when $\theta' < \pi - \varepsilon_0$). With interest attaching to the potential distribution at distances, r, from the bubble center which are much larger than the bubble radius, it is seen that the Green function, G_2, for $S_1 + S_2'$ can be replaced in Eq. (70) without any considerable error by the Green function, G, for the external Neiman problem for a spherical surface of radius a , the latter being given (for example in [11]) by

$$G = \frac{1}{4\pi}\left[\frac{2}{R} - \frac{1}{a}\ln\frac{a+R-r\cos\gamma}{(1-\cos\gamma)r}\right].$$

(71)

Here, R is the distance between an arbitrarily chosen point on the spherical surface (coordinates θ', φ', and an arbitrarily chosen point outside of the sphere (coordinates, r, θ, $\cos\gamma = \cos\theta\cos\theta' + \sin\theta\sin\theta'\cos(\varphi - \varphi')$. Since consideration has been limited to cases in which $r \gg a$, both function and logarithm can be developed as power series in the small parameter a/r, second power and higher terms in a/r neglected and the function G brought to the form

$$G = \frac{1}{r} + \frac{3a}{2r^2}[\cos\theta\cos\theta' + \sin\theta\sin\theta'\cos(\varphi - \varphi')]\frac{1}{4\pi}.$$

(71')

Equation (70) can then be rewritten as

$$\varphi(r,\theta) = -\frac{a^3}{2r^2}\left(\int_0^{\pi-\varepsilon}\frac{\partial\varphi}{\partial r}\sin\theta'd\theta' + \int_{\pi-\varepsilon}^{\pi}\frac{\partial\varphi^*}{\partial r}\sin\theta'd\theta'\right)$$
$$-\frac{3a^3\cos\theta}{4r^2}\left(\int_0^{\pi-\varepsilon}\frac{\partial\varphi}{\partial r}\cos\theta'\sin\theta'\,d\theta' + \int_{\pi-\varepsilon}^{\pi}\frac{\partial\varphi^*}{\partial r}\cos\theta'\sin\theta'\,d\theta'\right).$$

(70')

When the second term in the first parentheses is rewritten with the aid of Eq. (68), it is at once seen that the binomial in question here is equal to zero. Since $\cos\theta$ is approximately equal to unity over the interval from $\pi - \varepsilon$ to π, Eq. (58) can be drawn on and the expression in the second parentheses written as

$$\varphi^{(g)}(r,\theta) = \frac{3a^3\cos\theta}{4r^2}\int_0^{\pi}\frac{\partial\varphi}{\partial r}(a,\theta')(1-\cos\theta')\sin\theta'd\theta'.$$

(72)

The upper limit of integration has been replaced by π since $\varepsilon \ll \pi$ and the integral expression tends to zero as $\theta' \to \pi$.

We will now consider the evaluation of the normal component of the electric field, $\partial\varphi/\partial r(a, \theta')$, which figures in the preceding integrals.

The participation of the faces of the secondary double electric layer in the liquid motion gives rise to convective electric fluxes along the bubble surface. Since the charge distribution in the faces of this secondary layer can be calculated, these fluxes and their surface divergences can be evaluated by the procedures which were used in the case of the equilibrated double electric layer. There it was a matter of finding an expression for the normal component of the electric current at the external face of the equilibrated double electric layer, and here a matter of obtaining an expression for the angular dependence of the normal component of the electric current at the external face of the secondary double electric layer, $i_n[a + \delta(\theta), \theta]$, as it appears in (63). Realization of this plan meets with difficulty, however, just as surface conduction and diffusion had to be taken into account in calculating the electric field associated with the motion of the equilibrated layer. Actually, the approximate electrical neutrality of the secondary double electric layer entails almost complete compensation of the convective electric currents arising from movement of the external and internal faces of the layer. This fact requires that account be taken of supplementary contributions to the tangential currents arising from surface conduction and diffusion, as well as contributions arising from corrections to the electromigrational and diffusional components associated with $E_2(r, \theta)$, grad $C_2^+(r, \theta)$, and grad $C_2^-(r, \theta)$.

Thus Eq. (63) can be drawn on, and the normal component of the electric current written as

$$- E_n \left[a + \delta\left(\theta\right), \theta\right] = \varkappa_0^{-1} \operatorname{div}_S \left(I_S + I_\delta\right),$$

(73)

I_S being the surface current density and I_δ the tangential current per unit length of a curve $\theta = \text{const}$ on the bubble surface, this current arising from processes occurring in the external face of the secondary layer, as described by the expression

$$I_\delta = \int\limits_0^\infty i_0 \left(x, \theta\right) \frac{2\pi \left(a + x\right) \sin\theta}{2\pi a \sin\theta} \, dx.$$

(74)

The surface current density can be calculated through Eq. (13), replacing $C^\pm$ by $\Gamma^\pm$, and $D^\pm$ by the coefficient of surface diffusion, $D_S^\pm$. Second approximation terms are to be retained in setting up the expressions for the convective surface current from (3) and (4). These second approximation terms can be neglected, however, when setting up expressions for the currents arising from surface conduction and diffusion, these being different from zero, even in the first approximation. When account is taken of the component of the current, I_ξ, arising from the difference in the rates of movement of the faces of the double electric layer, one is led to the following expression:

$$I_S = I_\zeta + F \left(z^+\Gamma_2^+ - z^-\Gamma_2^-\right) v_\theta - Fa^{-1} \left(D_S^+ z^+ \frac{\partial\Gamma_1^+}{\partial\theta} - D_S^- z^- \frac{\partial\Gamma_1^-}{\partial\theta}\right)$$

$$+ F^2 \left(RT\right)^{-1} \left[D_S^+ \left(z^+\right)^2 \Gamma_1^+ + D_S^- \left(z^-\right)^2 \Gamma_1^-\right] E_{1\theta}.$$

(75)

A more useful form of this equation can be obtained by drawing on (58) to express $\Gamma_1^\pm$ in terms of the molar adsorption, Γ, and $E_{1\theta}$ in terms of $\partial C / \partial\theta$:

$$I_S = I_\zeta + F \left(z^+\Gamma_2^+ - z^-\Gamma_2^-\right) v\left(\theta, 0\right) + I_S'.$$

(76)

Here

$$I_S' = A \frac{\partial C}{\partial\theta} + B \sin\theta,$$

(77)

where

$$A = \frac{Fz^+ z^- \left(z^+ + z^-\right) \left(D_S^- D^+ - D_S^+ D^-\right)}{z^+ D^+ + z^- D^-} \frac{\Gamma_0}{C_0}, \quad B = \frac{g\Delta\rho z^+ z^- \varepsilon\zeta}{12\pi\mu} \frac{D_S^+ z^+ + D_S^- z^-}{D^+ z^+ + D^- z^-} \frac{\Gamma_0}{C_0}.$$

(78)

Before setting up the expression for I_δ, i_θ will be considered in more detail, drawing on Eq. (13) with $C_1^+ + C_2^+$, $C_1^- + C_2^-$, and $E_{1\theta} + E_{2\theta}$ substituted for C^+, C^-, and $\vec{E}$:

$$i_\theta = F \left(z^+ C_2^+ - z^- C_2^-\right) v_0 - F \left(D^+ z^+ \operatorname{grad}_\theta C_1^+ - D^- z^- \operatorname{grad}_\theta C_1^-\right)$$

$$+ \frac{F^2}{RT} \left[D^+ \left(z^+\right)^2 C_1^+ + D^- \left(z^-\right)^2 C_1^-\right] E_{1\theta} + i_\theta',$$

(79)

where

$$i_\theta' = - F \left(z^+ D^+ \operatorname{grad}_\theta C_2^+ - z^- D^- \operatorname{grad}_\theta C_2^-\right) + \frac{F^2}{RT} \left\{\left[D^+ \left(z^+\right)^2 C_1^+\right.\right.$$

$$\left.\left. + D^- \left(z^-\right)^2 C_1^-\right] E_{2\theta} + \left[D^+ \left(z^+\right)^2 C_2^+ + D^- \left(z^-\right)^2 C_2^-\right] E_{1\theta}\right\}.$$

(80)

The condition expressed by (3) has been used here and the $[D^+(z^+)^2 C_2^+ + D^-(z^-)^2 C_2^-] E_{2\theta}$ term omitted because its value is low. The second member of Eq. (79) will cancel one of the two terms which arise on substituting $\vec{E}_1 = - \operatorname{grad}\varphi_1$ into the third member, (58) showing this last to be a binomial expression. The third term component remaining after substitution of E_1 and cancellation is of the same order of magnitude as $I_\zeta(\delta/a)$, and can there-

68

fore be neglected in comparison with the terms of (75) representing the surface current density, I_S. From this it follows that the second and third terms of (79) can be dropped. The resulting simplified expression for i_θ is then substituted into (74), and (75) drawn on, to obtain the following expression for the total electric current in the secondary double electric layer in which the terms are regrouped for convenience:

$$I_S + I_\delta = I_\zeta + I_v + I_S' + I_\delta', \tag{81}$$

where

$$I_v = F \left\{ (z^+ \Gamma_2^+ - z^- \Gamma_2^-) v_\theta (\theta, 0) + \int_0^\delta (z^+ C_2^+ - z^- C_2^-) v_\theta (x) \left(1 + \frac{x}{a} \right) dx \right\}, \tag{82}$$

$$I_\delta' (\theta) = \int_0^\delta i' (x, \theta) \left(1 + \frac{x}{a} \right) dx. \tag{83}$$

The upper limit of integration has been changed from ∞ to δ, since the integrands of these expressions reduce to zero outside the diffuse boundary layer.

When the velocity field for the general case is written as the sum of two terms:

$$v_\theta (\theta, x) = v_\theta (\theta, 0) + \Delta v_\theta (\theta, x), \tag{84}$$

the one expressing the velocity distribution along the surface and the other the alteration in velocity in passing through the boundary layer, the expression for I_v is transformed to

$$I_v = I_{v1} + I_{v2},$$

where

$$I_{v1} = F v_\theta (\theta, 0) \left[(z^+ \Gamma_2^+ - z^- \Gamma_2^-) - \int_0^\infty (z^+ C_2^+ - z^- C_2^-) dx \right], \tag{85}$$

$$I_{v2} = F \int_0^\delta (z^+ C_2^+ - z^- C_2^-) \Delta v_\theta (\theta, x) \left(1 + \frac{x}{a} \right) dx. \tag{86}$$

Calculation of the Long-Range Electric Field for Re < 1, Pe >> 1, and Weak Surface Retardation

We now apply the effective method developed by V. G. Levich [7, §72] for handling the problem of convective diffusion when Pe >> 1, and Re < 1 to obtain the concentration distribution in the diffusion boundary layer under the condition of (18), namely

$$C_0 - C (\theta, x) = \frac{\mathscr{A}}{(4\pi \text{Pe})^{1/2}} \int_0^\theta \frac{\exp \left[-3x^2 \sin^4\theta \, \text{Pe} / 8 a^2 f (\theta, \theta') \right]}{f^{1/2} (\theta, \theta')} \cos \theta' \sin \theta' d\theta', \tag{87}$$

where $f(\theta, \theta') = \cos \theta' - \cos \theta - (\cos^3 \theta' - \cos^3 \theta)/3$, $\text{Pe} = a v_0 / D_e$,

$$v_0 = \frac{3}{2} u, \quad \mathscr{A} = \frac{2 v_0 \Gamma_0}{D_e}.$$

An expression for the alteration in the adsorption along the surface will be required in what follows and can be obtained from (87) if local adsorptional equilibrium is assumed:

$$\frac{\Gamma_0 - \Gamma (\theta)}{\Gamma_0} = \frac{C_0 - C (a, \theta)}{C_0} = \frac{\mathscr{A}}{C_0 (4\pi \text{Pe})^{1/2}} I (\theta), \tag{88}$$

where

$$I(\theta) = \int_0^\pi \frac{\cos\theta' \sin\theta' d\theta'}{f^{1/2}(\theta,\theta')}.$$

(89)

Equation (88) can be drawn on to calculate the retardation coefficient, γ, of (33); this proves to be a function of the angle θ which increases rapidly as $\theta \to \pi$. Thus it follows that the suggested method of solution will be acceptable when the retardation coefficient is small in comparison with μ over almost the entire drop surface.

Considering that

$$v_\theta(\theta, 0) = v_0 \sin\theta, \text{ and } \Delta v_\theta = 3\frac{x}{a}\frac{\mu'+\gamma}{\mu}\sin\theta,$$

according to Eq. (30), and drawing on (59), (60), and the well-known equations of vector analysis, it can be shown that

$$I_{v1} = \frac{\varepsilon v_0 \sin\theta}{4\pi} E_{1r}(a,\theta)\frac{\delta}{a}\left(\frac{\varepsilon'}{\varepsilon} + \frac{\delta}{a}\right),$$

(90)

$$I_{v2} = -\frac{\varepsilon v_0 \sin\theta}{4\pi a}\left(2 + 3\frac{\mu'+\gamma}{\mu}\right)\varphi_1(a,\theta).$$

(91)

With the approximation $E_1 \approx \varphi_1/\delta$, it can be readily proven that $|I_{v1}/I_{v2}| \approx \varepsilon'/\varepsilon + \delta/a \ll 1$. Thus, in view of (58), one can write

$$I_v \simeq I_{v2} \simeq Fv_0 \sin\theta\left(2 + 3\frac{\mu'+\gamma}{\mu}\right)\frac{\varepsilon RT}{4\pi F^2}\frac{D^+ - D^-}{z^+D^+ + z^-D^-}\frac{\Delta C(a,\theta)}{C_0} = -C'\Delta C(a,\theta),$$

(92)

where $\Delta C(a,\theta) = C_0 - C(a,\theta)$.

Though I_δ can be evaluated only with difficulty it can be shown to be negligible in comparison with I_v. Since it has been proved above that the three ratios C_2^+/C_1^+, C_2^-/C_1^-, and E_2/E_1 are of the same order of magnitude, it can be concluded that the same is true of all three groups of terms in (80); magnitude-wise, the first, and the whole expression for i'_θ, can, therefore be represented as

$$i_\theta \approx FD \operatorname{grad}_\theta(z^+C_2^+ - z^-C_2^-).$$

Transformations similar to those carried out above, lead finally to the approximation expressions

$$\frac{I'_\delta}{I_v} \approx \mathrm{Pe}^{-\frac{1}{2}}\frac{\partial}{\partial\theta}\ln\frac{\Delta C(a,\theta)}{C_0} \ll 1, \text{ where } (\theta < \pi - \varepsilon).$$

(93)

Thus a calculation of $\varphi^g(r,\theta)$ through Eq. (72) should take account of I_ζ, I_v, from Eq. (92), and I'_S; from Eq. (77)

$$\varphi^g(r,\theta) = \frac{3a\cos\theta}{4\varkappa r^2}\left[\frac{\mathscr{A}}{(4\pi \mathrm{Pe})^{1/2}}(2S_1 A + S_2 C') + \frac{4}{3}B\right],$$

(94)

where

$$S_1 = \int_0^\pi I(\theta)\cos\theta\sin\theta d\theta = 1.11, \text{ and } S_2 = \int_0^\pi I(\theta)\sin^3\theta d\theta = 0.745.$$

It has been pointed out repeatedly that the treatment of this article will show that $|\varphi_\zeta| \ll |\varphi_{D_1}|$. From this it follows that the B term can be dropped from the equation to give

$$\varphi^g(r, \theta) = \frac{3a^2 \cos \theta}{4\kappa r^2} \left[\frac{u}{a} \frac{\varepsilon \zeta}{2\pi} + \frac{1.5 \varepsilon v_0^2}{4\pi a} \frac{\Gamma_0}{C_0} \frac{RT}{F} \frac{D^+ - D^-}{z^+ D^+ + z^- D^-} \frac{1}{(4\pi \text{ Pe})^{1/2}} \right.$$

$$\left. + \frac{2.6F}{(4\pi \text{ Pe})^{1/2}} \frac{\Gamma_0^2 V_0}{C_0 a} \frac{D_s^- D^+ - D_s^+ D^-}{D^+ D^-} z^+ z^- \right].$$

(95)

The first member of this three-term equation gives the contribution to the potential of the moving droplet arising from the ζ-potential, the second the contribution from the secondary double layer, and the third the contribution from surface diffusion and conduction.

The Possibility of Calculating the ζ-Potential from Electrokinetic Measurements

The definitions of Γ^+ and ξ^+ can be combined with Eqs. (5) and (36) to show that

$$\xi \approx \frac{\Gamma}{C} d.$$

(96)

On this basis it can be readily proven that the ratio second member of the numerator of (42) to the first is of the same order of magnitude as d/a, and the former can therefore be neglected. The second term in the denominator can also be neglected, and for exactly the same reason. One thus obtains the expression

$$\gamma = \frac{RT}{3D_e} \left(\frac{\Gamma_0}{C_0} \right)^2 C_0.$$

(97)

It is readily seen that the second term in the denominator of (42) is less than the first, when the conditions are such that

$$\gamma/\mu \ll a/d,$$

(98)

γ being determined through Eq. (97). Application of (96) will show the ratio of the first term of (40) to the second (these terms have been designated by φ_ζ and φ_D, respectively) to be much less than unity when the conditions of (98) is satisfied and the retardation coefficient is given by (97)

$$|\varphi_\zeta/\varphi_D| \approx (d/a)(\gamma/\mu) \ll 1.$$

(99)

The contrary case is that in which the second term in the denominator of (42) is the larger, and the inequalities of (98) and (99) are not satisfied; here, the approximation expressions for the retardation coefficient and the φ_ζ/φ_D ratio are:

$$\gamma \approx (\Gamma_0/C_0)(\mu a/\xi) \approx \mu a/d,$$

(100)

$$|\varphi_\zeta/\varphi_D| \approx 1.$$

(101)

Just as in the case of $\Gamma_0/C_0 > a$, Eq. (96) can now be drawn on to show that the second term in the numerator of (53) is much less than the first. When the second term in the denominator is also neglected, one obtains

$$\gamma = \frac{RTa}{3D_{Se}} \Gamma_0.$$

(102)

It is then easy to see that the second term in the denominator is actually less than the first when the conditions are such that

$$\gamma/\mu \ll a/d,$$

(103)

and γ is determined from (102).

When the condition of (103) is satisfied and the retardation coefficient is given by Eq. (102), the approximation of (96) can be drawn on and an expression obtained for the ratio of the first term, φ_ζ, of (52) to the second, φ_D:

$$|\varphi_\zeta/\varphi_D| \approx (d/a)(\gamma/\mu) \ll 1.$$

(104)

It can be shown, in the same way, that

$$\gamma \sim (\Gamma_0/C_0)\,(\mu a/\xi) \sim \mu a/d;$$

(105)

$$|\varphi_\zeta/\varphi_D| \approx 1$$

(106)

when the second term of (53) is the larger.

By formulating the expression for the ratio of first and second terms in Eq. (95),

$$|\varphi_\zeta/\varphi_v| \sim \left(\delta_D \frac{C_0}{\Gamma_0}\, \zeta\right) \Big/ (RT/F),$$

(107)

it can be readily shown that the contribution of the ζ-potential to the Dorn potential is greater than the contribution from the secondary double layer when the ζ-potential is low and $\Gamma_0/C_0 \ll \delta$.

Conclusions

The Dorn potential arises from diffusional electric effects when Pe $\ll$ 1 and the surface activity and bulk concentration are both low. The diffusional electric field is localized within the secondary double electric layer, however, when Pe $\gg$ 1; and its significance is so reduced that it needs be taken into account only in a second approximation. Thus with low concentrations of a reagent of moderately low surface activity, measurement of the Dorn potential at Pe $\gg$ 1 and Re $<$ 1 permits determination of the ζ-potential from the simplest form of the Smoluchowski equation.

No account has been taken, either here or in [4], of the inverse effect of the electric field on the movement of the double electric layer faces. Thus both the formula for the Dorn potential of the solid particle which was derived in [4] as a more general version of the Smoluchowski formula, and the equations for the liquid drop which were derived in the present work, are applicable only when the parameter values are such that the electroosmotic rate of movement of the double layer faces is much less than the velocity of motion in the field of gravity. It is, unfortunately, difficult to obtain criteria which would apply here since this requires a knowledge of the deviation from spherical symmetry of the electric field within the moving double electric layer, and discussion has been limited to the part of the field lying outside of the layer.

V. G. Levich [7, §75] has treated the analogous problem of the potential difference in a column of falling mercury droplets, showing the potential difference, $\mathscr{E}$, in a column 1 cm high to be related to the potential difference between the upper and lower poles of the drop by the equation

$$\mathscr{E} = 2\pi n \delta\varphi.$$

(108)

The potential distribution over the drop surface is given by

$$\varphi = \Phi \frac{a^2}{r^2} \cos\theta,$$

(109)

so that

$$\delta\varphi = 2\Phi. \tag{110}$$

A similar potential distribution is set up in the neighborhood of a moving droplet or bubble such is of interest here, the only difference being in value of the coefficient Φ; calculation of $\mathscr{E}$ can therefore be made through the equation

$$\mathscr{E} = 4\pi n\Phi, \tag{111}$$

which follows from (108) and (110).

We will now follow Levich [7, § 103] and relate this potential difference, $\mathscr{E}$, with the measured current strength, I. If the circuit formed by the column of liquid of length L and cross section S is closed by an external resistance, Ω, the external potential will balance up in such a way that a current I, will flow, additively composed of the local currents arising from particle movements. Since the emf of this circuit is equal to the product $\mathscr{E}L$, it follows that

$$I = \frac{\mathscr{E}L}{\frac{L}{S\varkappa} + \Omega} = \frac{I_0 = \frac{L}{S\varkappa}}{\frac{L}{S\varkappa} + \Omega}, \tag{112}$$

I_0 being the current which would flow if $\Omega = 0$, that is to say, the current corresponding to the short-circuited column. According to (112) and (108)

$$I_0 = \mathscr{E}S\varkappa = 4\pi n\Phi S\varkappa. \tag{113}$$

Drawing on Eqs. (111) and (113) (which are open to experimental vertification), we can now summarize the expressions for Φ which are to be used with various parameter values:

$$\Phi = \frac{\Delta\rho g a}{3\,\mu}\left[\frac{FCz^+z^-}{aD_e\varkappa}(D^+\xi^- - D^-\xi^+) + \frac{1}{2 + 3\frac{\mu' + \gamma}{\mu}}\frac{RT}{FD_e}\frac{D^+ - I^-}{z^+D^+ + z^-D^-}\frac{\Gamma_0}{C_0}\right]$$

$$(\mathrm{Pe} \ll 1, \ \Gamma_0/C_0 \ll a),$$

$$\Phi = \frac{\Delta\rho g a^2}{3\mu\,(z^+ + z^-)}\left[\frac{FCz^+z^-}{aD_{Se}\varkappa}(D_S^+\xi^- - D_S^-\xi^+) + \frac{1}{2 + 3\frac{\mu' + \gamma}{\mu}}\frac{RT}{FD_{Se}}\frac{D_S^+ - D_S^-}{z^+D^+ + z^-D^-}\frac{\Gamma_0}{C_0}\right]$$

$$(\mathrm{Pe} \ll 1, \ \Gamma_0/C_0 \gg a),$$

$$\Phi = \frac{3}{4\varkappa}\left[\frac{u\varepsilon\zeta}{2\pi a} + \frac{1,5\varepsilon v_0^2\Gamma_0}{4\pi a C_0}\frac{RT}{F}\frac{1}{(4\pi\mathrm{Pe})^{1/2}}\frac{D^+ - D^-}{z^+D^+ + z^-D^-} + \frac{2,6F}{(4\pi\mathrm{Pe})^{1/2}}\frac{\Gamma_0^2 v_0}{C_0 a}\frac{D_S^- D^+ - D_S^+ D^-}{D^+D^-}z^+z^-\right]$$

$$(\mathrm{Pe} \gg 1, \ \ \mathrm{Re} \ll 1).$$

The author wishes to express his indebtedness to B. V. Deryagin for his valued advice and interest in this work.

Literature

1. M. Smoluchowski. Bull. Akad. Sci. Cracovie, Classe Sci. Math. Natur. 1:182 (1903).
2. A. N. Frumkin. Izv. Akad. Nauk SSSR, Ser. Khim. (1945), p. 223.
   V. G. Levich. Zh. Fiz. Khim. 21:689 (1947).
3. S. S. Dukhin and B. V. Deryagin. Dokl. Akad. Nauk SSSR 121:503 (1958); Kolloidn. Zh. 21:37 (1959).

4. B. V. Deryagin and S. S. Dukhin, Dokl. Akad. Nauk SSSR 129:1328 (1959).

5. B. V. Deryagin, S. S. Dukhin, and V. V. Lisichenko, Zh. Fiz. Khim. 33:2280 (1959); 34:524 (1960).

6. B. V. Deryagin and S. S. Dukhin, Trans. Mining Met. 70, Part 5:222 (1960-1961).

7. V. G. Levich, Physical Chemical Hydrodynamics, Moscow, Gosfizmatizdat (1961).

8. A. N. Frumkin and V. G. Levich. Zh. Fiz. Khim. 21:1183 (1947).

9. S. S. Dukhin and B. V. Deryagin. Kolloidn. Zh. 20:705 (1958).
   S. S. Dukhin. Zh. Fiz. Khim. 34:1053 (1960).

10. S. S. Dukhin. In collection: Research in Surface Forces, Moscow, Izd. Akad. Nauk SSSR (1961), p. 38.
    [English translation: Consultants Bureau, New York (1963).]

11. N. N. Koshlyakov, E. V. Gliner, and N. N. Smirnov, Basic Differential Equations and the Partial Derivatives of Mathematical Physics, Moscow, Gosfizmatizdat (1962).

THE STATISTICAL THEORY OF THE DOUBLE ELECTRIC LAYER
Part I
G. A. Martynov

The Physical Presuppositions and Limits of Applicability

of the Gouy — Stern Theory

I. With the present communication we begin the publication of a series of articles on the statistical theory of the double electric layer. We will attempt to treat those factors which have been either neglected (e.g., image forces), or only inadequately investigated (e.g., intrinsic ionic volumes in the Gouy—Chapman—Stern theory) working from the Gibbs canonical distribution, or, more exactly, from the Bogolyubov correlation functions which follow identically from the latter [1]. It is, however, in order to decide whether improvements are actually required in the current theory before passing to what would be essentially a refinement and generalization of it. This is all the more necessary in view of the widely prevailing belief that the Gouy—Stern theory gives a very exact account of the experimental data and stands in need of no correction.

To bring out the physical presuppositions inherent in the theory, we will first show that its fundamental equations are special instances of the more general Bogolyubov equations.* These latter will then be used as a basis for estimates which are necessary to a determination of the limits of applicability of the theory. We will finally consider the experimental results and show that our estimates of the applicability limits are actually confirmed.

II. It has been shown by N. N Bogolyubov [1] that the concentration, $G_a(z_a)$, of particles of type a at a distance z_a from the surface of a solution is fixed by the equation

$$\frac{\partial G_a(z_a)}{\partial z_a} + \frac{G_a(z_a)}{\theta}\frac{\partial U_a(z_a)}{\partial z_a} + \frac{1}{\theta}\int\limits_{z_1 \geqslant 0}\sum_b \nu_b G_{ab}(\vec{r}_a,\vec{r}_b)\frac{\partial \Phi_{ab}(\vec{r}_a,\vec{r}_b)}{\partial z_a}\,d\vec{r}_b = 0. \tag{1}$$

Here $\theta = kT$ is the temperature, measured on an energy scale, ν_a is the number of particles of type a per unit volume of solution, $U_a(z_a)$ is the energy of a type a particle in the external field, $\Phi_{ab}(\vec{r}_a, \vec{r}_b)$ is the energy of interaction of particles of types a and b, $\vec{r}_a$ is the radius vector of the a particle, and $G_{ab}(\vec{r}_a, \vec{r}_b)$ is a binary function satisfying the boundary conditions

$$G_{ab} \to G_a(z_a)\,G_b(z_b) \text{ when } r_{ab} = |\vec{r}_a - \vec{r}_b| \to \infty, \tag{2}$$

$$G_{ab} \to f(\vec{r}_a, \vec{r}_b)\exp[-\Phi_{ab}^{(s)}(r_{ab})/\theta] \to 0, \quad f \geqslant 0 \text{ when } r_{ab} \to 0,$$

$\Phi_{ab}^{(s)}$ being the energy of the Born repulsion forces which assure particle rigidity.

In view of (2), it is convenient to write the expression for G_{ab} as

$$G_{ab} = g_a g_b \exp[-\Phi_{ab}^{(s)}/\theta]\,(1+g_{ab}),$$

---

*This was first done by V. G. Levich and V. A. Kir'yanov [2].

where

$$G_a = g_a, \quad \lim_{r_{ab} \to \infty} g_{ab}(\vec{r}_a, \vec{r}_b) = 0.$$

Substitution of this expression into (1) leads to:

$$\frac{\partial \ln g_a}{\partial z_a} + \frac{1}{\theta}\frac{\partial U_a}{\partial z_a} + \frac{1}{\theta} \int\limits_{z_b \geqslant 0} \sum_b v_b g_b (1 + g_{ab}) e^{-\Phi_{ab}^{(s)}/\theta} \frac{\partial \Phi_{ab}}{\partial z_a} d\vec{r}_b = 0.$$

(3)

The summation of (3) extends over all types of particles present in the solution.

It is necessary to know the forms of U_a and Φ_{ab} in order to specify the system. The general expression for the energy of an ion in the neighborhood of the interface between two media is

$$U_a = \Phi_a^{(coul)} + \Phi_a^{(im)} + \Phi_a^{(ad)}.$$

(4)

Here $\Phi_a^{(coul)} = e_a[\Psi_0 - (2\pi \eta/\varepsilon)z_a]$ is the electrostatic energy of a charge e_a located at a distance z_a from a surface carrying a charge η, ε is the dielectric constant of the solvent located in the semi-space $z \geqslant 0$,

$$\Phi_a^{(im)} = \frac{e_a^2}{2\varepsilon}\frac{\varepsilon - \varepsilon'}{\varepsilon + \varepsilon'} \cdot \frac{1}{2z_a}$$

is the energy of interaction of an ion with its own image, as determined by the difference in polarizabilities of the solvent and the external medium (the dielectric constant of the latter is ε'), and $\Phi_a^{(ad)}$ is the ion-external medium interaction energy, resulting from the presence of nonelectrical forces. The ion-pair interaction energy is, on the other hand, expressed by

$$\Phi_{ab} = \Phi_{ab}^{(coul)} + \Phi_{ab}^{(im)} + \Phi_{ab}^{(s)},$$

(5)

$\Phi_{ab}^{(coul)} = (e_a e_b/\varepsilon)(1/r_{ab})$ being the energy of coulombic interaction of two point-charges, and

$$\Phi_{ab}^{(im)} = \frac{e_a e_b}{\varepsilon}\frac{\varepsilon - \varepsilon'}{\varepsilon + \varepsilon'}\frac{1}{r_{ab}^*}$$

the energy of interaction of ion a with the image of ion b.

Reference to the solvent in Eq. (3) is through the dielectric constant, ε, alone.

In order to obtain the Gouy–Stern system of equations from (3)-(5), we will use the relation $\ln g_a + \Phi_a^{(ad)} = -e_a\varphi(z)/\theta$ to introduce a new function, φ, with the dimensions of a potential. We now substitute this expression for φ into (3), set $\Phi_a^{(im)} = \Phi_{ab}^{(im)} = 0$; $g_{ab} = 0$; and $\Phi_{ab}^{(s)} = 0$, and then act on the resulting equation with the hamiltonion operator $\vec{\nabla}_a$ to obtain:

$$\varphi''(z) = -\frac{4\pi}{\varepsilon}\rho(z), \quad \rho(z) = \sum_a v_a e_a \exp\left[-\frac{\Phi_a^{(ad)}(z) + e_a\varphi(z)}{\theta}\right].$$

(6)

This equation can be simplified somewhat by noting that there must be a distance of closest approach, h, of the centers of the ions of finite dimensions to the interface (i.e., $\Phi_a^{(ad)} = +\infty$ over the interval $(0, h)$), and that $\Phi_a^{(ad)}(z)$ must fall off rapidly with increasing z when $z > h$. One finds the Gouy–Stern system of equations for the determination of φ [3] by assuming the region of nonzero (and different from $+\infty$) energy of specific adsorption, $\Phi_a^{(ad)}$, to be an infinitesimally thin layer (the second plate of the Stern molecular condenser) characterized by the mean value $\Phi_a^{(ad)} \simeq \Phi_a^{(0)}$ and an electrostatic potential ψ_1:

$$\varphi_1''(z) = 0, \text{ when } 0 \leqslant z \leqslant h;$$

(7)

$$\varphi''(z) = -\frac{4\pi}{\varepsilon} \rho(z), \quad \rho = \sum_a v e_a e^{-\frac{e_a \varphi(z)}{\theta}} \quad \text{when } h \leqslant z \leqslant \infty; \tag{8}$$

$$\varphi_1(0) = \psi_0, \quad \varphi_1(h) = \varphi(h) = \psi_1, \quad \varepsilon_1 \varphi_1(h) - \varepsilon \varphi'(h) = -4\pi\eta_1,$$

$$\varphi(\infty) = 0. \tag{9}$$

Here account has been taken of the possibility that the dielectric constant, ε_1, in the molecular condenser (i.e., the layer $(0, h)$) may not be equal to ε. The

$$\eta_1 = \sum_a v_a e_a \exp\left[-\frac{1}{\theta}(\Phi_a^{(0)} + e_a\psi_1)\right] \tag{10}$$

of Eq. (9) is a special case of a Stern isotherm for low surface coverage. It is not possible to obtain the complete Stern isotherm from (3).

III. We will now consider in more detail the presuppositions involved in passage from (3) to (7)-(10). These can be separated into two groups, the first containing assumptions related to the intrinsic volumes of the ions. The expression of (6) for the double layer charge could be obtained only by neglecting the repulsive forces which act between the ions (which is equivalent to setting $\Phi_{ab}^{(s)} = 0$). It is clear tha a point-ion model ceases to be acceptable at high concentrations. Since $g_a = v_a \exp\left[-(1/\theta)(\Phi_a^{(ad)} + e_a\varphi)\right]$, it follows that the concentration, g_a, in the double layer will be high when either v_a or the exponent is large, the latter being the case at high potentials. But is is exactly under these circumstances that one would expect breakdown of the Gouy–Stern description of the charge distribution through Eqs. (8) and (10). This by no means indicates, however, that all conclusions based on the theory will be in error at high concentrations. The finiteness of dimensions was taken into account in passing from (6) to (7)-(10) by assuming that the ions could not approach the interface too closely because of repulsive forces from the external medium. Thus it is to be anticipated that those conclusions of the Gouy–Stern theory which relate only to the presence of the molecular condenser would be valid at all concentrations.

The second group presuppositions trace back to the neglect of image forces. In estimating the error from this source, we note that the g_{ab} term of the integral of (3) allows for the screening of an ion by its own Debye atmosphere. Since this screening reduces the energy of interaction of the ion with particles lying outside its atmosphere, an equation exaggerating the effect of image forces would be obtained by setting $g_{ab} = 0$ in (3). Writing $\ln g_a + \Phi_a^{(im)} + \Phi_a^{(ad)} = -e_a \varphi/\theta$ and carrying out the same operations as before replaces (8) by

$$\varphi''(z) = -\frac{4\pi}{\varepsilon}, \quad \rho(z)\,\rho(z) = \sum_a v_a e_a \exp\left[-\frac{e_a \varphi(z)}{\theta} - k^2 \frac{\varepsilon - \varepsilon'}{\varepsilon + \varepsilon'} \frac{z^*}{2z}\right],$$

$$e_a = ke, \quad z^* = \frac{e^2}{2\varepsilon\theta}. \tag{11}$$

From this it is evident that:

1) The effect of the image forces is essentially dependent on the ratio of the dielectric constants of solvent and external medium. If the solvent is bounded by a vacuum ($\varepsilon' = 1$), $\varepsilon > \varepsilon'$ and the image forces give rise to negative adsorption; if it is bounded by a metal ($\varepsilon' = \infty$), the reverse is true, and the adsorption positive.

2) The effect of the image forces increases as the dielectric constant of the solvent diminishes. Thus the value of z^* for ethyl alcohol ($\varepsilon \cong 30$) is twice as large as for water (3.57 A);

3) The effect of the image forces is also closely dependent on the valence, k, of the ion. For most inorganic ions in aqueous solution, $2h \simeq z^*$. The maximum alteration in concentration ($v_{max} - v_0)/v_0$, resulting from the image forces is therefore approximately equal to $e^{\pm k^2}$, and the range over which this alteration is

significant (i.e., over which the image force energy is greater than θ) is $l \simeq k^2(z*/2)$. For tetravalent ions $(\nu_{max} - \nu_0)/\nu_0 \simeq e^{16}$, $l \simeq 20$ Å, and is several times larger for weak electrolytes.

4) The relative significance of the image forces diminishes as ψ_1 increases. A comparison of the two terms in the exponential of (11) shows that the image forces of univalent ions can be neglected when $e\psi_1 \gtrsim \theta$, but that the conditions must be such that $e\psi_1 \gtrsim 16\,\theta$ before the same can be done for tetravalent ions. Potentials of this order of magnitude are never realized in practice, and image forces should therefore be invariably taken into account in treating polyvalent electrolytes.

5) The relative significance of the image forces falls off at low concentrations where $1/\varkappa \gg l$ ($\varkappa = \sqrt{4\pi \Sigma \nu_a e_a^2}/\theta$ is the inverse of the Debye radius), l, the depth of the "image region," being independent of the concentration while, $1/\varkappa$, the double layer depth, diminishes as the concentration rises.

6) When the electrolyte is unsymmetrical, the difference in the ion-image interaction energies near the interface gives rise to a region [approximate depth, $l \simeq k^2(z*/2)$] of predominance of the ion of higher (or lower) valence, specific adsorption resulting from the differences in polarizability of the solvent and the external medium (the electrode). This effect is not accounted for in either the Gouy equation (8) or the Stern isotherm (10).

Though these estimates are obviously too high, they do clearly indicate limitations in the applicability of the Gouy—Stern theory to polyvalent solutions of strong electrolytes and to solutions of weak electrolytes.

IV. The belief that the Gouy—Stern theory gives a highly exact representation of the experimental data goes back to an experimental check of the solution of Eqs. (7)-(10) for the case of the uniform plane, ideally polarized, electrode. The present communication will therefore be largely limited to a consideration of this particular system, the more involved case of overlapping double layers being treated in less detail.

We now pass to the solution of (7)-(10), limiting the discussion, for simplicity, to a symmetrical electrolyte ($e_+ = -e_- = e$) which is not specifically adsorbed on the electrode ($\eta_1 = 0$). Then it is readily shown that

$$\psi_0 = \psi_1 + E_1 h, \quad E_1 h = \frac{1}{C_1} \sqrt{\frac{2\nu\,\theta\varepsilon}{\pi}}\, \text{sh}\,\frac{e\psi_1}{2\theta}, \quad C_1 = \frac{\varepsilon_1}{4\pi h} = \frac{\eta}{\psi_0 - \psi_1}, \tag{12}$$

$$\eta = \sqrt{\frac{2\nu\,\theta\varepsilon}{\pi}}\, \text{sh}\,\frac{e\psi_1}{2\theta}, \tag{13}$$

E_1 being the electric field strength inside the molecular condenser and C_1 the integral capacity of the condenser.

Elimination of ψ_1 from (12) and (13) yields a relation in η and ψ_0:

$$\psi_0 = \frac{\eta}{C_1} + \frac{2\theta}{e}\, \text{arsh}\left[\frac{e\eta}{2\theta}\,\bigg/\,\frac{\varepsilon\varkappa}{4\pi}\right]. \tag{14}$$

In addition to $\eta = \eta(\psi_0)$, two other parameters of the double layer can be directly estimated from experiment, namely, the total capacity of the double layer:

$$C = \frac{\partial\eta}{\partial\psi_0} = \frac{\partial\eta}{\partial\psi_1} \cdot \frac{1}{\partial\psi_0/\partial\psi_1}, \tag{15}$$

and the alteration of the surface tension, $\Delta\sigma$, at the interface between the electrolyte and the external medium:

$$\Delta\sigma = -\int_{\psi_0^*}^{\psi_0} \eta'(\psi_0')\,d\psi_0' = -\int_{\psi_1^*}^{\psi_1} \eta'(\psi_1')\,\frac{\partial\psi_0'(\psi_1')}{\partial\psi_1'}\,d\psi_1'. \tag{16}$$

Here the primed quantities represent intermediate values assumed by η, ψ_0, and ψ_1 as the double layer potential changes from $\psi_0' = \psi_0^*$, when $\Delta\sigma = 0$, to $\psi_0' = \psi_0$.

It is useful to introduce certain transformations in (15) and (16). In addition to the total double layer capacity, C, we now introduce the differential capacity of the molecular condenser*

$$C_{1df} = \frac{\partial \eta}{\partial (\psi_0 - \psi_1)} = \frac{\partial}{\partial (hE_1)} \left(\frac{\varepsilon_1 E_1}{4\pi} \right) = \frac{1}{4\pi} \frac{\varepsilon_1 + E_1 \dfrac{\partial \varepsilon_1}{\partial E_1}}{h + E_1 \dfrac{\partial h}{\partial E_1}},$$

(17)

and the capacity of the diffuse part of the double layer

$$C_2 = \frac{\partial \eta}{\partial \psi_1} = \frac{\varepsilon \varkappa}{4\pi} \, \mathrm{ch} \, \frac{e\psi_1}{2\theta}; \quad \varkappa^2 = \frac{8\pi v e^2}{\varepsilon \theta}.$$

(18)

Since (12) shows that

$$\frac{\partial \psi_0}{\partial \psi_1} = 1 + \frac{\partial \eta}{\partial \psi_1} \cdot \frac{\partial (hE_1)}{\partial \eta} = 1 + \frac{C_2}{C_{1df}},$$

(15) can be written as

$$C = \frac{C_{1df} C_2}{C_{1df} + C_2}, \quad \frac{1}{C} = \frac{1}{C_{1df}} + \frac{1}{C_2}.$$

(19)

We now pass to a consideration of the surface tension. By drawing on (12) and the fact that $\eta = D_1/4\pi$, where $D_1 = \varepsilon_1 E_1$ is the electromagnetic induction vector, (16) can be transformed to:

$$\Delta \sigma = \Delta \sigma_\eta + \Delta \sigma_c, \quad \Delta \sigma_\eta = - \int_{\psi_1^*}^{\psi_1} \eta'(\psi_1') \, d\psi_1', \quad \Delta \sigma_c = - h \int_{E_1''}^{E_1} \frac{D_1}{4\pi} \, dE_1.$$

(20)

Here $\Delta \sigma_\eta$ is the increase in the free energy per unit surface area resulting from the alteration of the charge (charging work), and $\Delta \sigma_c$ is the increase in the free energy resulting from the accumulation of electrostatic energy in the molecular condenser.

V. All the fundamental static properties of the double electric layer in the neighborhood of a uniform ideally polarized electrode are covered by Eqs. (12)-(20). We make the following points prior to checking these equations with experiment.

1. Equations (15) and (16) (which can be written as $C = \partial \eta / \partial \psi_0 = \partial^2 \sigma / \partial \psi_0^2$) have absolutely no connection with the double layer model proposed by Gouy and Stern. In fact these equations are not associated with any model, but follow from the laws of thermodynamics. Thus all attempts at developing the function $C = C(\psi_0)$ from data on $\eta = \eta(\psi_0)$, or $\sigma = \sigma(\psi_0)$, are tests of thermodynamic identities rather than the Gouy—Stern theory.

2. For the derivation of Eqs. (19) and (20) it must be assumed that there is a region, $\{0, h\}$, in the double layer which is free of bulk charges. Thus the values of C and $\Delta \sigma$ determined from these relations depend on the charge distribution in the diffuse part of the double layer (Gouy layer) only in so far as C_2 and $\Delta \sigma_\eta$ are fixed by (13) and (18). It follows that C and $\Delta \sigma$ will be independent of C_2 and $\Delta \sigma_\eta$ when $C_2 \gg C_{1df}$ (i.e., when the concentration is high and the value of $\varkappa$ large, or when the potential, ψ_0, is high and $e\psi_1/2\theta$ large),†

---

*It will be assumed for simplicity that h = const and $(\partial / \partial E_1)\{ E_1(\varepsilon_1 / 4\pi h)\} = C_1 + E_1(\partial C_1 / \partial E_1)$.

† The part of this statement concerning C is obvious from (19). The part concerning $\Delta \sigma$ follows from the fact that $\Delta \sigma \simeq - (C_1/2) (\psi_0 - \psi_1)^2$ when $\psi_1^* = 0$, as shown by (12) and (16). Then $C_2 \gg C_1$ implies that $\psi_0 \gg \psi_1$, and thus, $\Delta \sigma \simeq C_1 \psi_0^2 /2$.

and the charge structure then becomes independent of the double layer as well. It follows that measurements of the double layer capacity, or the surface tension, at high concentrations, or potentials, can confirm the presence of the molecular condenser, but are incapable of furnishing any indication whether the Gouy–Stern model correctly describes the charge distribution in the layer.

3. The single defining equation for $\eta = \eta(\psi_0)$, (14), reflects all the specific details of the Gouy–Stern model. Unfortunately, this equation also contains the capacity of the molecular condenser, C_1, which cannot be theoretically estimated. It is also true that C_1 cannot be determined from measurements on $C = C(\psi_0)$ or $\Delta\sigma = \Delta\sigma(\psi_0)$, it having been shown at the outset that these relations are identities following from $\eta = \eta(\psi_0)$. * In fact, there has, until recently, been no method for the direct measurement of C_1, and therefore no possibility of a direct test of the Gouy–Stern theory. B. M. Grafov and V. G. Levich [4] have, however, just proposed a method for determining ψ_1 from measurements on the passage of alternating current through the electrode-solution system. A knowledge of ψ_1 makes it possible to calculate the double layer charge from Eq. (13), and the value of η obtained in this way can then be compared with that resulting from measurements of the total double layer capacity, C. Such experiments have not yet been carried out, although they would be of great significance for the Gouy–Stern theory.

VI. Grahame [5] was the first to note that the relation between C, η, $\Delta\sigma$, and the bulk concentration, ν, of the electrolyte is the only consequence of the Gouy–Stern theory for the uniform ideally polarized electrode open to experimental check. A check of this type using measured double layer capacities for the mercury electrode–aqueous NaF solution system† showed brilliant agreement between experimental results and theoretical predictions. This comparison was inexact in two respects, however. First, C_1 was assumed to be constant and its value calculated by applying the Gouy theory to the results of experiments at $\nu = 0.916$ mole/liter. The theory is, however, known to be far from exact in this range of concentrations. Had the determination of C_1 been made from the results of experiments at $\nu = 0.001$ mole/liter, where the Gouy theory is more nearly exact, Grahame would have found no agreement at all with the experimental results (see the figure). ‡ Second, Grahame assumed C_1 to be a function of ψ_0, although it is obvious that the capacity of the molecular condenser must depend on $\psi_0 - \psi_1$, the potential drop between its plates, (or, what amounts to the same thing, on the internal field strength, E_1) rather than the total potential drop between electrode and solution. We have reworked Grahame's data in the light of these corrections, calculating C_1 from Eq. (14) and using the values of η and ψ_0 given in [5]. The results of these calculations are shown in the figure where C_1 has been plotted against $(\psi_0 - \psi_1)$ for four values $\nu = 0.001$; 0.01; 0.1; and 0.916 mole/liter.

The quantity $(\psi_0 - \psi_1)$ was chosen as the independent variable for the following reasons. C_1 is, in general, a function of ψ_0 and ν since $C_1 = f(\psi_0 - \psi_1, \nu_{max}^{(+)}, \nu_{max}^{(-)})$, where $\nu_{max}^{(\pm)} = \nu \exp[\pm e\psi_1/\theta]$ being the number of positive and negative ions on the external plate of the molecular condenser. But C_1 no longer depends on ν_{max} when ν_{max} is low, (for example, less than 0.1–1.0 mole/liter), for the molecular condenser then consists essentially of solvent molecules. Thus one has $C_1 \cong f(\psi_0 - \psi_1)$ when ν and $e\psi_1$ are small. The outer plate of the condenser is free of both positive and negative ions when $e\psi_1 \gg \theta$, and one then has $C_1 \cong f(\psi_0 - \psi_1, \nu e^{e\psi/\theta}) = f(\psi_0 - \psi_1, C_1^2(\psi_0 - \psi_1)^2) = f(\psi_0 - \psi_1)$, since (12) shows that $\nu e^{+e\psi_1/\theta} \simeq C_1^2(2\pi/\varepsilon\theta)(\psi_0 - \psi_1)^2$. Thus

---

*We will explain this statement by an example. Let it be supposed that theory has shown that $\eta = a\psi_0^n$, a being an unknown function of ψ_0. Experimental measurements on $\eta = \eta(\psi_0)$ lead to the form of the function $a = a(\psi_0)$. Let it be further assumed that other experiments have led to the form of the function $C = C(\psi_0)$, where $C = d\eta/d\psi_0$. It is now clear that these latter experiments cannot be used to check the relation $\eta = a\psi_0^n$, since $C = a'\psi_0^n + na\psi_0^{n-1}$. Substitution of $a = \eta/\psi_0^n$ and $a' = (d/d\psi_0)(\eta/\psi_0^n) = C/\psi_0^n - n\eta/\psi_0^{n+1}$, as found from the first experiments, leads to an identity, $C = C$, which would be satisfied by any form of the function $\eta = \eta(\psi_0)$, and, more specifically, for any value of n.

†Grahame has pointed out that specific adsorption does not occur in this system and that $\eta_1 = 0$. Equation (10) shows that a new unknown, $\Phi_a^{(0)}$, appears in $\eta_1 = \eta(\psi_0)$ when $\eta_1 \neq 0$, thus complicating the check still more.

‡This has been kindly pointed out by V. Krylov in a personal communication, for which the author is especially grateful.

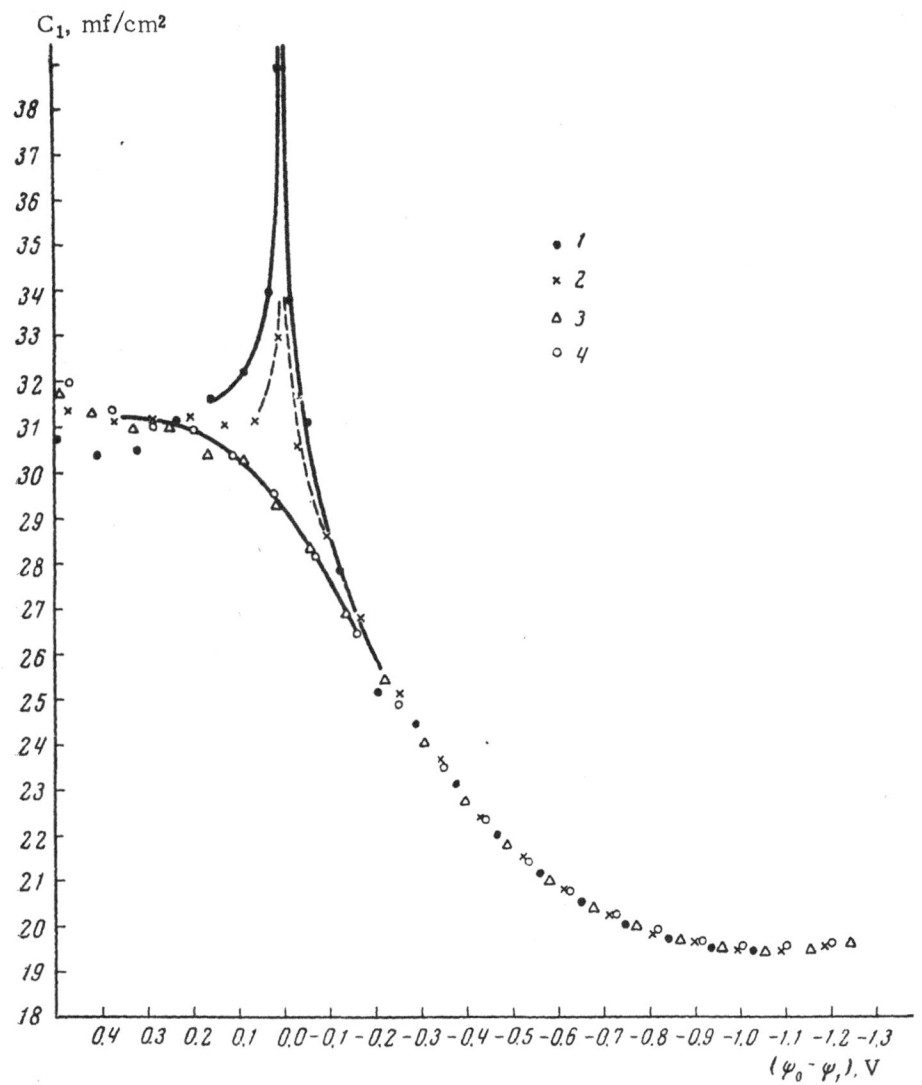

The relation between the capacity, C_1, of the molecular condenser and the potential drop, $\psi_0 - \psi_1$, between the condenser plates: 1) $\nu = 0.001$ mole/liter; 2) $\nu = 0.010$ mole/liter; 3) $\nu = 0.100$ mole/liter; 4) $\nu = 0.916$ mole/liter.

the ν dependence vanishes when C_1 is considered as a function of $\psi_0 - \psi_1$.* This, itself, could serve as indirect confirmation of the Gouy—Stern theory.

It is seen from the figure that while the points for the concentrations ν = 0.1 and 0.916 mole/liter fall on a single curve over the entire $(\psi_0 - \psi_1)$ interval, the points for ν = 0.001 and 0.01 mole/liter fall on this curve, only up to $(\psi_0 - \psi_1) \gtrsim 0.25$ V. Since the value of $\varepsilon \varkappa / 4\pi$ at ν 0.1 and 0.916 mole/liter is 71 and 214 mf/cm², respectively, $C_2 = \varepsilon \varkappa / 4\pi \, ch \, e\psi_1 / 2\theta$ is invariably considerably greater than $C_1 \simeq 20\text{-}30$ mf/cm² at these concentrations. It has been pointed out above that only the presence of the molecular condenser can be confirmed from results obtained with $C_2 \gg C_1$, judgments concerning the charge distribution in the double layer being beyond the scope of such work. Thus the fact that C_1 is independent of ν when $C_2 \gg C_1$, is more confirmation of the existence of the molecular condenser.

*It is obvious that this statement is not exact.

Simple calculations show that $C_2 \gtrsim 100$ mf/cm$^2 \gg C_1$ when $\nu = 0.001$ and 0.01 mole/liter, and $|\psi_0 - \psi|$ ≥ 0.5 V. Thus C_1 is also independent of ν when the concentration is low and the potential difference $(\psi_0 - \psi_1)$ high. The relation between C_1 and $(\psi_0 - \psi_1)$ is very close, however, when both the concentration and $(\psi_0 - \psi_1)$ are low. This is clear indication that the double layer charge distribution predicted by the Gouy–Stern theory is not exact, and the error actually amounts to as much as 30%. These departures must be presumed to arise from the failure of the Gouy–Stern theory to take image forces into account since intrinsic ionic volumes can be of no significance at such low concentrations.

The departure between the various curves diminishes as ν and $(\psi_0 - \psi_1)$ increase in value. This is not indication, however, that the Gouy–Stern theory gives a more exact description of the double layer charge distribution under these conditions; in fact, the above analysis would suggest that exactly the opposite is true. The diminishing divergence between the curves goes back to the fact that C_2 increases with ν and ψ_0, so that the experimental results become less and less sensitive to the charge structure. Moreover, calculation shows that ν_{max} varies from 1 to 20 mole/liter over the interval $(\psi_0 - \psi_1) \geq 0.25$ V for all four values of ν. Having the intrinsic ionic volumes, it is obvious that the Gouy–Stern theory is in no position to give a correct description of the double layer charge distribution at these high concentrations.

VI. Although a vast amount of work has been published on the determination of $C = C(\psi_0)$, $\eta = \eta(\psi_0)$, and $\Delta\sigma = \Delta\sigma(\psi_0)$, only the data of Grahame are such as to permit a comparison of theoretical predictions and experimental results. The preceding discussion has made it clear, however, that only limited conclusions can be reached from tests with uniform electrodes. Considerable interest attaches to systems involving overlapping double layers, where the double layer charge distribution can be studied at any ψ_0. The introduction of a secondary electrode unfortunately gives rise to a new, van der Waals, component of the splitting pressure and thus makes for difficulty in the interpretation of the experimental results. A. Sheludko and D. Ekserova [6] have, nevertheless, been able to show that the Deryagin–Landau theory of the electrostatic splitting pressure based on the Gouy Eq. (8) gives a good description of the experimental data in aqueous KCl solutions at $\nu \simeq 10^{-4}$ mole/liter.* Agreement with experiment can be maintained up to concentrations of the order of 10^{-3} mole/liter if the van der Waals component of the splitting pressure is assumed to be negligibly small at film depths ranging from 0.05 to 0.1 μ. These authors note, however, that this assumption is in contradiction with conclusions based on their own work with KCl solutions of higher concentrations [8]. The results of [6] can therefore serve as indication that certain deviations between theory and experiment set in at $\nu \approx 10^{-3}$ mole/liter, a conclusion consistent with the above analysis of the Grahame data.

Considerable interest attaches to studies of hydrophobic colloid stability by V. I. Barboi and Yu. M. Glazman [9] which show the existence of definite deviations between the experimental data and the theory as presently formulated. The source of these deviations is not yet clear; they could arise from the inadequacy of the Gouy model of the double layer, or from failure to allow for specific adsorption in the Stern layer in the stability calculations, or from the fact that the alteration of the van der Waals component of the splitting pressure with distance has not yet been thoroughly studied.

Conclusions

1. Comparison of the predictions of the Gouy–Stern theory with the experimental data has confirmed the existence of the molecular condenser over the entire range of concentrations and potentials.

2. Confirmation of the Gouy model for the double layer charge distribution has been obtained at very low concentrations ($\nu \lesssim 10^{-4}$ mole/liter) and potentials. The error in the theory increases as the concentration rises, reaching 20-30% at $\nu \simeq 10^{-2} \div 10^{-3}$ mole/liter. None of the data confirm the validity of this aspect of the theory at high concentrations.

3. Theoretical analysis shows that the errors inherent in the Gouy–Stern theory will increase with the concentration and potential, and rise with passage to solutions of weak or polyvalent electrolytes.

---

*It should be emphasized that it is here a matter of a numerical check of the Deryagin–Landau theory; the qualitative correctness of the theory is well supported.

Literature

1. N. N. Bogolyubov. Problems of Dynamic Theory and Statistical Physics, Moscow, Gostekhizdat (1946).
2. V. G. Levich and V. A. Kir'yanov, Dokl. Akad. Nauk SSSR 131(5):1134 (1960).
3. H. R. Kruyt, The Science of Colloids [Russian translation], Moscow, IL (1955).
4. B. M. Grafov and V. G. Levich, Dokl. Akad. Nauk SSSR 146(6):1372 (1962).
5. D. C. Grahame. J. Am. Chem. Soc., 76(19):4819 (1954).
6. A. Sheludko and D. Ekserova. Dokl. Akad. Nauk SSSR, 127(1):149 (1959).
7. B. V. Deryagin and L. D. Landau. Zh. Eksperim. i Teor. Fiz. 11:802 (1941); Acta Physica URSS 14:633 (1941).
8. A. Sheludko and D. Ekserova. Izv. Bolgar. Akad. Nauk, Khim. Inst. (1959), p. 105.
9. V. M. Barboi and Yu. M. Glazman, Dokl. Akad. Nauk SSSR, 138(1):139 (1961); Kolloidn. Zh. 23(4):376 (1961); 24(4):438 (1962).

THE STATISTICAL THEORY OF THE DOUBLE ELECTRIC LAYER
Part II
G. A. Martynov

Institute of Physical Chemistry, Acad. Sci., USSR
Laboratory for Surface Phenomena

The Effect of Image Forces

It is a well-known fact that the Gouy–Stern theory takes no account of either Debye atmospheres or image forces in fixing the charge distribution in the double electric layer. The simple calculations of Communication I have shown that these factors can indeed be neglected when dealing with aqueous solutions of univalent electrolytes at very low concentrations. The significance of the Debye atmospheres and image forces increases markedly, however, in passing to higher concentrations, especially when the ions are of higher valency and the dielectric constant of the solvent low. The present communication will attempt to treat these effects more rigorously.

Formulation of the Problem

We will start, as in Communication I, from the N. N. Bogolubov expression [1] for the distribution functions:

$$\frac{\partial G_{a_1,\ldots,a_s}}{\partial r_1^\alpha} + \frac{G_{a_1,\ldots,a}}{\theta} \frac{\partial U_{a_1,\ldots,a_s}}{dr_1^\alpha} + \frac{1}{\theta} \int \sum_{1 \leqslant a_{s+1} \leqslant M} \nu_{a_{s+1}} G_{a_1,\ldots,a_{s+1}} \frac{\partial \Phi_{a_1 a_{s+1}}}{\partial r_1^\alpha} \, dr_{s+1} = 0. \tag{1}$$

Here $\theta = kT$ is the temperature, measured on the energy scale, $G_{a_1, \ldots, a_s}$ is the distribution function for s particles, $\nu_a = N_a/V$ is the number of particles of type a per unit volume, M is the number of particles of the various types, $dr_i = dx_i dy_i dz_i$, and $U_{a_1, \ldots, a_s}$ is the interaction energy in the complex consisting of the particles $a_1, \ldots, a_s$,

$$U_{a_1,\ldots,a_s} = \sum_{1 \leqslant i \leqslant s} U_{a_i}(\vec{r_i}) + \sum_{1 \leqslant i < j \leqslant s} \Phi_{a_i a_j}(\vec{r_i}, \vec{r_j}), \quad \Phi_{a_i a_j} = \Phi_{a_i a_j}^{(s)} + \Phi_{a_i a_j}^{(e)}. \tag{2}$$

U_{a_j} is the energy of the particle a_j in the external field, $\Phi_{a_i a_j}^{(S)}$ is the short-range component of this energy, and $\Phi_{a_i a_j}^{(e)}$ the electrostatic component of the interaction of particles a_i, a_j.

The $G_{a_1, \ldots, a_s}$ function must be normalized and symmetrical and show diminishing correlation with approach to infinity:

$$G_{a_1,\ldots,a_s} \to \prod_{1 \leqslant i < j \leqslant s} G_a \text{ when } r_{ij} = |\vec{r_i} - \vec{r_j}| \to \infty, \ 1 \leqslant i < j \leqslant s. \tag{3}$$

It must also satisfy an impermeability condition

$$G_{a_1,\ldots,a_s} \to f(\vec{r_1}, \ldots, \vec{r_s}) e^{-\Phi_{a_k, a_l}^{(s)}/\theta} \to 0, \ f \geqslant 0 \text{ when } r_{kl} \to 0, \ 1 \leqslant k < l \leqslant s. \tag{4}$$

It will be useful for what follows to pass from the distribution functions $G_{a_1, \ldots, a_s}$ to the correlation functions $g_{a_1, \ldots, a_s}$ through the relations: $G_a = g_a$, $G_{ab} = g_a g_b \gamma_{ab}(1 + g_{ab})$, $G_{abc} = g_a g_b g_c \, \gamma_{ab} \gamma_{ac} \gamma_{bc} \, (1 + g_{ab} + g_{ac} + g_{bc} + g_{abc})$, etc., where $\gamma_{ab} = \exp[-\Phi_{ab}^{(s)}/\theta]$.

It is obvious that $G_{a_1, \ldots, a_s}$ functions of this type will satisfy the conditions of symmetry and impenentrability, (4); with the following limitations:

$$g_{a_1, \ldots, a_s} \to 0 \text{ when } r_{ij} \to \infty, \quad 1 \leqslant i < j \leqslant s, \; s \geqslant 2; \tag{5}$$

they will also satisfy the condition of (3).

Substitution of (4) into (1) leads to:

$$\frac{\partial \ln g_a}{\partial r_a^\alpha} + \frac{1}{\theta} \frac{\partial U_a}{\partial r_a^\alpha} + \frac{1}{\theta} \int\limits_{z_c \geqslant 0} \sum_{1 \leqslant c \leqslant M} \nu_c g_c \gamma_{ac} (1 + g_{ac}) \frac{\partial \Phi_{ac}}{\partial r_a^\alpha} dr_c = 0, \tag{6}$$

$$\frac{\partial g_{ab}}{\partial r_a^\alpha} + \frac{1 + g_{ab}}{\theta} \frac{\partial \Phi_{ab}}{\partial r_a^\alpha} + \frac{1}{\theta} \int\limits_{z_c \geqslant 0} \sum_{1 \leqslant c \leqslant M} \nu_c g_c \gamma_{ac} \, [\gamma_{bc} (1 + g_{ab} + g_{ac} + g_{bc})$$

$$- (1 + g_{ab} + g_{ac}) + (\gamma_{bc} g_{abc} - g_{ab} g_{ac})] \frac{\partial \Phi_{ac}}{\partial r_a^\alpha} dr_c = 0. \tag{7}$$

We will now specify the forms of the functions U_a and Φ_{ab} in (6) and (7). It has been shown in the previous communication that the general case is such that $U_a = \Phi_a^{(coul)} + \Phi_a^{(im)} + \Phi_a^{(ad)}$, where $\Phi_a^{(coul)} = e_a[\psi_0 - (2\pi\eta/\varepsilon)z_a]$ is the coulombic energy of an a ion of charge e_a in the field of a plate carrying a charge at potential ψ_0, $\Phi_a^{(im)} = (e_a^2/2\varepsilon)(\beta/2z_a)$; $\beta = (\varepsilon - \varepsilon')/(\varepsilon + \varepsilon')$, Φ_a (im) is the energy of interaction of the ion with its own image (ε and ε' are the dielectric constants of solvent and external medium, respectively), and $\Phi_a^{(ad)}$ is the interaction energy of the ion and external medium resulting from forces of nonelectrostatic origin.

We pause here to go into some detail concerning the physical meanings to be attached to η and ε', since the treatment of these parameters is no longer trivial when it is a matter of calculating image forces. A double electric layer can be formed either by selective adsorption of ions of fixed charge on a neutral electrode in a neutral electrolyte, or by the introduction of a charged electrode into a neutral electrolyte. We will consider the first of these cases, supposing for simplicity that the electrode is an infinite plane surface. It will be assumed that $N_a^{(ad)}$ ions of type a have passed from the solution and been adsorbed on the electrode surface, so that this surface carries a charge

$$Q^{(ad)} = \sum_a e_a N_a^{(ad)}.$$

Since the adsorption points are scattered randomly over the electrode surface, this charge will be considered to be "smeared out" to a mean density $\eta^{ad} = Q^{(ad)}/S$. The process of adsorption will clearly leave the electrolyte with a charge $Q^{(E)} = -Q^{(ad)}$. The fact that the electrode acquires an image charge $Q^{(im)} = \beta Q$ from the adsorbed ions and the ions remaining in solution cannot lead to any misunderstanding here since summation over all ion charges in solutions shows that $Q = Q^{(ad)} + Q^{(e)} = 0$.

The situation is more involved when a charged electrode is immersed in an electrolyte. Such situation can obviously arise when two infinite plane-parallel plates are put into solution. Let it be supposed that an external emf has been applied and the charge $Q = \eta S$ carried from one of these plates to the other. The result of this transfer is to form a double layer of bulk charge $-Q$ in the neighborhood of plate I and a double layer of bulk charge $+Q$ in the neighborhood of plate II. The distance, H, between the plates being much greater than the depth of the double layer, the electric field of plate I will have no effect at all on the double layer charge distribution in the neighborhood of plate II, and vice versa. The interaction energy for an ion in layer II and the image of this ion in plate I will also be neglibibly small when the value of H is quite large. It is therefore convenient to replace the system of two plates and the included electrolyte by a half-system by removing the plane H/2 to infinity, and this will be done in what follows.

This half-system will not be neutral, however. Actually, the external charge Q will be completely neutralized by the double layer charge, $Q^{(e)} = -Q$. At the same time, $Q^{(e)}$ will form an image charge, $Q^{(im)} = \beta Q^{(e)} = -\beta Q$, on the plane surface. Thus the surface charge density is not $\eta = Q/S$, but $\eta + \eta^{(im)} = (Q + Q^{(im)})/S = (1 - \beta)\eta$. This absurd conclusion results from working with the half-system rather than the complete system of two planes plus the intervening electrolyte. When it is remembered that the second double layer charge, $-Q^{(e)} = Q$, will produce an image charge, $Q^{(im)} = -\beta Q^{(e)} = \beta Q = \beta \eta S$,* on the first plate and thus neutralize the image charge resulting from the first layer, it becomes clear that η must be treated as the external charge alone in a correct formulation of the problem, the image charge, $\eta^{(im)}$, disappearing when account is taken of the ions at infinity, which is to say of the ions located in the neighborhood of the second plate.

We now pass to a discussion of the physical meaning of ε', the dielectric constant of the external medium. The charges adsorbed in the Stern layer will obviously retain a certain degree of mobility in the plane of this layer. From this it follows that the approach of a dissolved ion to the solution interface will somewhat displace the adsorbed ions from their equilibrium positions, the result being that the interface acquires a "bound" adsorption charge that opposes the penetration of the electric field of the "bulk" ion into the external medium. It is possible to allow for this effect by replacing the actual dielectric constant, ε', of the external medium by a certain effective value, ε'_{eff}. The field of the "bulk" ion will not penetrate the external medium at all in the special case in which the mobility of the adsorbed ion is much greater than the mobility of the ion in solution (a situation which can arise at the solution—air interace), and one will then have $\varepsilon'_{eff} = \infty$, even though the true value of ε' for air is unity. Proceeding in this way, we will understand ε' to designate this effective value in what follows.

The final step in the formulation of the problem is to fix the function $\Phi_{ab}^{(e)}$ of (6) and (7). It has been shown in Communication I that $\Phi_{ab}^{(e)} = \Phi_{ab}^{(coul)} + \Phi_{ab}^{(im)}$, where $\Phi_{ab}^{(coul)} = (e_a e_b/\varepsilon)(1/r_{ab})$ being the coulombic interaction energy of two point charges, e_a and e_b, and $\Phi_{ab}^{(im)} = (e_a e_b/\varepsilon)(\beta/r_{ab}^*)$ the energy of interaction of ion a with the image of ion b. It will be assumed for simplicity that the ions are point structures, so that $\Phi_{ab}^{(s)} = 0$ and $\gamma_{ab} \equiv 1$. It is clear that this assumption would be justified only in the case of rather dilute solutions.

The Equation for the Unary Function

We will transform the system of Eqs. (6)–(7) to a somewhat more suitable form before attempting its solution. New functions

$$\frac{\partial \chi_a(z_a)}{\partial z_a} = \frac{\partial \Phi_a^{(im)}}{\partial z_a} + \int\limits_{z_c \geqslant 0} \sum\limits_{1 \leqslant c \leqslant M} v_c g_c g_{ac} \frac{\partial \Phi_{ac}}{\partial z_a} dr_c, \tag{8}$$

$$w_a = \chi_a(z_a) + e_a \varphi(z_a) = e_a \varphi(z_a) + \left[\Phi_a^{(im)}(z_a) + \int\limits_{z_c \geqslant 0} \sum\limits_{1 \leqslant c \leqslant M} v_c g_c \left(\int\limits_{\infty}^{z_a} g_{ac} \frac{\partial \Phi_{ac}}{\partial z_a} dz_a \right) dr_c \right],$$

$$g_a = \exp\left[-\frac{1}{\theta}(\chi_a + \Phi_a^{(ad)} + e_a \varphi) \right] = \exp\left[-\frac{1}{\theta}(\omega_a + \Phi_a^{(ad)}) \right]. \tag{9}$$

will first be introduced into (6).

---

*Although the energy of interaction of a charge e with its own image tends to zero as the distance from the charge to the plate increases, it is to be recalled that the image charge which it imposes on this plate will remain constant at βe.

Substitution of (8) and (9) into (6) leads to

$$\frac{\partial \varphi}{\partial z_a} + \frac{2\pi\eta}{\varepsilon} - \frac{1}{\varepsilon} \int\limits_{z_c > 0} \sum_{1 \leqslant c \leqslant M} v_c e_c e^{-\frac{w_c + \Phi_c^{(ad)}}{\theta}} \frac{\partial f^{(ac)}}{\partial z_a} \, dr_c = 0,$$

(10)

where

$$\Phi_{ab} = \Phi_{ab}^{(coul)} + \Phi_{ab}^{(im)} = \frac{e_a e_b}{\varepsilon} \left(\frac{1}{r_{ab}} + \frac{\beta}{r_{ab}^*} \right) = \frac{e_a e_b}{\varepsilon} f(\vec{r}_a, \vec{r}_b) = \frac{e_a e_b}{\varepsilon} f^{(ab)}.$$

(11)

Equation (10) can be readily integrated with respect to z_a. Application of the Laplace operator, Δ_a, to the resulting integral equation with

$$\Delta_a f^{(ab)} = - 4\pi \{\delta(r_{ab}) - \beta\delta \ r_{ab}^*)\},$$

(12)

leads Eq. (10) to the form *

$$\varphi''(z) = - \frac{4\pi}{\varepsilon} \sum_a e_a v_a \exp\left[- \frac{1}{\theta} (\Phi_a^{(ad)}(z) + \chi_a(z) + e_a\varphi(z)) \right].$$

(13)

We now draw on the fact that

$$\Phi_a^{(ad)}(z) = \begin{cases} + \infty, & 0 \leqslant z < h, \\ \Phi_a^{(0)}, & h \leqslant z \leqslant h + \lambda, \ \lambda \simeq 0 \\ 0, & h + \lambda < z \leqslant \infty \end{cases}$$

(14)

The result obtained here is

$$\varphi_1^{..}(z) = 0, \ 0 \leqslant z < h;$$

(15)

$$\varphi''(z) = - \frac{4\pi}{\varepsilon} \sum_a e_a v_a \exp\left[- \frac{1}{\theta} (\chi_a + e_a\varphi) \right], \ h + \lambda \simeq h < z \leqslant \infty.$$

(16)

It is to be noted that the condition $g_a(\infty) = 1$ implies that

$$\varphi(\infty) = 0.$$

(17)

The other boundary conditions for the function φ are found by multiplying (13) by $\partial f^{(ac)}/\partial z_c$ and then integrating over $z_a \geq 0$. Here the integral of the right-hand member of (13) proves to be identically equal to the integral in Eq. (10). Thus full equivalence of (10) and (13) requires that

$$\frac{\partial \varphi}{\partial z} + \frac{2\pi\eta}{\varepsilon} = - \frac{1}{4\pi} \int\limits_{z_c \geqslant 0} \Delta_a \varphi \, \frac{\partial f^{(ac)}}{\partial z_c} \, dr_a.$$

Evaluating this integral through the Green equation and taking account of the fact that the image charge, $\beta\eta$, is neutralized by the ion charges at infinity, one obtains

$$z = 0, \ \varphi_1(0) = \psi_0,$$

(18)

---

* The second δ-function of (12) makes no contribution to (13) since integration of (10) is over the half-space $z_c \geq 0$ in which r_{ab}^* does not go to zero.

$$z = h, \quad \varphi_1(h) = \varphi(h) = \psi_1, \quad \varepsilon_1 \varphi_1'(h) - \varepsilon \varphi'(h) = -4\pi \eta^{(ad)},$$

$$\eta^{(ad)} = \sum_{1 \leqslant a \leqslant M} e_a \nu_a \exp\left[-\frac{1}{\theta} \left(\Phi_a^{(0)} + e_a \psi_1 + \chi(h) \right) \right]. \tag{19}$$

Here ε_1, the dielectric constant of the layer $(0, h)$, is different from ε. It is clear from these equations that φ is an electrostatic potential, (15) and (16) being forms of the Poisson Equation and (17)-(19) the usual electrostatic boundary conditions. It is to be noted that (16) is not a differential equation, χ_a being related to φ through the integral equation (8).

The Equation for the Binary Function

We now pass to Eq. (7) for the binary function, neglecting from the very outset the nonlinear terms, $\gamma_{bc} g_{abc} - g_{ab} g_{ac}$, which are second order in comparison with terms proportional to g_{ab}. For the case in which

$$\varkappa r_0 \ll 1 \quad \left(\text{where} \quad \varkappa^2 = \frac{4\pi \Sigma \nu_a e_a^2}{\varepsilon \theta}, \quad r_0 = \frac{e_a^2}{\varepsilon \theta} \right) \tag{20}$$

one can assume $g_{ab} = (e_a e_b / \varepsilon) q_1 + q_2$ and pass to dimensionless variables to show that the $g_{ab}(\partial / \partial r_a^\alpha / (\Phi_{ab}^{(e)} / \theta))$ term of (7) is much smaller than the other terms and can be neglected for this reason. For aqueous solutions of symmetrical electrolytes, the condition of (20) takes the form

$$\nu_0 \ll \frac{0.01}{k^6}, \tag{21}$$

$k = e_a / e$ being the valence of ion a and ν_0 the concentration of the electrolyte in mole/liter. The condition of (20) will be assumed satisfied in what follows. Equation (7) then reduces to

$$\frac{\partial g_{ab}}{\partial r_a^\alpha} + \frac{1}{\theta} \frac{\partial \Phi_{ab}^{(e)}}{\partial r_a^\alpha} + \frac{1}{\theta} \int\limits_{z_c \geqslant 0} \sum_{1 \leqslant c \leqslant M} \nu_c \, g_c \, g_{bc} \frac{\partial \Phi_{ac}^{(e)}}{\partial r_a^\alpha} \, dr_c = 0. \tag{22}$$

We now seek a solution of (22) which would have the form

$$g_{ab} = -\frac{e_a}{\varepsilon \theta} \varphi_b^{(ab)}, \tag{23}$$

the superscripts here indicating the independent variables for the function $\varphi_b^{(ab)}$, just as in (11), and the subscript the ion whose charge is proportional to $\varphi_b^{(ab)}$.

Substitution of (23) into (22) followed by passage to vector notation leads to

$$\vec{\nabla}_a \varphi_b^{(ab)} - \frac{e_b}{\varepsilon} \vec{\nabla}_a f^{(ab)} + \frac{1}{\varepsilon \theta} \int\limits_{z_c \geqslant 0} \sum_{1 \leqslant c \leqslant M} \nu_c e^{-w_c/\theta} \varphi_b^{(bc)} \vec{\nabla}_a f^{(ac)} \, dr_c = 0 \tag{24}$$

By acting on (24) with the hamiltonian operator, $\vec{\nabla}_a$, and drawing on (12), (24) can be carried over to

$$\Delta_a \varphi_a^{(ab)} - \varphi_b^{(ab)} \left(\frac{4\pi}{\varepsilon \theta} \sum_{1 \leqslant a \leqslant M} \nu_a e_a^2 e^{-w_a(z_a)/\theta} \right) = -\frac{4\pi e_b}{\varepsilon} \delta(r_{ab}). \tag{25}$$

One boundary condition for $\varphi_b^{(ab)}$ is obtained from (5), namely

$$\varphi_b^{(ab)} \to 0 \quad \text{when} \quad r_{ab} \to \infty. \tag{26}$$

The boundary condition at the electrolyte-external medium interface is obtained as above, multiplying (25) by $\nabla_c f^{(ac)}$ and integrating over $z_a \geq 0$, neglecting for simplicity the layer $\{0, h\}$ in which $\varepsilon_1 \neq \varepsilon$ and $g_a = 0$. The equivalence of (24) and (25) combined with the Green formula leads to

$$\int_{-\infty}^{\infty} \left\{ f^{(ac)} \frac{d\varphi_b^{(bc)}}{dz_c} - \varphi_b^{(bc)} \frac{df^{(ac)}}{dz_c} \right\}_{z_c=0} dx_c \, dy_c = 0. \tag{27}$$

Since $r_{ac} = r_{ac}^*$ and $\partial/\partial z_c(1/r_{ac}) = -\partial/\partial z_c(1/r_{ac}^*)$, when $z_c = 0$ it follows that

$$f^{(ac)}\Big|_{z_c=0} = (1+\beta)/r_{ac}, \quad \frac{\partial f^{ac}}{\partial z_c}\Big|_{z_c=0} = \frac{\partial}{\partial z_c}\left(\frac{1-\beta}{r_{ac}}\right).$$

Remembering, finally, that $(1+\beta) = 2\varepsilon'/(\varepsilon+\varepsilon')$, and $(1-\beta) = 2\varepsilon/(\varepsilon+\varepsilon')$, (27) can be written as

$$\iint_{-\infty}^{+\infty} \left\{ \varphi_b^{(bc)} \frac{\partial}{\partial z_c}\left(\frac{1}{r_{ac}}\right) - \frac{1}{r_{ac}}\left[\frac{\varepsilon}{\varepsilon'} \frac{\partial \varphi_b^{(bc)}}{\partial z_c}\right] \right\}_{z_c=0} dx_c \, dy_c = 0. \tag{28}$$

It is readily shown that a solution of (25) with the boundary conditions of (26) and (28) is equivalent to a solution of the system

$$\Delta_a \varphi_b^{(ab)} = -\frac{4\pi}{\varepsilon}\left\{ e_b \delta(r_{ab}) - \varphi_b^{(ab)}\left[\frac{4\pi}{\varepsilon\theta} \sum_{1\leq a\leq M} v_a e_a^2 \exp\left(-\frac{w_a}{\theta}\right)\right]\right\}, \quad z_a, \ z_b \geq 0; \tag{29}$$

$$\Delta_a \psi_b^{(ab)} = 0, \qquad z_a < 0, \qquad z_b > 0; \tag{29'}$$

$$z_a = 0, \qquad \varphi_b^{(ab)} = \psi_b^{(ab)}, \qquad \varepsilon \frac{\partial \varphi_b^{(ab)}}{\partial z_a} = \varepsilon' \frac{\partial \psi_b^{(ab)}}{\partial z_a}, \tag{29''}$$

$$r_{ab} = \infty, \qquad \varphi_b^{(ab)} = \psi_b^{(ab)} = 0. \tag{29'''}$$

In order to prove as much, we multiply (29") by r_{ab}^{-1} and integrate over the half-space $z_a \leq 0$. The resultant integral can be transformed with the aid of the Green formula to obtain

$$\int_{z_a\leq 0} \frac{1}{r_{ab}} \Delta_a \psi_b^{(ab)} \, dr_a = \int_{z_a\leq 0} \psi_b^{(ab)} \Delta_a \left(\frac{1}{r_{ab}}\right) dr_a + \iint_{-\infty}^{+\infty}\left[\frac{1}{r_{ab}} \frac{\partial \psi_b^{(ab)}}{\partial z_a} - \psi_b^{(ab)} \frac{\partial}{\partial z_a} \times \left(\frac{1}{r_{ab}}\right)\right]_{z_a=0} dx_a \, dy_a = 0. \tag{30}$$

The volume integral in the right-hand side of this expression vanishes since $\Delta_a(1/r_{ab}) = -4\pi \delta(r_{ab})$ and r_{ab} does not go to zero in the region $z_a \leq 0$. Equation (30) can be transformed to (28) and the identity of the two systems thus proven by replacing the dummy index, a, of the surface integral by c and drawing on (29''').

The physical meaning of (29) is obvious — this is a system of electrostatic equations describing the field of a point charge immersed in a medium of bulk charge ρ, the magnitude of this bulk charge being fixed by the potential $\varphi_b^{(ab)}$ at point r_a. A peculiarity of (29) is that η does not enter into its boundary condition, (29'''), even though the external charge is located on the surface $z = 0$.

The Solution of the Equations

We will solve the system of Eqs. (15), (16), (18), (19), and (29) by successive approximations, assuming for simplicity that the electrolyte is symmetrical $(e_+ = -e_- = e, \ e_a^2 = e^2)$ and the potential, ψ_1, so low that

$$e\psi_1 \ll \theta. \tag{31}$$

We will also assume that it is the potential ψ_1 in the plane $z = h$ which has been given rather than the charges η, $\eta^{(ad)}$. The zeroth approximations for $\varphi(z)$ and $\varphi_b^{(ab)}$ will be obtained from (16) and (29''') by setting $\exp(-\chi_a/\theta) = 1$. It is clear that this procedure is justified when $\chi = \chi_a \ll \theta$. It follows from (8), however, that $\chi \simeq (e^2/2\varepsilon)(1/2z)$. Thus the requirement $\chi \ll \theta$ is not satisfied in the interval $h \leq z \leq (1/2)z^* = e^2/4\varepsilon\theta$ where χ is actually greater than θ. The approximation $\chi = 0$ is quite good, however, when $\varkappa z^* \ll 1$.* This can be most readily seen from the study of the original integral equations, (10) and (24). Both of these equations contain χ, but only under the sign of integration and in the form $[-\chi/\theta]$. Let us now break each of these integrals down into two, one over the interval $\{h, z^*\}$ and the other over the interval $\{z^*, \infty\}$. The quantity $\exp[-\chi/\theta]$ differs markedly from unity in the first interval and is essentially equal to unity in the second. If, however, the value of the integral over $\{h, z^*\}$ is much less than its value over the interval $\{z^*, \infty\}$, replacing $\exp[-\chi/\theta]$ by unity in the first interval will not markedly alter the value of the entire integral and the error resulting from the assumption $\chi = 0$ will not be large. The last inequality will clearly be satisfied if the depth of the double layer is many times greater than z^*, that is to say, if $\varkappa z^* \ll 1$. Since the integral Eqs. (10) and (24) are fully equivalent to (16) and (29), replacement of $\exp[-\chi/\theta]$ in the latter by unity with $\varkappa z^* \ll 1$ will again not lead to a large error.

Expressing the $\exp[-\chi/\theta]$ term of (16) as a series and breaking this series off with the first nonzero member leads to the Gouy–Chapman equation as the zeroth approximation for $\varphi(z)$:

$$\varphi'' - \varkappa^2 \varphi = 0, \quad \varphi(h) = \psi_1, \quad \varphi(\infty) = 0, \tag{32}$$

ψ_1 being obtained from (19).

The solution of (32) being $\varphi = \psi_1 e^{-\varkappa z}$, it follows that

$$g_a = \exp\left[-\frac{e_a \psi_1}{\theta} e^{-\varkappa z}\right] \simeq 1 - \frac{e_a \psi_1}{\theta} e^{-\varkappa z} \simeq 1. \tag{33}$$

The zeroth approximation for $\varphi_b^{(ab)}$ is obtained by expressing the $e^{-w_a/\theta} = e^{-e_a \varphi/\theta}$ term of (29) as a series and breaking this series off with the first nonzero member (i.e., unity):

$$\left.\begin{aligned}
&\Delta_a \varphi_b^{(ab)} - \varkappa^2 \varphi_b^{(ab)} = -\frac{4\pi e_b}{\varepsilon} \delta(r_{ab}), \; z_a \geqslant 0, \; z_b \geqslant 0; \\
&\Delta_a \psi_b^{(ab)} = 0, \; z_a \leqslant 0, \; z_b > 0; \\
&z_a = 0, \; \psi_b^{(ab)} = \varphi_b^{(ab)}, \; \varepsilon' \frac{\partial \psi_b^{(ab)}}{\partial z_a} = \varepsilon \frac{\partial \varphi_b^{(ab)}}{\partial z_a}; \\
&r_{ab} \to \infty, \; \psi_b^{(ab)} = \varphi_b^{(ab)} = 0.
\end{aligned}\right\} \tag{34}$$

This system of equations is identical with that obtained from the model studies of [2, 3, 4].

It has been shown in [3] that the solution of (34) has the form

$$\varphi_b^{(ab)} = \frac{e_b}{\varepsilon} \left\{ \frac{e^{-\varkappa r_{ab}}}{r_{ab}} + \int_0^\infty \frac{\lambda \mathcal{J}_0(\lambda R)}{\sqrt{\varkappa^2 + \lambda^2}} \frac{\varepsilon\sqrt{\varkappa^2 + \lambda^2} - \varepsilon'\lambda}{\varepsilon\sqrt{\varkappa^2 + \lambda^2} + \varepsilon'\lambda} \exp\left[-(z_a + z_b)\sqrt{\varkappa^2 + \lambda^2}\right] d\lambda \right\}, \tag{35}$$

where $R = \sqrt{(x_a - x_b)^2 + (y_a - y_b)^2}$ and $\mathcal{J}_0$ is the Bessel function.

---

*It is to be recalled that $z^* = 3.57 \cdot k^2 \cdot 10^{-8}$ cm, for aqueous solutions of electrolytes, k being the valence of the ion.

Equations (33) and (35) give the zeroth approximations for φ and $\varphi_b^{(ab)}$. The integral

$$\int\limits_{z_c \geqslant 0} \sum_c v_c\, g_c\, g_{ac} \frac{\partial \Phi_{ac}}{\partial z_a}\, dr_c$$

of Eq. (8) must be evaluated in order to obtain the first approximations. Here we draw on (22), from which it follows that

$$\int\limits_{z_c \geqslant 0} \sum_{1 \leqslant c \leqslant M} v_c g_c g_{bc} \frac{\partial \Phi_{ac}}{\partial z_a}\, dr_c = -\theta \frac{\partial g_{ab}}{\partial z_a} - \frac{\partial \Phi_{ab}^{(e)}}{\partial z_a} = \frac{e_a}{\varepsilon} \frac{\partial \varphi_b^{(ab)}}{\partial z_a} - \frac{e_a\, e_b}{\varepsilon} \frac{\partial f^{ab}}{\partial z_a}. \tag{36}$$

Substitution of the expression obtained for $\varphi_b^{(ab)}$ from (35) leads to

$$\int\limits_{z_c \geqslant 0} \sum_{1 \leqslant c \leqslant M} v_c g_c g_{bc} \frac{\partial \Phi_{ac}}{\partial z_a}\, dr_c = -\frac{e_a\, e_b}{\varepsilon} \left\{ \left[\frac{(1 + \varkappa r_{ab}) e^{-\varkappa r_{ab}} - 1}{r_{ab}^2} \cdot \frac{z_a - z_b}{r_{ab}} \right] + \right.$$

$$\left. + \int\limits_0^\infty \lambda \mathcal{J}_0(\lambda R) \frac{\varepsilon \sqrt{\varkappa^2 + \lambda^2} - \varepsilon'\lambda}{\varepsilon \sqrt{\varkappa^2 + \lambda^2} + \varepsilon'\lambda} \exp\left[-(z_a + z_b)\sqrt{\varkappa^2 + \lambda^2} \right] d\lambda - \beta \frac{z_a + z_b}{[R^2 + (z_a + z_b)^2]^{3/2}} \right\}. \tag{37}$$

The only difference between the integral of the left-hand member of (37) and the integral of (8) is that the one contains $g_{ac}(\vec{r}_a, \vec{r}_c)$ instead of $g_{bc}(\vec{r}_b, \vec{r}_c)$. The integral of interest here can therefore be obtained from (37) by replacing e_b by e_a and allowing $\vec{r}_b \to \vec{r}_a$. The expression in the square brackets then disappears, and one has

$$\int\limits_{z_c \geqslant 0} \sum_{1 \leqslant c \leqslant M} v_c g_c g_{ac} \frac{\partial \Phi_{ac}}{\partial z_a}\, dr_c = -\frac{\partial}{\partial z_a} \left(\frac{\beta e^2}{4\varepsilon z_a} \right) +$$

$$+ \frac{\partial}{\partial z_a} \frac{e^2}{2\varepsilon} \int\limits_0^\infty \frac{\exp\left[-2z_a \sqrt{\varkappa^2 + \lambda^2} \right]}{\sqrt{\varkappa^2 + \lambda^2}} \cdot \frac{\varepsilon \sqrt{\varkappa^2 + \lambda^2} - \varepsilon'\lambda}{\varepsilon \sqrt{\varkappa^2 + \lambda^2} + \varepsilon'\lambda} \lambda d\lambda.$$

Substitution of this expression into (8) finally leads to

$$\chi(z) = \frac{e^2}{2\varepsilon} \int\limits_0^\infty \frac{\exp\left[-2z \sqrt{\varkappa^2 + \lambda^2} \right]}{\sqrt{\varkappa^2 + \lambda^2}} \cdot \frac{\varepsilon \sqrt{\varkappa^2 + \lambda^2} - \varepsilon'\lambda}{\varepsilon \sqrt{\varkappa^2 + \lambda^2} + \varepsilon'\lambda} \lambda\, d\lambda. \tag{38}$$

This integral can be expressed in terms of known functions only if $\varepsilon'/\varepsilon = 0$, a situation approximated by an aqueous solution of electrolyte bounded by air where $\varepsilon \simeq 80 \gg \varepsilon' \simeq 1$:

$$\chi(z) = \theta \frac{z^*}{2z} e^{-2\varkappa z}, \quad z^* = \frac{e^2}{2\varepsilon\theta}. \tag{39'}$$

When $\varepsilon = \varepsilon'$ (i.e., the electrolyte is bounded by a medium having the same dielectric constant as the solvent),

$$\chi(z) = \theta \frac{z^*}{2z} e^{-2\varkappa z} \left\{ \left(1 + \frac{1}{\varkappa z} \right)^2 - 2e^{2\varkappa z} K_2(2\varkappa z) \right\}, \tag{39''}$$

ν_0, mole/liter	C, μ/cm^2	ε	$C_2^{(\text{exp})}/C_2^{(\text{theo})}$	
			$B=1$	$B\neq1$
0.01	13.06	78.44	0.906	0.923
0.10	20.66	77.1	0.689	0.794
0.66	24.83	71.3	0.467	0.705
0.916	25.71	69.3	0.470	0.794

K_2 being a Bessel function of dummy argument; when $\varepsilon'/\varepsilon = \infty$ (i.e., the electrolyte is bounded by a metal),

$$\chi(z) = -\, 0\, \frac{z^*}{2z}\, e^{-2\varkappa z}.$$

(39''')

The final first approximation expression for φ is obtained by substituting the expression for $\chi(z)$ into (16) and developing the exponential as a power series in $e\varphi/\theta$:

$$\varphi''(z) - \varkappa^2 e^{-\frac{\chi(z)}{\theta}} \varphi(z) = 0;\quad \varphi(h) = \psi_1, \varphi(\infty) = 0.$$

(40)

This equation differs from the Gouy–Chapman equation, (32), only in so far as $\varkappa^2 = (8\pi e^2/\varepsilon\theta)\,\nu$ has been replaced by $\varkappa^2(z) = \varkappa^2 \exp[-\chi(z)/\theta]$. Here $\chi(z)$ is the energy of coulombic correlation of the particles in the double layer, which is to say, the energy of interaction of an ion with its own image plus the energy of interaction of this same ion with its own distorted atmosphere.

Discussion of Results

Equation (40) shows that $\varphi(z)$ will be equal to zero when there is zero potential drop at the interface between the two media, i.e., when $\psi_0 = 0$. Coulombic correlation will at the same time require that $g_a(z) = \exp[-\chi(z)/\theta]$ and that the adsorption

$$\Gamma_a = \nu_0 \int\limits_0^\infty (g_a - 1)\,dz$$

be different from zero, as can be seen from (9) and (38). We will spend no more time on this matter since the expression obtained for $\chi(z)$ is identical with that found from the model studies of [4, 5]. We merely note that the adsorption is negative and quite low when there are no image forces, $\varepsilon = \varepsilon'$, and the change in $g_a(z)$ is entirely due to Debye atmosphere distortion. By expressing the $\chi(z)$ of (39''') as a power series in $\varkappa Z$ and remembering that $\varkappa z^* \ll 1$, it can be shown that $\chi/\theta \simeq \varkappa z^*[1/3 + (1/8)\ln \varkappa z^*] \ll 1$ when z is small and $\chi(z)$ at a maximum.

We now pass to the case in which $\psi_0 \neq 0$ and $\varphi(z)$ is determined through (40). Unfortunately, only numerical solutions can be obtained for this equation. One such solution has been obtained in [5] and will now be used for setting up a comparison with the experimental data. If we designate the double layer charge calculated from the Gouy–Chapman equation, (32), by $\eta = (\varepsilon\varkappa/4\pi)\psi_1$ and the charge calculated through Eq. (40) by $\psi_0 = 0$, it follows that

$$C = \frac{\partial\eta}{\partial\psi_0} = \frac{C_1 C_2}{C_1 + C_2}, \quad C_1 = \frac{\varepsilon_1}{4\pi h}, \quad C_2 = \frac{\partial\eta}{\partial\Psi_1} = \frac{\varepsilon\varkappa}{4\pi}B,$$

(41)

at $\eta = (\varepsilon\varkappa/4\pi)\psi_1 \cdot B = \eta_0 B$, C being the total capacity of the double layer, C_1 the capacity of the molecular condenser, and C_2 the capacity of the diffuse portion of the double layer.

The experimental data on the $C = C(\nu)$ relation, and the known values of C and C_2 at $\nu = 0.001$ mole/liter for the aqueous NaF solution — mercury electrode system were used and the value of C_1 for this system calculated from Eq. (41). The values of C and C_1 were then used to calculate $C_2^{(exp)} = CC_1/(C_1 - C)$ at other concentrations, ν, from the assumption $C_1 = \text{const}$ and comparison finally made with the results obtained from the theoretical equation $C_2^{(theo)} = (\varepsilon \varkappa /4\pi)B$.* The accompanying table shows the results of such calculations, first for $B = 1$, no account being taken of image forces, and then for $B \neq 1$. The data on $\varepsilon = \varepsilon(\nu_0)$ were taken from [7]. Though the introduction of the image forces essentially reduces the errors of the theory, it is seen that the agreement with experiment remains something less than perfect. It must be remembered, however, that this is not an exact method of comparison, first because the capacity, C_1, of the molecular condenser can vary with ν, and second because the calculations for the table applied Eq. (40) over a range of concentrations in which it is known to be only an approximation. We have, unfortunately, found no data other than these [6] which could be used for a check of this type.

Conclusions

1. An equation has been derived in which account is taken of the effect of image forces on the potential distribution in the double electric layer formed at the interface between an electrolyte and an external medium. This equation follows from the Gibbs canonical distribution. It is significant that no "model" considerations are involved in this derivation.

2. This equation has been solved for the case of a weak solution of a symmetrical electrolyte, and proof thereby given that allowance for the image forces introduces a correction term into the Gouy—Chapman equation, the magnitude of this term rising sharply with increasing valence of the electrolyte.

3. An exact statistical basis has been developed for the Onsager theory of electrolyte surface tensions.

Literature

1. N. N. Bogolyubov. Problems of Dynamic Theory and Statistical Physics, Moscow, Gostekhizdat (1946).
2. L. Onsager and N. N. T. Samaras. J. Chem. Phys. 2:528 (1934).
3. V. E. Bravina. Vestn. Mosk. Gas. Univ. Ser. Mat-Mekh., No. 2:85 (1958).
4. F. H. Stillinger. J. Chem. Phys., 35(5):1584 (1961).
5. G. A. Martynov, V. G. Melamed, and P. T. Ali-Zade. Dokl. Akad. Nauk SSSR, 151(3):601 (1963).
6. D. C. Grahame. J. Am. Chem. Soc. 76(19):4819 (1954).
7. G. H. Haggis, J. B. Hasted, and T. J. Buchanan. J. Chem. Phys., 20(9):1952 (1952).

---

*This type of check was kindly suggested by V. Krylov, to whom the author is especially indebted.

THE STATISTICAL THEORY OF THE DOUBLE ELECTRIC LAYER
Part III

G. A. Martynov

Institute of Physical Chemistry, Acad. Sci., USSR
Laboratory for Surface Phenomena

The Effect of the Intrinsic Ionic Volumes

It is a well-known fact that the Gouy—Stern theory of the double electric layer calculates the mutual interionic interactions by considering the ions to be point masses (see Communication I), and, in fact, allows for intrinsic volumes only in dealing with the interactions between ions and the electrolyte — external medium interface. It is clear that such a model is not consistent. The present communication will attempt to give a more exact treatment of the intrinsic ionic volumes. Here it will be shown that the Gouy—Stern description of double layer processes is neither qualitatively nor quantitatively correct at high concentrations.

Formulation of the Problem

The present work is based on the modified form of the Bogolubov equation for the unary distribution function, g_a, which was used in the two earlier communications on the statistical theory of the double electric layer, namely

$$\theta \frac{\partial \ln g_a}{\partial z_a} + \frac{\partial U_a}{\partial z_a} + \int_V \sum_c \nu_c g_c \gamma_{ac} (1 + g_{ac}) \frac{\partial \Phi_{ac}}{\partial z_a} \, dr_c = 0. \tag{1}$$

The notation here is as before.

In order to obtain a description of processes in the double layer, the binary correlation function, g_{ac}, and the functions U_a and Φ_{ac} which fix the energies of particle interaction with the external field and with one another must be specified. It is to be remembered that in general it is so that $U_a = \Phi_a^{(coul)} + \Phi_a^{(ad)} + \Phi_a^{(im)}$, $\Phi_a^{(coul)}$ being the energy of ion a in the field of the charged surface, $\Phi_a^{(ad)}$ the energy due to the specific (i.e., non-electrostatic) adsorption forces, $\Phi_a^{(im)}$ the energy of interaction of ion a with its own image; furthermore $\Phi_{ac} = \Phi_{ac}^{(coul)} + \Phi_{ac}^{(s)} + \Phi_{ac}^{(im)}$, where $\Phi_{ac}^{(coul)}$ is the energy of coulombic interaction of ions a and c, $\Phi_{ac}^{(s)}$ is the energy associated with short-range forces, and, $\Phi_{ac}^{(im)}$ is the energy of interaction of ion a with the image of ion c.

It has been shown in Communication II that the Debye atmosphere distortion (expressed by the g_{ac}) invariably leads to negative particle adsorption in the neighborhood of the electrode. On the other hand, positive adsorption can result from the image forces which are established when the dielectric constant, ε', of the external medium is greater than the dielectric constant, ε, of the solvent. The image force effect will overcompensate the Debye atmosphere effect when the solution is bounded by a metal (i.e., when $\varepsilon' = \infty$). Thus it is clear that there will be a certain value, ε', in the interval $\varepsilon < \varepsilon' < \infty$, at which the image force effect will exactly neutralize the Debye atmosphere effect. Calculation shows that the processes occurring at ε' are essentially those which would be expected in a system free of both image forces and Debye atmosphere. All of the calculations of Communication II were based on the assumption that the concentrations were low, but there is no reason to anticipate any change in passing over to higher concentrations where the effect of intrinsic ionic volumes would become significant. It will therefore be assumed that $\Phi_a^{(im)} = \Phi_{ac}^{(im)} = 0$; $g_{ac} = 0$, the system being bounded by an external medium whose dielectric constant is such that the image force effect will just neutralize the Debye atmosphere effect. This is indeed an "exotic" case, but one which brings out in pure form the conditions necessitating the introduction of intrinsic ionic volumes in double electric layer theory.

It will be assumed, as before, that $\Phi_a^{(coul)} = (2\pi \eta e_a / \varepsilon) \, z_a$, and $\Phi_{ac}^{(coul)} = (e_a e_c / \varepsilon) \cdot 1/r_{ac}$. Moreover, we will consider that the ions are rigid spheres, which is to say that

$$\Phi_{ac}^{(s)} = \begin{cases} \infty, & 0 \leqslant r_{ac} < r_0 \\ 0, & r_0 \leqslant r_{ac} \leqslant \infty \end{cases}, \quad \gamma_{ac} = \exp\left[-\frac{\Phi_{ac}^{(s)}}{\theta}\right] = \begin{cases} 0, & 0 \leqslant r_{ac} < r_0 \\ 1, & r_0 \leqslant r_{ac} \leqslant \infty \end{cases}. \tag{2}$$

Finally the function, $\Phi_a^{(ad)}$, which fixes the nonelectrostatic component of the energy of interaction of particle and surface will be written as

$$\Phi_a^{(ad)} = \begin{cases} +\infty, & -h \leqslant z_a < 0 \\ \dot{\Phi}_a^*(z_a), & 0 \leqslant z_a \leqslant \infty \end{cases},$$

the fact that the function is infinite over the interval $(-h, 0)$ simply indicating that the closest approach of the particle to the interfacial surface is $h \simeq r_0/2$.

We will now set $\theta \ln g_a = -[e_a\varphi(z_a) + \Phi_a^*(z_a)]$, $\varphi(z_a)$ being a new unknown potential function. As a result, (1) takes the form

$$\varphi'(z) + \frac{2\pi\eta}{\varepsilon} - \frac{1}{\varepsilon} \int\limits_{z_c \geqslant 0} \sum_c v_c e \gamma_{ac} \exp\left[-\frac{e_c\varphi(z_c) + \Phi^*(z_c)}{\theta}\right] \frac{d}{dz_a}\left(\frac{1}{r_{ac}}\right) dr_c$$

$$+ \frac{\theta}{e_a} \int\limits_{z_c \geqslant 0} \sum_c v_c \exp\left[-\frac{e_c\varphi(z_c) + \Phi_c^*(z_c)}{\theta}\right] \frac{d\gamma_{ac}}{dz_a} dr_c = 0. \tag{3}$$

Here, the ratio of the second integral to the first is proportional to $(e^2/\varepsilon r_0\theta)^{-1} \ll 1$, as can be readily seen by passing to dimensionless variables. The second integral of (3) can therefore be omitted without any considerable error. Equation (3) can be further simplified by writing the remaining integral as

$$\int\limits_V' \vec{\nabla}_c Q(z_c) \cdot \vec{\nabla}_c\left(\frac{1}{r_{ac}}\right) dr_c,$$

where

$$\rho = \sum_c v_c e_c g_c(z), \quad Q(z) = \int\limits_z^\infty \rho(z)\,dz = \int\limits_z^\infty \sum_c v_c e_c \times \exp\left[-\frac{e_c\varphi(z_c) + \Phi_c(z_c)}{\theta}\right] dz_c.$$

This last expression covers the double layer charge over the half-space (z, ∞), the prime attached to the integral indicating that integration is over all the semispace $z_c \geqslant 0$ which lies outside a sphere of radius r_0 drawn around the particle a. This integral can be readily evaluated through the Green formula, the result obtained depending on whether the sphere of radius r_0 does or does not intersect the surface $z = 0$. We will first consider the region $z \geqslant r_0$ in which there can be no intersection. By carrying out the required calculations, one finds that

$$r_0 \leqslant z \leqslant \infty, \quad \varphi'(z) + \frac{2\pi}{\varepsilon}\{\eta + Q(0)\} - \frac{4\pi}{\varepsilon}\frac{1}{2r_0}\int\limits_{z-r_0}^{z+r_0} Q(t)\,dt = 0. \tag{4}$$

If the interfacial perturbation is assumed to fall off to zero as one moves away from the interface (i.e., if $g_a \to 1$, $\varphi \to 0$, $\varphi' \to 0$), the requirement that (4) be valid for any arbitrary large value of z can be satisfied only if

$$\eta + Q(0) = \eta + \int_0^\infty \rho(z)\,dz = 0, \quad \sum_c \nu_c e_c = 0.$$

(5)

The physical meaning of these relations is obvious: the one implies neutrality in the solution — electrode system and the other neutrality of the electrolyte at points far removed from the interface.

For the region $0 \leq z \leq r_0$, (5) implies that

$$\varphi'(z) = \frac{2\pi\eta}{\varepsilon}\left(1 - \frac{z}{r_0}\right) - \frac{4\pi}{\varepsilon}\frac{1}{2r_0}\int_0^{z+r_0} Q(t)\,dt = 0.$$

(6)

The function $\varphi(z)$ defined by Eqs. (4) and (6) is **n o t t h e d o u b l e l a y e r e l e c t r o s t a t i c p o t e n - t i a l**, ψ. This can be readily seen by differentiating (4) to obtain

$$\varphi''(z) = -\frac{4\pi}{\varepsilon}\frac{1}{2r_0}\int_{z-r_0}^{z+r_0}\sum_c \nu_c e_c \exp\left[-\frac{e_c\varphi(z_c) + \Phi_c(z_c)}{\theta}\right] dz_c,$$

(7)

where $r_0 \leq z \leq \infty$.

This result is quite similar to the Poisson Equation; its right-hand member does not, however, contain the true z plane charge, which is

$$\rho = \sum_c \nu_c e_c g_c(z),$$

but a certain mean value, namely,

$$\bar{\rho} = \frac{1}{2r_0}\int_{z-r_0}^{z+r_0} \rho(z_c)\,dz_c.$$

The corresponding boundary condition for $\varphi(z)$ is obtained by setting $z = 0$ in Eq. (6):

$$\varphi'(0) = -\frac{4\pi}{\varepsilon}\left[\eta + \frac{1}{2r_0}\int_0^{r_0} dt \int_0^t \rho(\tau)\,d\tau\right];$$

(8)

this is again different from the boundary condition for the true electrostatic potential $[\varphi'(0) = -(4\pi/\varepsilon)\eta]$. Equations (7) and (8) pass over to the usual self-consistent equations of the Gouy—Stern theory, and $\varphi(z)$ becomes the familiar electrostatic potential, only at the limit $r_0 \to 0$.

It is clear from what has been said that the relation between the macroscopic double layer parameters η and $\psi_0 = \psi(0)$ cannot be developed simply by determining $\varphi(z)$ from (6) and (7). The Poisson Equation

$$\psi''(z) = -\frac{4\pi}{\varepsilon}\rho(z), \quad \psi(z) = \frac{4\pi}{\varepsilon}\int_z^\infty (z - \zeta)\rho(\zeta)\,d\zeta$$

(9)

must be first applied to $\rho = \rho(\varphi, z)$ as a known function of $\varphi(z)$, the potential distribution, $\psi(z)$, found, and the latter then used to develop the relation between η and ψ_0.

Certain conclusions concerning $\varphi(z)$ can be drawn from the general form of the above equations. Thus Eq. (9) shows that this function cannot have points of discontinuity since the free energy of the system could not otherwise be a minimum, the electric field strength going to infinity at a point of discontinuity. Equations (4) and (6) indicate that $\varphi(z)$, being continuous over the entire interval $z \geq 0$, must also have first and second derivatives which are continuous over this interval and a third derivative showing a discontinuity at the plane $z = r_0$.

This last result is solely due to the specific details of the assumed solid sphere model. The function $\varphi(z)$ would have derivatives of every order if the Born repulsive forces were assumed to fall off continuously with the distance.

We will now consider the simplest case of a symmetrical electrolyte (in which $e_+ = e_- = e$, $\nu_+ = \nu_- = \nu$) and no specific adsorption (i.e., in which $\Phi_\pm \equiv 0$). Here, $\rho = -2\nu e \, \mathrm{sh}[e\varphi(z)/\theta]$, and (4) and (6) become

$$0 \leqslant 1, \; \varphi'(\xi) = -\frac{2\pi\eta r_0}{\varepsilon}(1-\xi) - \frac{\theta}{e}\frac{(\varkappa r_0)^2}{2}\int\limits_0^{\xi+1} dt \int\limits_t^\infty \mathrm{sh}\,\frac{e\varphi(\tau)}{\theta}\,d\tau, \tag{10}$$

$$1 \leqslant \xi \leqslant \infty, \; \varphi''(\xi) = -\frac{\theta}{l}\frac{(\varkappa r_0)^2}{2}\int\limits_{\xi-1}^{\xi+1} dt \int\limits_t^\infty \mathrm{sh}\,\frac{e\varphi(\tau)}{\theta}\,d\tau.$$

Here

$$\xi = z/r_0, \quad \varkappa^2 = \frac{8\pi\nu\,e^2}{\varepsilon\theta}. \tag{11}$$

The expression of (11) passes over to the Gouy equation whose solution, ψ, satisfies the condition $\psi(z) \geq 0$ everywhere when $\varkappa r_0 \to 0$. Since the solution of the system (10)-(11) varies continuously with $\varkappa r_0$ when the value of this parameter is low, it follows that the function $\varphi(z)$ will also satisfy the condition $\varphi \geq 0$ for any $z \geq 0$ for low but finite values of $\varkappa r_0$. Equations (10)-(11) show that one will always have $\varphi'(z) \leq 0$ and $\varphi''(z) \geq 0$ for these values of $\varkappa r_0$, which is to say that the curve $\varphi = \varphi(z)$ will have neither extrema nor inflection points but fall off monotonically with increasing distance. Thus the maximum on the curve must be at the boundary where $z = 0$.

One of the fundamental defects of the Gouy–Stern theory is that the calculated values of the concentration, $\nu(0)$, in the plane $z = 0$ are too large, even at very low $\varkappa r_0$, when the potential drop, ψ_0, at the boundary of the diffuse part of the double layer is greater than 100-200 mv.*

We will now show that the values of $\nu(0)$ obtained from Eqs. (10) and (11) are lower than those given by the Gouy–Stern theory. For this purpose, let us designate the solution of the Gouy equation by $\psi(z)$, and the solution of Eqs. (10)-(11) by $\varphi(z)$. It follows from (5), the neutrality condition, that each of these solutions must satisfy

$$\eta = e\nu \int\limits_0^\infty \mathrm{sh}\,\frac{e\psi}{\theta}\,dz = e\nu \int\limits_0^\infty \mathrm{sh}\,\frac{e\varphi}{\theta}\,dz$$

for given η. At the same time, $\psi(z)$ must satisfy the boundary conditions

$$\psi'(0) = -\frac{4\pi}{\varepsilon}\eta, \text{ and } \varphi'(0) = -\frac{4\pi}{\varepsilon}\left\{\eta - \frac{\theta}{e}\frac{(\varkappa r_0)^2}{2r_0}\int\limits_0^{r_0} dt \int\limits_0^t \mathrm{sh}\,\frac{e\varphi}{\theta}\,d\tau\right\},$$

$$\text{i.e., } |\psi'(0)| > |\varphi'(0)|.$$

---

* Thus the experimental data on the capacity of 0.001 N aqueous NaF on mercury [1] (for which $\varkappa r_0 \simeq 0.03$) yield ν_1 values of 2-12 mole/liter for 190 mv $\leq \psi_0 \leq$ 250 mv when analyzed by this theory. It is clear that the point-ion model on which the Gouy–Stern theory is based is no longer applicable at such high concentrations.

Since both $\psi(z)$ and $\varphi(z)$ are continuous monotonically decreasing functions, the condition

$$\int\limits_0^\infty \mathrm{sh}\,\frac{e\varphi}{\theta}\,dz = \int\limits_0^\infty \mathrm{sh}\,\frac{e\psi}{\theta}\,dz,$$

requires that $\psi(0) > \varphi(0).$ * Since, furthermore, $\nu(0) \simeq \nu \exp(e\psi_0/\theta)$ for the Gouy case and $\nu(0) \simeq \nu \exp(e\varphi_0/\theta)$ for the case of Eqs. (10)–(11), low values of the exponent correspond to low values of $\nu(0)$, as was to be proved. Thus the introduction of the intrinsic ionic volumes eliminates this weak point in the Gouy theory.

The Case of High Concentrations and Low Potentials

Analytic solutions of the system (10), (11) can be obtained only when the potential is low, $\psi_0 = \psi(0)$. (Here, and what follows, it will be assumed that ψ_0 is given and $\eta = \eta(\psi_0)$.) In fact, (9) shows that $e\varphi/\theta$ will be low when the value of

$$e\,\psi_0/\theta = (\varkappa r_0)^2 \int\limits_0^\infty \xi\,\mathrm{sh}\,\frac{e\varphi(\xi)}{\theta}\,d\xi$$

is not too high, and a good approximation is therefore obtained by setting $\mathrm{sh}(e\varphi/\theta) \simeq e\varphi/\theta.$† Equations (10) and (11) then become, respectively,

$$0 \leqslant \xi \leqslant 1,\ \ \varphi'(\xi) = -\frac{2\pi\eta r_0}{\varepsilon}(1-\xi) - \frac{(\varkappa r_0)^2}{2}\int\limits_0^{\xi+1} dt \int\limits_t^\infty \varphi(\tau)\,d\tau, \tag{12}$$

$$1 \leqslant \xi \leqslant \infty,\ \ \varphi'''(\xi) = \frac{(\varkappa r_0)^2}{2}\{\varphi(\xi+1) - \varphi(\xi-1)\}. \tag{13}$$

Direct substitution readily shows that $e^{-w\zeta}$ will satisfy the condition of (13) when w is a solution of the transcendental equation

$$w^3 = (\varkappa r_0)^2\,\mathrm{sh}\,w. \tag{14}$$

The general solution of (13) can therefore be written as

$$\varphi(\xi) = \psi_0 \sum_{k=1}^\infty A_k e^{-w_k \xi}, \tag{15}$$

the constant multiplying factor, ψ_0, being taken out from under the summation sign for convenience. The A_k of (15) are to be determined from Eq. (12). Substitution of (15) into (12), followed by application of

$$1 = \sum_{k=1}^\infty A_k \left(\frac{\varkappa r_0}{w_k}\right)^2,\ \ \eta = \psi_0\,\frac{\varepsilon\varkappa}{4\pi}\sum_{k=1}^\infty A_k\,\frac{\varkappa r_0}{w_k}. \tag{16}$$

---

* The reasoning proceeds in exactly the same manner when

$$\psi_0 = \frac{\theta}{e}\,\frac{(\varkappa r_0)^2}{r_0^2}\int\limits_0^\infty z\,\mathrm{sh}\,\frac{e\varphi(z)}{\theta}\,dz.$$

is given.

† It should be noted that the larger $\varkappa r_0$ (i.e., the higher the concentration), the wider the ψ_0 interval over which $\mathrm{sh}(e\varphi/\theta) \simeq e\varphi/\theta$.

(these relations follow from the Poisson Equation, (9), with $z = 0$, and the neutrality condition (5)) one obtains

$$0 \leqslant \xi \leqslant 1, \quad (\varkappa r_0)^{-2} = \sum_{k=1}^{\infty} A_k \frac{e^{w_k (1-\xi)} - w_k (1 - \xi)}{w_k^2} \,.$$

(17)

This equation is a series expansion of $(\varkappa r_0)^{-2}$ in the functions $e^{w_k(1-\xi)} - w_k(1 - \xi)$. Each A_k can, theoretically, be evaluated from (17). The actual determinations are difficult, however, since the $e^{w_k(1-\xi)} - w_k(1 - \xi)$ from a complete system of independent functions, but are not orthonormal. On the other hand, conclusions concerning the properties of the double layer can be drawn from (14)–(17) even though the A_k have not been determined.

Equation (14) has an infinite number of roots, just as all other transcendental equations.* It can be shown to have only complex-conjugate roots (i.e., $w_k = \lambda_k \pm i w_k$*), when $\varkappa r_0 \geq 1.6$, and these plus two additional real roots, λ*, and λ**, when $0 \leq \varkappa r_0 \leq 1.6$. These λ_k satisfy the condition (see Appendix)

$$0 \leqslant \lambda^* \leqslant \lambda^{**} < \lambda_2 < \dots < \lambda_k < \dots, \text{where } \lambda_k \to \ln \frac{2 (k\pi)^3}{(\varkappa r_0)^2} \text{ when } k \to \infty.$$

(18)

Thus the properties of $\varphi = \varphi(\xi)$ will be essentially determined by the first two terms in the expansion of (15), at least when the value of ξ is large, and the study of the solution simplified accordingly.

When $\varkappa r_0 \to 0$, $\lambda^* \to \varkappa r_0$, since $\mathrm{sh}\,\lambda \simeq \lambda$ for low values of the argument: on the other hand, λ^{**} and all the remaining λ_k, $(k = 1, 2, \dots)$, tend to infinity. Thus for low values of $\varkappa r_0$ one can, with considerable accuracy, set $\varphi \simeq \psi_0 e^{-\lambda^* \xi} = \psi_0 e^{-\varkappa z}$, which is the familiar result first obtained by Gouy for highly dilute solutions.

The data presented below indicate that λ^* and λ^{**} tend to 3.0 as $\varkappa r_0 \to 1.6$, λ^* remaining in the interval $(0; 3.0)$ and λ^{**} in the interval $(3.0; \infty)$:

$\varkappa r^0$	0.100	0.300	0.600	0.800	1.000	1.200	1.400	1.642
λ^*	0.100	0.302	0.618	0.837	1.097	1.399	1.781	2.984
λ^{**}	12.99	10.01	7.93	6.96	6.14	5.37	4.60	2.984
h	1.000	1.005	1.030	1.045	1.087	1.14	1.20	1.31
$\varkappa r_0$	1.642	2.016	2.642	3.941	7.407	9.167	—	—
λ	2.984	2.721	2.365	1.814	1.000	0.000	—	—
ω	0	1.571	2.365	3.142	3.808	4.078	—	—

Thus the first two members of (15) are significant even at $\varkappa r_0 \simeq 1.6$.

The real roots, λ^* and λ^{**}, converge at $\varkappa r_0 = 1.6$ to become two complex-conjugate roots $\lambda \pm i\omega$, but the remaining roots undergo no change at this point. The value of λ diminishes from 3.0 to 0 as $\varkappa r_0$ increases from 1.6 to 9.2, and ω increases from 0 to 4.1 (see the table). This indicates that the effective depth of the double layer, $L \simeq r_0/\lambda$, increases from $L \simeq 0.33\, r_0$ to infinity as $\varkappa r_0$ passes through this interval. On the other hand, the alteration in L is appreciable only when $\varkappa r_0 \gtrsim 7.5$. Double layer expansion cannot generally be observed in electrolytes, the solubility being too low to give concentrations such that $7.5 \leq \varkappa r_0 \leq 9.2$. It is possible that an increase in the depth of the double layer could be detected in aqueous solutions of such highly soluble salts as $AgNO_3$, or in molten salts.

Not only does the double layer depth begin to increase when $\varkappa r_0 > 1.6$, but there is, at the same time, an alteration in the concentration and potential distributions within the layer, this as a result of the fact that $\varphi \simeq A e^{-\lambda \xi} \cos(\omega \xi + \omega_0)$. Oscillations cannot be observed when $\varkappa r_0 \lesssim 7.5$, the effective depth, $L \simeq r_0/\lambda$, of the double layer being less than the oscillation period $T = 2\pi r_0/\omega$. These oscillations are quite real in the interval

---

*In general, (14) will have four roots, $w_k = \pm \lambda_k \pm i\omega_k$, for each pair of values, λ_k, and ω_k. On the other hand, all of the A_k's which multiply terms of (15) in $\exp (+ \lambda_k \xi)$ must be set equal to zero, in order that the equation satisfy the condition at infinity.

$7.5 \lessgtr \varkappa r_0 \lessgtr 9.2$ since $1.64 \, r_0 \gtrless T \gtrless 1.54 \, r_0$ and $L > T$ here. This is indication that the double layer has a crystal-like structure in highly concentrated systems (the structure is gaseous when $0 \lessgtr \varkappa r_0 \lessgtr 1.6$).

Thus the solution, (15), of the system (12)–(13) has two singular points, one at $\varkappa r_0 = 1.6$ and the other at $\varkappa r_0 = 9.2$. There is, however, no discontinuity in the electrostatic parameters of the double layer at the first of these points, $|w_{1,2}| = \sqrt{\lambda^2 + \omega^2} = f(\varkappa r_0)$ being a continuous function of $\varkappa r_0$. On the other hand, $\varkappa r_0 = 9.2$ is an essential singularity of the solution, where the roots $\lambda \pm i\omega$ and $-\lambda \pm i\omega$ (the rejected roots) merge and disappear. Working at the same level of approximation as in the derivation of Eqs. (12) and (13), the binary distribution function, G_{ab}, for a homogeneous system of hard spheres interacting according to the Coulomb law can be shown to have the form $\rho_{ab} = r_{ab}/r_0 \geq 1$ when

$$G_{ab} = 1 - \frac{c_a e_b}{\varepsilon \theta r_0} \sum_{k=1}^{\infty} A \, \frac{e^{-w_k \rho_{ab}}}{\rho_{ab}},$$

w_k being roots of the transcendental equation (14), just as before. The first member of this series takes the form $e^{\pm i\omega \rho_{ab}}/\rho_{ab}$ when $\varkappa r_0 = 9.2$, and the square fluctuation of the number of particles

$$\Delta N_{ab}^2 \approx \int_0^\infty (G_{ab} - 1)\, \rho_{ab}^2 \, d\rho_{ab}$$

tends to infinity. This explains the fact that any small external perturbation (the charge η or the potential ψ_0 in the double layer) will induce an internal perturbation which will, in turn, propagate further and further away from the point at which the perturbation originated. In the case of the double layer, this result follows from the fact that the concentration, $g_a \simeq 1 - e_a \varphi / \theta$ at the point $\varkappa r_0 = 9.2$, regardless of how small the values of η and ψ_0 may be. The stability limits of the system are fixed by the temperature and concentration [2] at which the exponent of the first term in the series of (15) becomes a pure imaginary quantity. The indication is that a system of solid spheres with coulombic interaction such as is under discussion here can exist as a liquid phase * only when

$$\varkappa r_0 < 9.2, \text{where} \quad \frac{\nu\, e^2 r_0^2}{\varepsilon \theta} < \frac{9.2^2}{8\pi} \simeq 3.3.$$

(19)

The liquid phase becomes absolutely unstable, at higher values of $\varkappa r_0$ and the system can then exist only in the ionic crystal form.

All of these conclusions rest on the assumption that the potential drop at the electrolyte — external medium interface is small, so that the function $\mathrm{sh}\, e\varphi/\theta$ can be developed as a power series in $e\varphi/\theta$ and the series broken off with the first member, regardless of the value of ξ. It is impossible to proceed in this manner, however, when ψ_0 is large. But since $\varphi(\xi) \to 0$ as $\xi \to \infty$, $\mathrm{sh}\, e\varphi/\theta$ can always be expressed as a power series at the larger distances, and the previous conclusions concerning $\varphi(\xi)$ therefore apply to all values of ψ_0 provided ξ is large. It is quite likely that one can also formulate a stronger proposition to the effect that $\varphi = \varphi(\xi)$ is a monotonically diminishing function over the entire interval $0 \leq \xi \leq \infty$†; for all values of ψ_0 when $\varkappa r_0 < 1.6$, or conversely, that function then falls off with increasing distance, and changes sign an infinite number of times, when $\varkappa r_0 > 1.6$ regardless of the value of ψ_0.

---

*More exactly, as a phase showing this order type.

†It is possible that the function will change sign several times in the neighborhood of the electrode when $\varkappa r_0 < 1.6$ and the value of ψ_0 is large. Study of Eqs. (10) and (11) show that this can happen only if φ alters radically over the interval $\Delta z < r_0/2$, a rather unlikely situation.

We now return to the case of low ψ_0 values, and consider the double layer capacity $C = \eta / \psi_0$. It follows from the second equation of (16) that

$$C = C_0 h\,(\varkappa r_0), \text{where} \, h = \sum_{k=1}^{\infty} A_k\,(\varkappa r_0)\,\frac{\varkappa r_0}{w_k\,(\varkappa r_0)}\,.$$

$$(20)$$

Here, $C_0 = \varepsilon \varkappa / 4\pi$ is the double layer capacity calculated from the assumption that $r_0 = 0$.

The intrinsic ionic volume correction, h, is always finite but greater than unity. The second of these statements can be readily confirmed by writing the expression for h in the form

$$h = \sum_{k=1}^{\infty} A_k \left(\frac{\varkappa r_0}{w_k} \right)^2 \frac{w_k}{\varkappa r_0}\,.$$

This differs from the first series of (16) only in so far as each of the terms is here multiplied by $w_k / \varkappa r_0$, a quantity whose modulous is greater than unity (for all w_k when $0 < \varkappa r_0 \leq 3.8$, and for all w_k, with the exception of $\lambda_1 \pm i w_1$, when $3.8 \leq \varkappa r_0 < 9.2$). Since

$$\sum_{k=1}^{\infty} A_k \left(\frac{\varkappa r_0}{w_k} \right)^2 = 1,$$

one always has $h > 1$.

The fact that an upper limit is imposed on h follows from the boundary condition (17). In fact, (16) and (20) can be drawn on, and (17) written as

$$1 + (1 - \xi)\,h = \sum_{k=1}^{\infty} A_k \left(\frac{\varkappa r_0}{w_k} \right) e^{w_k\,(1-\xi)}.$$

From this it is seen that the boundary condition (17) cannot be satisfied if $h = \infty$, and the constants, A_k, of (15) would then be undefined. Since, however, the system of functions $e^{w_k(1-\xi)} - w_k(1 - \xi)$ remains complete and linearly independent when $0 < \varkappa r_0 < 9.2$, development of the constant $(\varkappa r_0)^{-2}$ as a series in these functions is possible at any point in this interval if it is possible at one point. From this it at once follows that h must be finite throughout.

The boundary condition of (17) is equivalent to an infinite system of algebraic equations for the infinite sequence of coefficients $A_k = B_k \pm i C_k$. This can be readily shown by multiplying both right and left members of (17) by $e^{im\pi\xi}$ ($m = 0, 1, 2, \ldots$) and then integrating both parts of the equation with respect to ξ over the interval from 0 to 1. It is possible to limit consideration to the first two terms since the sequence of (15) converges quite rapidly, especially for small values of $\varkappa r_0$. The result is that when $\varkappa r_0 < 1.6$,

$$\left.\begin{array}{l} \varphi = \dfrac{\psi_0}{(\varkappa r_0)^2} \dfrac{[(e^{\lambda_2} - 1) - \lambda_2]\,\lambda_1^2 e^{-\lambda_1 \xi} - [(e^{\lambda_1} - 1) - \lambda_1]\,\lambda_2^2 e^{-\lambda_2 \xi}}{(e^{\lambda_2} - e^{\lambda_1}) - (\lambda_2 - \lambda_1)}, \\[3mm] h = \dfrac{1}{\varkappa r_0}\,\dfrac{\lambda_1\,(e^{\lambda_2} - 1) - \lambda_2\,(e^{\lambda_1} - 1)}{(e^{\lambda_2} - e^{\lambda_1}) - (\lambda_2 - \lambda_1)}, \end{array}\right\}$$

$$(21)$$

where $\lambda_1 = \lambda^*$ and $\lambda_2 = \lambda^{**}$.

The constants A^* and A^{**} of this expression for $\varphi(\xi)$ are determined directly from the two roots of (17), the one obtained by setting $\xi = 0$ and the other by setting $\xi = 1$. Values of h calculated from (21) are shown in the table on p. 99.

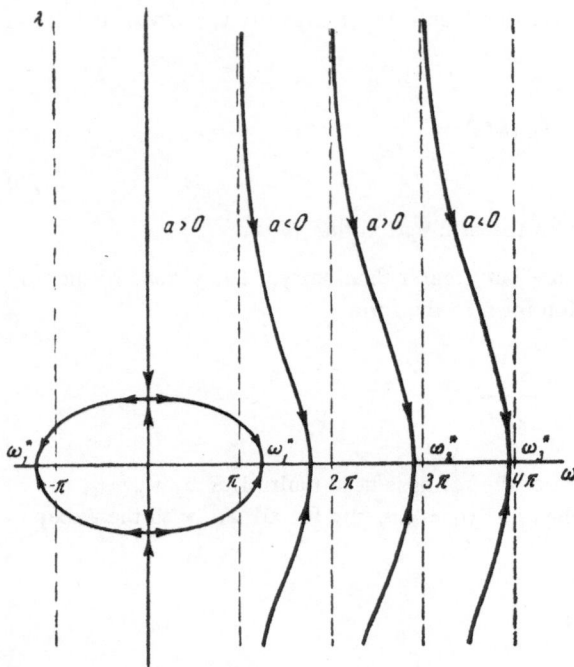

Distribution of the roots of Eq. (A)

It should be noted that it is impossible to limit the calculation of h to the first two members of (15) when $\varkappa r_0 \gtrsim 1.6$. This is indicated by the fact that calculation of h in this interval from these two members leads to

$$h = [e^\lambda (\lambda \sin \omega - \omega \cos \omega) + \omega]/[e^\lambda \sin \omega - \omega],$$

from which it follows that the value of h will approach zero when $\varkappa r_0 \simeq 3$, $\lambda \simeq 2.1$, $\omega \simeq 2.8$, and $e^\lambda \sin \omega - \omega = 0$, thus contradicting the previous conclusions. *

Conclusions

1. An equation for the potential distribution in a double electric layer of charged solid spheres has been obtained from the Gibbs cannonical distribution. This is the simplest model for the concentrated electrolytic solution and the ionic crystal melt.

2. This equation has been solved for the case of low potentials to show that there can be a marked double electric layer expansion in melts and dilute solutions.

Appendix

Since (14) has complex roots, it is best studied through the expressions for $w_k = \lambda_k \pm i\omega_k$, and λ, ω:

$$\frac{\text{th } \lambda}{\lambda} = \frac{\text{tg } \omega}{\omega} \frac{\lambda^2 - 3\omega^2}{3\lambda^2 - \omega^2}, \tag{A}$$

$$a = \frac{\omega \text{ ch } \lambda}{(\lambda^2 + \omega^2)^2} \left\{ \frac{\lambda}{\omega} \frac{\lambda^2 - 3\omega^2}{\lambda^2 + \omega^2} \text{ th } \lambda \cos \omega + \frac{3\lambda^2 - \omega^2}{\lambda^2 + \omega^2} \sin \omega \right\}, \quad a = (\varkappa r_0)^{-2}. \tag{B}$$

The distribution of the roots of Eq. (A) is shown in the figure, where

$$\omega_k^* = (k + 1) \pi + \Omega_k^* \quad \text{and} \quad \text{tg}\Omega_k^* = \frac{1}{3} \left\{ (k+1) \pi + \Omega_k^* \right\}.$$

Roots lying in the $(k\pi, \omega_k^*)$ bands with odd k lead to negative values of a in Eq. (B) and must therefore be discarded since $(\varkappa r_0)^{-2} = a > 0$ in the case under study. Roots lying in the bands with even k correspond to positive values of a in Eq. (B). Two additional roots are found on the axis of reals, when $0 \leq \varkappa r_0 \leq 1.642$; these merge when $\varkappa r_0 = 1.642$ to become complex-conjugate.

---

*The relation $h = \infty$ was first obtained in [3]. Clearly deciding this to be a meaningless result of the assumptions involved in the derivation of the original integral equations, the authors of [3] went no further in their study of the region of high $\varkappa r_0$ values. They thus failed to note the increase in double layer depth at high concentrations which follows directly from their equations. The fact that h goes to infinity, both here and in [3], is a consequence of having based the calculations on the first two members of the series rather than the entire expression.

x

102

The actual values λ_k and ω_k depend on a as a parameter. When $a \to \infty$, $\lambda^* \to \varkappa r_0$, and $\lambda^{**} \to \infty$. At the same time, the real part, λ_k, of each complex root tends to infinity and the imaginary part, ω_k, to $k\pi$. On the other hand, when

$$a \to -\frac{\sin \Omega^*_k}{[(k+1)+\Omega^*_k]^3}, \quad \omega_k \to \omega^*_k \text{ and } \lambda_k \to 0.$$

The value of a at which λ_1 goes to zero corresponds to $\varkappa r_0 = 9.167$. The higher the order number, k, of the root, the larger the value of $\varkappa r_0$ at which λ_k goes to zero.

The expression standing in brackets in Eq. (B) is approximately equal to unity for all λ_k and ω_k. Thus $[a \simeq -\omega \, \text{ch} \, \lambda/(\lambda^2 + \omega^2)^2]$. Since for large k, $\omega_k \simeq k\pi$, it at once follows that the higher k at fixed a, the higher λ_k, a relation which finds expression in Eq. (18).

Literature

1. D. C. Grahame. J. Am. Chem. Soc. 76(19):4819 (1954).
2. I. Z. Fisher. Statistical Theory of Liquid, Moscow, Gosfizmatizdat (1961).
3. F. H. Stillinger and J. G. Kirkwood. J. Chem. Phys., 33(5):1282 (1960).

THE THEORY OF THE ELECTROSTATIC INTERACTION OF MACROIONS CARRYING DISCRETE SURFACE CHARGES IN SOLUTIONS OF ELECTROLYTES

Yu. I. Yalamov

Institute of Physical Chemistry, Acad. Sci., USSR
Laboratory for Surface Phenomena

Introduction

The interaction between surface charged macroions in electrolytic solutions containing several types of microions is of considerable significance for lyophobic sol coagulation and other processes in colloidal solutions [1-10] and for various aspects of the behavior of protein globules in solutions of polyelectrolytes [11]. It is a well-known fact that double diffusion layers of electrolyte microions form around the macroions. The force of interaction between the macroions is then compounded of an electrostatic force, modified by the ion clouds, and a van der Waals force of attraction.

The theory of the interaction of massive macroions of fixed potential in electrolytic solutions was first carefully developed by B. V. Deryagin [1, 5, 6, 8-9] in his work on colloidal solutions. The interaction of plane surfaces carrying uniform surface charges at fixed potential has also been studied [2-4]. Mathematical calculations led to the basic Debye—Huckel equation, both here and in the work of Deryagin. It should be noted that macroions carrying fixed, uniformly distributed, charges were also studied in the work of Bergman, Low—Beer, and Zocher [2].

The present paper deals with the theory of the electrostatic interaction of macroions and the effect of discreteness of surface charge. The effect of this factor on macroions is of great interest in cases where the mean charge separation is comparable with, or larger than, the mean diameter of the Debye ionic atmosphere around the macroion. The force of charge interaction can be quite sensitive to the fact that the charge is discretely distributed when the distance of approach of the macroions is comparable with the diameter of the ionic atmosphere.

It should be pointed out that the force of macroion interaction depends, not only on the dielectric constant of the electrolytic solution (ε_2, in our notation), but also on the dielectric constant (ε_1) of the interior of the macroion when the surface charge is distributed discretely. This conclusion follows from the analytic expression for the interaction force developed in the present work.

The force of interaction between macroions of small surface curvature carrying a two-dimensional net of charges of different magnitude is also treated here, the assumption being made that the surface charge density is sufficiently low so that the potential distribution in solution can be described by the linearized Debye—Huckel equation. The fact that the fixed surface charge is discretely distributed presupposes an appreciable screening from the atmosphere of mobile electrolyte ions.

The integral equation for the force of interaction between macroions carrying discrete surface charges of different magnitude is compared with the force of interaction for uniform surface charge distribution. This equation passes over to the earlier relation for the uniform charge [1, 5] when the force is averaged over the macroion surface (which is physically equivalent to passage to a uniform charge).

The Electrostatic Interaction of Macroions with Discrete Surface Charges
of Different Magnitude

We will consider the interaction of two macroions of dielectric constant ε_1 separated by an electrolyte of dielectric constant ε_2. The distance between the macroions will be designated by 2R. The macroions will be assumed to carry a two-dimensional network of surface charges, in which the mean distance of charge separation is d. These charges are of different magnitude. Each charge is of magnitude q_1 on the first macroion and of magnitude q_2 on the second. The radius of the macroion will be assumed so large in comparison with the radius of the electrolyte macroions, and with the mean diameter, $1/\varkappa$, of the Debye ionic atmosphere, that the macroion—electrolyte interface is probably plane. The charges, q_1 and q_2, will, in turn, be assumed so small in comparison with $1/\varkappa$ that they can be treated as point charges. The preceding remarks suggest the use of the following system as a working model. We divide the space into three subregions distinguished in terms of the coordinate z.

I. The region z < 0, which represents the interior of the first macroion with dielectric constant ε_1. The two-dimensional net of charges q_1, with mean mutual separation, d, is assumed located in the plane z = − h.

II. The region bounded by the planes z = 0 and z = 2R which is assumed to be filled with the electrolyte of dielectric constant ε_2.

III. The region z > 2R, which represents the interior of the second macroion with dielectric constant ε_1. The two-dimensional net of point charges, q_2, with mean mutual separation d is assumed located in the plane z = 2R + h. The net of charges has been assumed located at a distance h within the macroions for ease of mathematical manipulation. The final equations for the interaction of macroions carrying surface charges is obtained by allowing h to approach zero in the derived relations. The potential distributions in these three regions must be determined before the force of interaction between the macroions can be estimated.

In region I ($-\infty < z \leq 0$), one has

$$\Delta\psi_1 = -\frac{4\pi}{\varepsilon_1} q_1 \sum_{m=-\infty}^{\infty} \sum_{n=-\infty}^{\infty} \delta(x - md)\,\delta(y - nd)\,\delta(z + h). \tag{1}$$

In region II ($0 \leq z \leq 2R$)

$$\Delta\varphi = -4\pi\rho/\varepsilon_2, \tag{2}$$

where

$$\rho = \sum_{i=1}^{s} n_i\,|e|\,z_i \exp(-|e|\,z_i\varphi/k_B T). \tag{3}$$

Linearizing (3), one obtains

$$\Delta\varphi = \varkappa^2\varphi, \tag{4}$$

where

$$\varkappa^2 = \frac{4\pi e^2}{\varepsilon_2 k_B T} \sum_{i=1}^{s} n_i z_i^2. \tag{5}$$

In region III ($2R \leq z < \infty$)

$$\Delta\psi_2 = -\frac{4\pi q_2}{\varepsilon_1} \sum_{m=-\infty}^{\infty} \sum_{n=-\infty}^{\infty} \delta(x - a - md)\,\delta(y - b - nd)\,\delta(z + h). \tag{6}$$

The right-hand member of Eq. (6) shows the charge net on the second macroion to be displaced relative to the charge net on the first microion, the respective distances of displacement along the x and y axes being a and b.

105

The boundary conditions for the potentials ψ_1, φ, and ψ_2 are:

$$\varepsilon_1 \frac{d\psi_1}{dz}\bigg|_{z=0} = \varepsilon_2 \frac{d\varphi}{dz}\bigg|_{z=0}, \quad \psi_1\big|_{z=0} = \varphi\big|_{z=0},$$

$$\varphi\big|_{z=2R} = \psi_2\big|_{z=2R},$$

$$\varepsilon_2 \frac{d\varphi}{dz}\bigg|_{z=2R} = \varepsilon_1 \frac{d\psi_2}{dz}\bigg|_{z=2R},$$

$$\psi_1 \to 0, \qquad \psi_2 \to 0,$$
$$z \to -\infty, \qquad z \to \infty.$$

$$(7)$$

We will abbreviate in what follows by setting

$$\sum_{m=-\infty}^{\infty} \sum_{n=-\infty}^{\infty} = \sum_{m,\,n} {}^*$$

Representing the solutions of (1), (4), and (6) as Fourier integrals, one has

$$\psi_1 = \iint_{-\infty}^{\infty} \psi_1(k,\ z) \sum_{m,\,n} \exp\left[i\left\{k_1(x-md)+k_2(y-nd)\right\}\right] dk_1\, dk_2,$$

$$\varphi = \iint_{-\infty}^{\infty} \varphi(k,\ z) \sum_{m,\,n} \exp\left[i\left(k_1(x-md)+k_2(y-nd)\right)\right] dk_1\, dk_2,$$

$$\psi_2 = \iint_{-\infty}^{\infty} \psi_2(k,\ z) \sum_{m,\,n} \exp\left[i\left(k_1(x-a-md)+k_2(y-b-nd)\right)\right] dk_1\, dk_2.$$

$$(8)$$

k is here the variable of integration.

Substituting (8) into (1), (4), and (6), and expressing the delta functions of the right-hand members of (1) and (4) in terms of Fourier integrals, one obtains:

1) from Eq. (1)

$$d^2\psi_1(k,\ z)/dz^2 - k^2\psi_1(k,\ z) = 0,$$

$$(9)$$

with the supplementary conditions

$$\frac{d\psi_1(k,\ z)}{dz}\bigg|_{-h^-}^{-h^+} = -\frac{q_1}{\varepsilon_1 \pi}, \quad \psi_1(k,\ z)\bigg|_{-h^-}^{-h^+} = 0;$$

$$(10)$$

2) from Eq. (4)

$$d^2\varphi(k,\ z)/dz^2 - (k^2 + \varkappa^2)\varphi(k,\ z) = 0,$$

$$(11)$$

---

*The summation actually ranges over a finite number of terms.

with the boundary conditions,

$$\varepsilon_1 \frac{d\psi_1(k,\,z)}{dz}\bigg|_{z=0} = \varepsilon_2 \frac{d\varphi(k,\,z)}{dz}\bigg|_{z=0},$$

$$\psi_1(k,\,z)\big|_{z=0} = \varphi(k,\,z)\big|_{z=0},$$

$$\varepsilon_2 \frac{d\varphi(k,\,z)}{dz}\bigg|_{z=2R} = \varepsilon_1 \frac{d\psi_2(k,\,z)}{dz}\bigg|_{z=2R} \cdot \exp\left[i\,(k_1a + k_2b)\right],$$

$$\varphi(k,\,z)\big|_{z=2R} = \psi_2(k,\,z)\big|_{z=2R} \cdot \exp\left[-i\,(k_1a + k_2b)\right]; \qquad (12)$$

3) from Eq. (6)

$$d^2\psi_2(k,\,z)/dz^2 - k^2\psi_2(k,\,z) = 0, \qquad (13)$$

with the supplementary conditions,

$$\frac{d\psi_2(k,\,z)}{dz}\bigg|_{z=(2R+h)^+}^{z=(2R+h)^-} = -\frac{q_2}{\varepsilon_1\pi},$$

$$\psi_2(k,\,z)\bigg|_{z=(2R+h)^+}^{z=(2R+h)^-} = 0. \qquad (14)$$

The supplementary conditions (10) and (14) cover the discontinuities in the derivatives of the Fourier components of the potentials ψ_1 and ψ_2 resulting from the presence of the point charges in the planes $z = -h$ and $z = 2R+h$.

The solutions of (9), (11), and (13) are such that

$$\psi_1(k,\,z) = \begin{cases} C_1(k)\,\mathrm{sh}\,(kz) + C_2(k)\,\mathrm{ch}\,(kz)\;\text{where}\,(-h \leqslant z \leqslant 0), \\ D(k)\,\exp(kz)\;\text{where}\,(-\infty < z \leqslant -h), \end{cases} \qquad (15)$$

$$\varphi = A(k)\,\mathrm{sh}\,(\sqrt{k^2 + \varkappa^2}z) + B(k)\,\mathrm{ch}\,(\sqrt{k^2 + \varkappa^2}z)\;\text{where}\;0 \leqslant z \leqslant 2R, \qquad (16)$$

$$\psi_2(k,\,z) = \begin{cases} F_1(k)\,\mathrm{sh}\,(kz) + F_2(k)\,\mathrm{ch}\,(kz)\;\text{where}\,(2R \leqslant z \leqslant 2R+h), \\ Q(k)\,\exp(-kz)\;\text{where}\,(2R+h \leqslant z < \infty). \end{cases} \qquad (17)$$

The last two boundary conditions of (7) for $+\infty$ and $-\infty$ were used in developing the solutions (15) and (17).

The constants C_1; C_2; D; A; B; F_1; F_2; Q can be evaluated through quite complicated calculations based on (10), (12), (14), and the boundary and supplementary conditions. An analytic expression for the potential (16) will be required in what follows. The constants $A(k)$ and $B(k)$ have the form

$$A(k) = \frac{e^{-kh}}{\pi}\left[\frac{q_2\mu\varepsilon_1 k - q_1\,(\varepsilon_2\lambda\,\mathrm{sh}\,2\lambda R + \varepsilon_1 k\,\mathrm{ch}\,2\lambda R)}{g}\right], \qquad (18)$$

$$B(k) = \frac{e^{-kh}}{\pi}\left[\frac{\varepsilon_2\lambda q_2\mu + q_1\,(\varepsilon_1 k\,\mathrm{sh}\,2\lambda R + \varepsilon_2\lambda\,\mathrm{ch}\,2\lambda R)}{g}\right], \qquad (19)$$

where

$$\mu = \exp\left[-i\,(k_1a + k_2b)\right], \qquad (20)$$

$$g = 2\mathrm{ch}^2\,(\lambda R)\,(\varepsilon_2\lambda\,\mathrm{th}\,\lambda R + \varepsilon_1 k)\,(\varepsilon_2\lambda + \varepsilon_1 k\,\mathrm{th}\,\lambda R), \qquad (21)$$

$$\lambda = \sqrt{k^2 + \varkappa^2}. \qquad (22)$$

On the basis of (8), (16), (18), and (19), the potential, φ, is seen to take the following form in the region occupied by the electrode when $h \to 0$:

$$\varphi = \iint\limits_{-\infty}^{\infty} R_1 \sum_{m, n} \exp\{i\,[k_1\,(x - md) + k_2\,(y - nd)]\} + R_2 \times$$

$$\times \sum_{m, n} \exp\{i\,[k_1\,(x - a - md) + k_2\,(y - b - nd)]\}\, dk_1 dk_2, \tag{23}$$

where

$$R_1 = \frac{q_1}{\pi g}\,[\varepsilon_1 k \,\mathrm{sh}\, \lambda\,(2R - z) + \varepsilon_2 \lambda \,\mathrm{ch}\, \lambda\,(2R - z)], \tag{24}$$

$$R_2 = \frac{q_2}{\pi g}\,[\varepsilon_1 k \,\mathrm{sh}\, \lambda z + \varepsilon_2 \lambda \,\mathrm{ch}\, \lambda z]. \tag{25}$$

The next step is the calculation of the change in electrical free energy which results in the system of the two macroions and electrolyte when the ion charges are raised from zero to the respective finite values q_1 and q_2. This has been shown [1, 2, 5, 6, 12] to be the only component of the free energy that depends on the distance between the ions, and it is this component which therefore fixes the force of particle interaction. Determination of this component of the free energy supposes a knowledge of the potentials, φ_{01} and φ_{02}, at q_1 and q_2 due to all of the other charges in the system. The intrinsic potentials of the latter are determined by the expressions

$$\varphi_c{}^{(1)} = 2q_1/(\varepsilon_1 + \varepsilon_2)\,r_1, \tag{26}$$

where

$$r_1 = \sqrt{x^2 + y^2 + z^2}$$

and

$$\varphi_c{}^{(2)} = 2q_2/(\varepsilon_1 + \varepsilon_2)\,r_2, \tag{27}$$

where

$$r_2 = \sqrt{(x - a)^2 + (y - b)^2 + (z - 2R)^2}.$$

Here we have selected charges for which $m = n = 0$. The potentials arising at $(0, 0, 0)$ and $(a, b, 2R)$ from all of the system charges except q_1 and q_2 can be readily determined by subtracting the potentials of (26) and (27) from (23) and then making use of the relations

$$1/r_1 = \int_0^{\infty} \mathcal{J}_0\,(k\rho) \exp\,(-kz)\, dk, \tag{28}$$

$$1/r_2 = \int_0^{\infty} \mathcal{J}_0\,(k\rho') \exp\,[-k\,(z - 2R)]\, dk, \tag{29}$$

$$\mathcal{J}_0\,(k\rho) = \frac{1}{2\pi} \int_{-\pi}^{\pi} \exp\,(ik\rho \cos \beta)\, d\beta, \tag{30}$$

$$\mathcal{J}_0 = (k\rho') = \frac{1}{2\pi} \int_{-\pi}^{\pi} \exp\,(ik\rho' \cos \beta')\, d\beta', \tag{31}$$

$$\left.\begin{array}{l} \rho = \sqrt{x^2 + y^2}, \quad \rho' = \sqrt{(x - a)^2 + (y - b)^2}, \\ \cos \beta = (k_1 x + k_2 y)/k\rho, \\ \cos \beta' = [k_1\,(x - a) + k_2\,(y - b)]/k\rho'. \end{array}\right\} \tag{32}$$

One then has:

$$\varphi_{01} = \lim \left(\varphi - \varphi_c^{(1)}\right) \tag{33}$$

as

$$r_1 \to 0.$$

$$\varphi_{02} = \lim \left(\varphi - \varphi_c^{(2)}\right) \tag{34}$$

as

$$r_2 \to 0.$$

The explicit forms of (33) and (34) are

$$\varphi_{01} = 2\left(\frac{\varepsilon_2}{\varepsilon_1 + \varepsilon_2}\right) \int_0^\infty \frac{[q_2(\varepsilon_1 + \varepsilon_2)\mathscr{J}_0(k\sqrt{a^2 + b^2}) + q_1 u]\,dk}{g} + 4\int_0^\infty \frac{\varepsilon_2 \lambda q_2}{g} \omega_{m.n}^{(2)} k\,dk + 4\int_0^\infty \frac{q_1 v}{g} \omega_{m.n}^{(1)} k\,dk, \tag{35}$$

$$\varphi_{02} = 2\left(\frac{\varepsilon_2}{\varepsilon_1 + \varepsilon_2}\right) \int_0^\infty \frac{[q_1(\varepsilon_1 + \varepsilon_2)\mathscr{J}_0(k\sqrt{a^2 + b^2}) + q_2 u]\,dk}{g} +$$

$$+ 4\int_0^\infty \frac{q_2 v}{g} \omega_{m.n}^{(1)} k\,dk + 4\int_0^\infty \frac{q_1 \varepsilon_2 \lambda}{g} \omega_{m.n}^{(2)} k\,dk. \tag{36}$$

The expressions for the u, v, $\omega_{m,n}^{(1)}$, $\omega_{m,n}^{(2)}$ of (35) and (36) are

$$u = (\varepsilon_1 k^2 - \varepsilon_2 \lambda^2)\,\text{sh}\,2\lambda R + (\varepsilon_2 - \varepsilon_1)\,\lambda k\,\text{ch}\,2\lambda R, \tag{37}$$

$$v = \varepsilon_1 k\,\text{sh}\,2\lambda R + \varepsilon_2 \lambda\,\text{ch}\,2\lambda R, \tag{38}$$

$$\omega_{m,n}^{(1)} = \sum_{m=0}^\infty \sum_{n=1}^\infty \mathscr{J}_0(kd\sqrt{m_2 + n_2}) + \sum_{m=1}^\infty \mathscr{J}_0(kdm), \tag{39}$$

$$\omega_{m,n}^{(2)} = \sum_{m=0}^\infty \sum_{n=1}^\infty \mathscr{J}_0(k\sqrt{(a+md)^2 + (b+nd)^2}) + \sum_{m=1}^\infty \mathscr{J}_0(k\sqrt{(a+md)^2 + b^2}). \tag{40}$$

The alteration of electrical free energy, $\Delta\Phi_{el}$, accompanying macroion surface charging from zero to the final value can be calculated from the relation

$$\Delta\Phi_{el} = q_1 N \int_0^1 \varphi_{01}(c'q_1, c'q_2)\,dc' = q_2 N \int_0^1 \varphi_{02}(c'q_1, c'q_2)\,dc', \tag{41}$$

c' being a parameter which varies from zero to unity, and N the number of charges per unit surface area.

Equation (41) can be obtained through the method proposed by B. V. Deryagin in [5]. The same result can also be obtained from the statistical mechanics of systems consisting of macroions and ordinary electrolyte microions by the Ikeda method [12]. It should be emphasized that (41) was derived for a discrete point charge on the macroion surface, which is somewhat different from the earlier case. This does not require alteration of the general Deryagin equation (see the first citation of [5]) if the expression for the alteration of the electrical free energy of the system is successively derived for simultaneous charging of all the macroions from zero to their final values.

The parameter c' is introduced to allow for the difference between q_1 and q_2, and represents the ratio of the instantaneous to the final charge values.

The force of macroion interaction per unit surface area is given by

$$f = - \partial (\Delta\Phi_{el})/\partial (2R). \tag{42}$$

The expression $\Delta\Phi_{el}$ can be obtained by substituting (35) and (36) into (41) and integrating over c'; application of (37) then leads to

$$f = N\varepsilon_2 \int_0^\infty \frac{k\lambda^2 \left[q_1 q_2 Q_1 \, v_{m,n}^{(1)} + \left(\frac{q_1^2 + q_2^2}{2} \right) Q_2 \, v_{m,n}^{(2)} \right] dk}{2\,\mathrm{ch}^4\, \lambda R \, [\varepsilon_2\lambda + \varepsilon_1 k \,\mathrm{th}\,\lambda R]^2 \, [\varepsilon_2\lambda\,\mathrm{th}\,\lambda R + \varepsilon_1 k]^2}. \tag{43}$$

Here

$$Q_1 = [(\varepsilon_2^2\,\lambda^2 + \varepsilon_1^2\,k^2)\,\mathrm{ch}\,2\lambda R + 2\varepsilon_1\varepsilon_2\lambda k\,\mathrm{sh}\,2\lambda R],$$

$$Q_2 = [(\varepsilon_2^2 - \varepsilon_1^2)\,k^2 + \varkappa^2\,\varepsilon_2^2],$$

$$v_{m,n}^{(1)} = \sum_{m,n} \mathcal{J}_0 (k\sqrt{(a+md)^2 + (b+nd)^2}),$$

$$v_{m,n}^{(2)} = \sum_{m,n} \mathcal{J}_0 (kd\sqrt{m^2+n^2}).$$

The final result is quite complex, but can be satisfactorily studied in certain special cases.

Equation (43) will now be analyzed for the case in which q_1 and q_2 have the same sign and $d \gg 2/\varkappa$, which is to say pronounced charge separation on each surface. The essential interaction is then due to the charges q_1 and q_2 at the points $r_1 = 0$ and $r_2 = 0$. Here one should set $v_{m,n}^{(2)} = 1$ and $v_{m,n}^{(1)} = \mathcal{J}_0 (k\sqrt{a^2+b^2})$ in Eq. (43), the only significant members of the sum being those for which $m = n = 0$. It will also be assumed that the dielectric constant of the interior of the macroion is considerably less than the dielectric constant of the electrolytic solution. Situations of this kind are usually met in aqueous solutions where $\varepsilon_2 = 80$. We can then set $\varepsilon_2 \gg \varepsilon_1$ in (43). These simplifications lead to

$$f = \frac{2N}{\varepsilon_2} \int_0^\infty \frac{k \left\{ q_1 q_2 \,\mathrm{ch}\,(2\lambda R)\, \mathcal{J}_0 (k\sqrt{a^2+b^2}) + \left(\frac{q_1^2+q_2^2}{2} \right) \right\} dk}{\mathrm{sh}^2\,2\lambda R}. \tag{44}$$

The force, f, will be greater than zero if the area under a $\mathrm{ch}(2\lambda R)\, \mathcal{J}_0 (k\sqrt{a^2+b^2})$ plot is no less than unity, the integrand being positive under these conditions. Actually $(q_1^2 + q_2^2)/2 \ge q_1 q_2$ and multiplication of the charge product, $q_1 q_2$, by a factor less than unity gives a negative sign to the numerator of the integrand. The denominator of this integrand is always positive, and has no effect on the sign of the integral. When $q_1 \gg q_2$, $(q_1^2 + q_2^2)/2 \gg q_1 q_2$, and the first member of the numerator needs be taken into account only if the area under the $\mathcal{J}_0 = (k\sqrt{a^2+b^2})\,\mathrm{ch}\,2\lambda R$ graph is much greater than unity in absolute value. The magnitude of this area depends on $\varkappa R$, and on a and b as well, since $\lambda = \sqrt{k^2 + \varkappa^2}$. We now return to the general expression for the interaction force, (43). The sign of this force can be negative when $\varepsilon_1 \gg \varepsilon_2$ and $q_2 \gg q_1$. Here a reduction of R leads to a diminution of the multiplying factor of $q_1 q_2$, and there can be a change in the sign of the numerator of the integrand at a certain value of R. The presence of the series $v_{m,n}^{(1)}$ and $v_{m,n}^{(2)}$ in the periodic Bessel functions makes for difficulty in analysis. The term in $(q_1^2 + q_2^2)/2$ being negative, it follows from the properties of zero-order Bessel functions that the probability of a change in the sign of the interaction force at a certain low value of R will increase when $v_{m,n}^{(1)} > v_{m,n}^{(2)}$. Conditions are, however, usually such that the dielectric constant, ε_2, of the solution is higher than the dielectric constant of the densely packed globule, whether this be a colloid, or a polymeric macroion of protein or synthetic material.

TABLE 1

t	$\varepsilon_2/\varepsilon_1$				
	1	2	3	4	5
1.0	0.730	0.182	0.081	0.045	0.0073
1.1	0.266	0.097	0.049	0.027	0.0056
1.2	0.194	0.075	0.040	0.025	0.0047
1.3	0.045	0.056	0.032	0.0194	0.0038
1.4	0.119	0.049	0.026	0.0164	0.00324
1.5	0.100	0.042	0.022	0.0144	0.00287
1.6	0.084	0.035	0.019	0.0122	0.00243
1.7	0.070	0.030	0.016	0.0104	0.00210
1.8	0.0535	0.022	0.012	0.0076	0.00178
1.9	0.0505	0.021	0.0108	0.0075	0.00151
2	0.0425	0.0181	0.0101	0.0063	0.00126
2.1	0.0356	0.0152	0.0084	0.0054	0.00109
2.2	0.0301	0.013	0.0072	0.0046	0.00093
2.3	0.0256	0.011	0.0062	0.0039	0.00079
2.4	0.0222	0.0094	0.0052	0.0033	0.000675
2.5	0.0195	0.0084	0.0046	0.0029	0.000575
2.6	0.0155	0.0067	0.0038	0.0024	0.000485

Thus the sign of the interaction force can change only if the periodic functions under the sign of integration, $v_{m,n}^{(1)}$ and $v_{m,n}^{(2)}$, are multiplied by a monotonically increasing function. These functions are quite complex and we will therefore not go into a general discussion of the sign of the interaction force. It is, in principle, possible to graph the integrand for fixed values of $\varkappa$, d, a, b, q_1, and q_2, determine the area under the graph, and thus automatically deduce the sign of the interaction force as a function of macroion separation for various values of the electrolyte concentration. We will, however, pause on the displacement of the charge net, more specifically, on the determination of most probable values of a and b. The most probable values are $a = b = d/2$ and the electrostatic interaction energy a minimum when the microion charges, q_1 and q_2, are of the same sign but different magnitude and d is the same for each.

Here, the substitution of $a = b = d/2$ in Eq. (43) will somewhat simplify the result. The most probable values of a and b are zero when the microion charges differ in sign ($q_1 > 0$; $q_2 < 0$). This can be investigated in the following manner.

Setting $a = b = 0$ in (43) leads to

$$f = N\varepsilon_2 \int_0^\infty \frac{k\lambda^2 \left[q_1 q_2 Q_1 + \left(\frac{q_1^2 + q_2^2}{2} \right) Q_2 \right] v_{m,n}^{(2)} \, dk}{2 \, \mathrm{ch}^4 \, \lambda R \, [\varepsilon_2\lambda + \varepsilon_1 k \, \mathrm{th} \, \lambda R]^2 \, [\varepsilon_2\lambda \, \mathrm{th} \, \lambda R + \varepsilon_1 k]^2} .$$

(45)

There will be a force of attraction between the macroions when $v_{m,n}^{(1)} = 1$, i.e., $d \gg 2/\varkappa$, and $\varepsilon_2 \ll \varepsilon_1$. The sign of the force of interaction will depend on the relative values of the oppositely signed terms $\varepsilon_2 > \varepsilon_1$ and $v_{m,n}^{(1)}$ when $q_1 q_2 Q_1$ and $(q_1^2 + q_2^2) Q_2/2$. The sign of the force must be determined by the introduction of the actual values of the charge and $\varkappa$ when the distance between the macroions is small, $Q_1 \gtrsim Q_2$ and $|(q_1^2 + q_2^2)/2| > |q_1 q_2|$. At greater distances, $Q_1 \gg Q_2$. Thus

$$|q_1 q_2 Q_1| \gg (q_1^2 + q_2^2) Q_2/2.$$

The sign of the interaction force will be negative when $|q_1| \approx |q_2|$.

TABLE 2

t	$\varepsilon_2/\varepsilon_1$					
	1	2	3	4	10	20
1.0	100	25	11.1	6.25	1.070	0.266
1.1	3.88	2.72	1.98	1.50	0.475	0.159
1.2	2.67	1.93	1.45	1.16	0.386	0.136
1.3	2.19	1.59	1.21	0.95	0.340	0.122
1.4	1.96	1.44	1.11	0.87	0.312	0.112
1.5	1.84	1.36	1.03	0.81	0.292	0.105
1.6	1.780	1.270	0.990	0.780	0.264	0.093
1.7	1.745	1.264	0.960	0.750	0.263	0.092
1.8	1.740	1.260	0.954	0.740	0.253	0.091
1.9	1.708	1.230	0.925	0.720	0.248	0.087
2.0	1.709	1.216	0.910	0.700	0.238	0.0835
2.1	1.720	1.214	0.880	0.698	0.225	0.0775
2.2	1.716	1.212	0.875	0.693	0.224	0.0770
2.3	1.715	1.198	0.874	0.670	0.218	0.0740
2.4	1.730	1.195	0.860	0.660	0.212	0.0715
2.5	1.740	1.188	0.855	0.645	0.206	0.0690
2.6	1.751	1.184	0.853	0.643	0.203	0.0675
5.0	1.890	1.092	0.702	0.492	0.125	0.0375
10.0	1.370	0.727	0.395	0.259	0.0565	0.0160

The general case, $|q_1| = |q_2|$, can be analyzed in more detail. The expression for the force of interaction corresponding to $q_1 = - q_2$ takes the following form when the substitution $a = b = 0$ is made in (43):

$$ f = Nq^2 \sum_{m,\,n} \int_0^\infty \frac{\varepsilon_2 \, k\lambda^2 \, \mathcal{J}_0 \,(kd \, \sqrt{m^2 + n^2}) \, dk}{\operatorname{ch}^2 (\lambda R) \, [\varepsilon_2\lambda \operatorname{th} \lambda R + \varepsilon_1 k]^2} . $$

(46)

Determination of the sign of the interaction force requires a study of the integral

$$ I = \int_0^\infty \frac{\varepsilon_2 \, k\lambda^2 \, \mathcal{J}_0 \,(kd \, \sqrt{m^2 + n^2}) \, dk}{\operatorname{ch}^2 (\lambda R) \, [\varepsilon_2\lambda \operatorname{th} \lambda R + \varepsilon_1 k]^2} , $$

(47)

this being the general form of the terms of (46). Changing the variable of integration to $\lambda = \sqrt{k^2 + \varkappa^2}$ carries (47) to

$$ I = \int_\varkappa^\infty \frac{\varepsilon_2\lambda^3 \, \mathcal{J}_0 \,(d\sqrt{\lambda^2 - \varkappa^2} \cdot \sqrt{m^2 + n^2}) \, d\lambda}{\operatorname{ch}^2 (\lambda R) \, [\varepsilon_2\lambda \operatorname{th} (\lambda R) + \varepsilon_1 \, \sqrt{\lambda^2 - \varkappa^2}]^2} . $$

(48)

The function $\mathcal{J}_0$ is periodic and damped; it furnishes the principal contribution to the integral when the argument ranges between zero and the first root of the zero-order Bessel function. Passage to the dimensionless variable $t = \lambda/\varkappa$ makes it possible to determine the maximum interval of values corresponding to the first root of the zero-order Bessel function for members of the series of (46) from the condition

$$ \varkappa d \, \sqrt{t^2 - 1} \cdot \sqrt{m^2 + n^2} \approx 2.4. $$

(49)

The minimum nonzero value of the term $\sqrt{m^2 + n^2}$ in this relation is unity while the maximum tends to infinity. The member of (46) corresponding to $\sqrt{m^2 + n^2} = 0$ does not contain a Bessel function and is positive, the integrand being positive. The condition of (49) implies that the variation of t is over the interval $1 \lesssim t \lesssim 2.6$ when $\varkappa d \simeq 1$. The case $\varkappa d \gg 1$ is that in which there is zero surface charge interaction because of screening, while the case $\varkappa d \ll 1$ is essentially that of uniform surface charge distribution.

We will now study the behavior of the function

$$g_0(\lambda) = \frac{\varepsilon_2 \lambda^3}{\text{ch}^2(\lambda R)\, [\varepsilon_2 \lambda\, \text{th}\, \lambda R + \varepsilon_1 \sqrt{\lambda^2 - \varkappa^2}]^2}. \tag{50}$$

When $\varkappa \le \lambda < \infty$, one has $g_0(\lambda) \le p(\lambda)$, where

$$p(\lambda) = \frac{\varepsilon_2 \lambda^3}{\text{ch}^2(\lambda R)\, [\varepsilon_2 \lambda\, \text{th}\, \lambda R]^2} = \frac{\lambda}{\varepsilon_2\, \text{sh}^2(\lambda R)}. \tag{51}$$

The function $p(\lambda)$ falls off with increasing λ, its derivative being less than, or equal to, zero

$$p'(\lambda) = \frac{\text{ch}\,\lambda R}{\varepsilon_2\, \text{ch}^3\,\lambda R}\, (\text{th}\,\lambda R - 2\lambda R) \leqslant 0. \tag{52}$$

The fact that $\text{ch}(\lambda R)/\varepsilon_2\, \text{sh}^3\,\lambda R \geq 0$ and $(\text{th}\,\lambda R - 2\lambda R) \le 0$ for all real values such that $\lambda \ge 0$ implies that the function $g_0(\lambda)$ will also decrease over a wide λ interval, remaining less than $p(\lambda)$. We, however, need a monotonically decreasing $g_0(\lambda)$ function, and it is not obvious that such can be obtained over narrow λ (or $t = \lambda/\varkappa$) intervals. The expression for the derivative, $g_0'(\lambda)$, is rather cumbersome and its sign cannot, in general, be predicted. The above remarks would, however, indicate that both $g_0(\lambda)$ and the majorant, $p(\lambda)$, diminish monotonically in those physically important cases where $\varepsilon_2/\varepsilon_1$ is large. We now change over to the variable t in $g_0(\lambda)$. The result is

$$g_0(t) = \varepsilon_2 \varkappa u(t), \tag{53}$$

where

$$u(t) = \frac{t^3}{\text{ch}^2(\varkappa R t)\, [(\varepsilon_2 t)\, \text{th}\,(\varkappa R t) + \varepsilon_1 \sqrt{t^2 - 1}]^2}. \tag{54}$$

We will construct a graph of the function $u(t)$ in order to study the behavior of the latter over the interval $1 \le t \le 2.6$ where the members of (46) make the principal contribution to f. The desired accuracy can be obtained here by working over t intervals of 0.1. The results of these calculations are shown in Table 1 (for $\varkappa R = 1$) and Table 2 (for $\varkappa R = 0.1$). The case $\varkappa R = 1$ corresponds to the interaction of two plane surfaces separated by distances comparable to the distance of double layer screening. It is seen from these tables that the function $u(t)$ falls off monotonically even for low values of the $\varepsilon_2/\varepsilon_1$ ratio. In actual cases, the dielectric constant of the solution will be such that $\varepsilon_2 \approx 80$, while $\varepsilon_1 \approx 2\text{-}5$ for weakly swollen media, so that the interface between the electrolyte and the polymeric macromolecule or colloidal particle can be considered as plane. The monotonic fall-off of the function $u(t)$ is more pronounced when $\varepsilon_2/\varepsilon_1 \approx 10\text{-}20$. Especial interest attaches to the case in which $\varkappa R = 0.1$; this corresponds to $R \approx 10^{-7}\text{-}10^{-8}$ cm for concentrated electrolytes which follow the Debye–Huckel theory. The tables again show that there is an insignificant breakdown in the monotonic fall-off of the function $u(t)$ when $\varepsilon_2/\varepsilon_1 = 1$. The function diminishes montonically in all other cases, remaining positive throughout. Calculations of the function $u(t)$ were also carried out in the neighborhood of the second root of the Bessel function with $t = 5$. The function continues to fall off for all $\varepsilon_2/\varepsilon_1 = 1$ values with the exception of unity, a case without physical interest. An analysis based on (53) shows the function $g(t)$ to be monotonically diminishing, at least over that interval of t values in which the principal contribution is from the integral of (47). The product of $g(t)$, a positive, monotonically diminishing, function and $\mathcal{J}_0$, a diminishing function with a maximum at $t = 1$, gives a function with positive integral.

The interaction force calculated from (46) will then be negative over the entire R interval from infinity to $0.1/\varkappa$, which is to say, up to atomic distances. This is indication of electrostatic attraction at all physically realizable distances of approach for coarse macroions carrying discrete charges of the same magnitude but opposite sign in a solution of electrolyte.

The charge potential distribution in the electrolyte will be determined by the simple Debye–Huckel equation

$$\frac{d^2\varphi}{dx^2} = \varkappa^2 \varphi \tag{55}$$

with the boundary condition

$$\frac{d\varphi}{dx}\bigg|_{x=0} = -\frac{4\pi\sigma_1}{\varepsilon_2}, \quad \frac{d\varphi}{dx}\bigg|_{x=2R} = \frac{4\pi\sigma_2}{\varepsilon_2} \tag{56}$$

when the macroion surface charges are assumed to be constant and uniformly distributed at the respective densities σ_1 and σ_2. A solution of (55) satisfying the boundary conditions of (56) has the form

$$\varphi = \frac{4\pi\,[\sigma_1 \operatorname{ch} \varkappa\,(x - 2R) + \sigma_2 \operatorname{ch} \varkappa x]}{\varkappa\varepsilon_2 \operatorname{sh} 2\varkappa R}. \tag{57}$$

The data of [1, 5] and Eq. (41) show the possibility of obtaining the free energy change required for the calculation of the interaction force from

$$\Delta\Phi_{el} = \sigma_1 \int_0^1 \varphi_{x=0}(C'\sigma_1, C'\sigma_2)\, dC' + \sigma_2 \int_0^1 \varphi_{x=2R}(C'\sigma_1, C'\sigma_2)\, dC'. \tag{58}$$

Substitution of (57) into (58) with x = 0 and x = 2R gives

$$\Delta\Phi_{el} = \frac{2\pi\,[2\sigma_1\sigma_2 + (\sigma_1^2 + \sigma_2^2) \operatorname{ch} 2\varkappa R]}{\varepsilon_2 \varkappa \operatorname{sh} 2\varkappa R}. \tag{59}$$

By differentiating (59) with respect to 2R and drawing on (47), one obtains

$$f = \frac{2\pi\,[2\sigma_1\sigma_2 \operatorname{ch} 2\varkappa R + (\sigma_1^2 + \sigma_2^2)]}{\varepsilon_2 \operatorname{sh}^2 2\varkappa R}. \tag{60}$$

This result is positive for any value of R. Thus the present model is such that the interaction force between two surfaces carrying fixed, uniformly distributed charges becomes the splitting pressure when the signs of σ_1 and σ_2 are identical. Averaging the expression of (43) for the force over the interfacial surface must naturally give the result of (60). Integration over m and n will, in fact lead to the delta functions, $\delta(k_1)$ and $\delta(k_2)$. The final result is obtained by subsequent passage to the limit:

$$f = \frac{1}{N_s} \int\!\!\!\int_{-\infty}^{\infty} f\, dm\, dn = \frac{4\pi N \varepsilon_2}{2\pi d^2} \times \lim_{k \to 0} \left\{ \frac{\lambda^2 \left[q_1 q_2 Q_1 + \left(\frac{q_1^2 + q_2^2}{2} \right) Q_2 \right]}{2 \operatorname{ch}^4(\lambda R)\,[\varepsilon_2\lambda + \varepsilon_1 k \operatorname{th} \lambda R]^2\,[\varepsilon_2\lambda \operatorname{th} \lambda R + \varepsilon_1 k]^2} \right\} =$$

$$= \frac{2\pi N}{\varepsilon_2 d^2} [2q_1 q_2 \operatorname{ch} 2\varkappa R + (q_1^2 + q_2^2)]/\operatorname{sh}^2 2\varkappa R. \tag{61}$$

Since $q_1/d^2 = Nq_1 = \sigma_1$, and $q_2/d^2 = Nq_2 = \sigma_2$ in the averaging, it follows that (61) is identically (60).

Conclusions

1. The potential distribution has been developed for a system consisting of two macroions carrying two-dimensional charge nets and a separating microionic electrolytic solution. The radius of curvature of the macroion was assumed large in comparison with the radius of the surrounding ion atmosphere so that the interface between macroion and electrolyte could be treated as plane.

2. Calculation of the force of macroion interaction has been carried out for the general case of different ion charges. The magnitude of this force depends on both the dielectric constant of the electrolyte and the dielectric constant of the interior of the macroion.

3. The sign of the force of interaction has been studied for the various special cases met in actual colloidal and polymeric solutions.

There is a certain distance of separation at which the sign of the force of interaction becomes negative, when the macroion charges are of the same sign but different magnitude and the dielectric constant, ε_2, of the solution is different from the dielectric constant, ε_1, of the interior of the ion. This distance is closely dependent on the ε_2, ε_1 ratio.

This conclusion traces back to the fact that the microion surface charges are discretely distributed, the potential distribution in the electrolyte depending on both ε_1 and ε_2. It is to be noted that neither the interaction force nor the potential distribution depends on ε_1 when the charge distribution is uniform.

The mathematically simpler case of the interaction of oppositely charged macroions has been studied in detail. This is the situation which arises in mixtures of various macroions. For $|q_1| \neq |q_2|$, the analysis of the interaction force is possible only if the charge distribution is markedly discrete and the interaction between charges on different macroions rather than the effect of neighboring microion charges is the essential factor at distances of approach such that $2R < 2/\varkappa$. Here the force is always an attraction when $d \gg 2/\varkappa$. The macroions attract at the larger distances of separation and repel at the smaller distances, when $\varepsilon_2 > \varepsilon_1$ and $|q_1| \sim |q_2|$, and the concentration of the electrolyte is fixed.

The analysis is most detailed for the case in which $q_1 = -q_2$ and $|q_1| = |q_2| = q$. Here one has to do with the interaction of macroions, each carrying a charge net, the charges in the two nets being equal in absolute magnitude but opposite in sign. Interaction between neighboring macroion charges must now be taken into account since it is a matter of treating the most general case with $d \approx 2/\varkappa$. The force of interaction proves to be negative over the entire R interval from infinity to atomic dimensions. This is indication of electrostatic attraction at all physically realizable distances of separation between macroions carrying charges which are equal in magnitude but opposite in sign.

It has been shown that averaging the force of interaction over the macroion—electrolyte interface (this is physically equivalent to a passage to the uniformly distributed surface charge) quickly leads to the same result as a direct study of the interaction of surfaces of uniform charge or fixed potential separated by an electrolytic solution of low potential ($|ez_i\psi/kT)| \ll 1$) [1, 2, 5].

5. The conclusions of 3 above are in essential agreement with those obtained by B. V. Deryagin [5 and 9], by B. V. Deryagin and V. G. Levich [8], and by B. V. Deryagin and L. D. Landau [6]. The discreteness of charge distribution leads to new results which are reflected in the fact that the sign of the force of interaction depends on both ε_1 and ε_2.

6. Although the general equation, (43), for the force of macroion interaction is complex, it can be used for calculations on the various observed interactions.

The author would like to express his deep thanks to B. V. Deryagin and V. G. Levich, Corresponding Members, Acad. Sci., USSR, for discussion of the results obtained in the course of this work.

Literature

1. B. V. Deryagin. Izv. Akad. Nauk SSSR, Ser. Khim., No. 5:1153 (1937); Acta Physicochim. URSS 10:333 (1939).
2. P. Bergmann, P. Löw—Beer, and H. Zocher, Z. Phys. Chem. 181:301 (1938).
3. A. N. Frumkin and A. Gorodetskaya, Acta Physicochim. URSS 9:327 (1938).
4. S. Langmuir. Science 88:430 (1938); J. Chem. Phys. 6:873 (1938).
5. B. V. Deryagin. Kolloidn. Zh. 6(4):291 (1940); Trans. Faraday Soc. 36:203 (1940); 36:730 (1940).
6. B. V. Deryagin and L. D. Landau, Acta Physicochim. URSS 14(6):633 (1941). Zh. Éksperim. i Teor. Fiz. 15:663 (1945).

7. E. Verwey and I. Overbeek, Theory of the Stability of Liophobic Colloids, New York (1948).

8. B. V. Deryagin and V. G. Levich, Dokl. Akad. Nauk SSSR 98:985 (1954).

9. B. V. Deryagin, Discussions Faraday Soc., No. 18:85 (1954).

10. H. R. Kruyt, The Science of Colloids, Chapter IV [Russian translation] Moscow, IL (1955).

11. S. E. Bresler, Biokhimiya 14:180 (1949).

12. J. Ikeda, J. Phys. Soc. Japan, 8:49 (1953).

SECTION III

EXPERIMENTAL STUDIES OF THE PROPERTIES OF THIN FILMS

THE DOUBLE REFRACTION OF THIN LIQUID FILMS IN SWOLLEN MONTMORILLONITE

R. Green-Kelley* and B. V. Deryagin

Institute of Physical Chemistry, Acad. Sci., USSR
Laboratory for Surface Phenomena
Rothamsted Experimental Station (England)

Introduction

There are numerous indications that thin liquid films have a special anisotropic structure, but it is quite difficult to prove as much because of the minuteness of the measured effects.† These difficulties can be surmounted by study of the intracrystalline swelling of oriented aggregates of montmorillonite particles, using x-ray diffraction to measure the liquid film depth. These same aggregates can be studied optically by measuring the double refraction of the liquid films between the crystalline planes, the parallel arrangement of the planes rendering coherent optical effects which depend on path differences. We have used this method to study films of water and various organic liquids with depths ranging from 30 to several hundred angstroms.

Experimental Methods

The principal material used in these studies was a montmorillonite from Wyoming (USA). The properties of this material had been determined at the Rothamsted Experimental Station where quartz was found to be the principal contaminating material. This was removed by dilute sol sedimentation in the superultracentrifuge at Sharply, clay particles less than 0.2 μ in diameter being obtained. The product was washed with sodium chloride solution to sodium ion saturation and then freed of chloride ions by treatment with water—alcohol mixtures.

A new method was developed for preparing well oriented aggregates suitable for optical study. The clay particles were moistened with water to form a fluid sol and the latter then evaporated through a cellophane membrane in such way that aggregates would form on the membrane itself.

By preventing evaporation from the gel surface, it was possible to avoid the formation and eventual buckling of a surface scum. This considerably improved the orientation of the aggregate and, at the same time, facilitated removal of the aggregate from the base, an operation which proves to be quite difficult when the gel is evaporated from a glass surface. Small parallelepipeds, 0.1-0.2 mm in cross section, were then cut out of the clay sheet with a safetly razor blade and certain of these selected on the basis of microscopic observations. The selected samples were also studied under the polarizing microscope (Fig. 1). The clay block, b, was set into the depression of the object glass, G, in such way that the c axis lay in the plane of the latter; the immersion liquid was then added and a cover glass, k, carefully put into place in order to prevent evaporation. Swelling was allowed to proceed to completion and the path different, Δ, then measured for the sodium D line with a Senarmon compensator, the double refraction, B, being finally calculated from the known thickness of the sample.

---

*England, Rothamsted Experimental Station.

†N. N. Fedyakin [1] has recently shown that the characteristic anomalous thermal expansion does not appear in water in microcapillaries of ~200 A radius. This is direct confirmation that the structure of such films is different from that of bulk water, but not by any means proof that this structure is anisotropic.

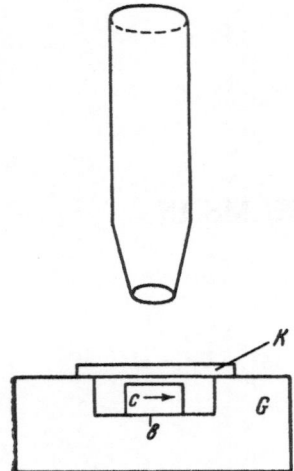

Fig. 1. Schematic representation of the apparatus for measuring the optical path difference along the c axis of the clay aggregate.

Interpretation

The structural double refraction must be taken into account in calculating the double refraction of the intracrystalline films from the overall double refraction of the swollen clay aggregates. Wiener [2] has shown that a system of parallel isotropic plates of nonuniform dielectric constant must be optically anisotropic, the mean value of the dielectric constant being given by the relations

$$\left.\begin{aligned}\varepsilon_0 &= \delta_1\varepsilon_1 + \delta_2\varepsilon_2 + \ldots\ldots, \\ \frac{1}{\varepsilon_a} &= \frac{\delta_1}{\varepsilon_1} + \frac{\delta_2}{\varepsilon_2} + \ldots\ldots\end{aligned}\right\} \tag{1}$$

Here, ε_0 is the mean (effective) dielectric constant of the parallel sheets, ε_a is the mean dielectric constant of the perpendicular sheets, and δ_x is the volume fraction of sheets with dielectric constant ε_x. Applicability of the Wiener equation requires that the thickness of each of the parallel sheets be much less than the wave length of the incident light. This condition is fulfilled for sodium light when the light film depth is less than 300 A. Applicability of the Wiener equation is further limited in the present instance by the fact that the polarizability of the silicate sheets cannot be measured by the ordinary bulk dielectric constants, these sheets being only ~10 A thick. The values adopted for $\varepsilon_1^|$ and $\varepsilon_1^"$ are, however, close to those for the bulk phase.

It will be shown that this element of uncertainty has no qualitative effect on the final results. It should be noted that it has not yet been possible to subject the Wiener equation to experimental check, although qualitative proof of the existence of the effect in question was obtained by Ambronn and Frey in 1926 [3].

The intrinsic double refraction of the silicate sheets in a system of pyrophillite-like sheets (each 10 A thick) separated by liquid films, is of approximately the same magnitude as for pyrophillite itself, namely 0.04 negative. The double refraction of the dehydrated montmorillonite aggregates is somewhat lower (from 0.02 to 0.03 negative), possibly because of incomplete orientation of the constituent crystallites, and an orientation coefficient of 0.25/0.4 = 0.6 was accordingly used to obtain the double refraction of the aggregate from the double refraction of a single crystal. The form of Wiener equation used here was

$$n_0^2 = \delta_1\,(n_1^{\|})^2 + \delta_2\,n_2^2, \tag{2}$$

$$n_a^2 = \frac{(n^{\perp})^2\,n_2^2}{\delta_1\,n_2^2 + \delta_2\,(n_1^{\perp})^2}, \tag{3}$$

$$B = 0.6\,(n_a - n_0).$$

Here n_0 is the effective index of refraction of the aggregate of parallel sheets, n_a is the effective index of refraction of the perpendicular sheets, $n_1^{\|}$ and $n_1^{\perp}$ are the indices of refraction of the silicate sheets (pyrophillite), n_2 is the index of refraction of the liquid films, assumed to be isotropic, and B is the double refraction of the aggregates.

The macroswelling, S, defined as the ratio of the c axis depth after and prior to swelling (see Fig. 1), was determined microscopically with the aid of a measuring eyepiece. Since the porosity of the block was low, it could be assumed that

$$S = \frac{1}{\delta_1}.$$

Water Films

Norrish has used x-ray diffraction [4] to show that sodium saturated montmorillonites undergo intracrystalline swelling in aqueous NaCl, the magnitude of this effect increasing to several hundred angstroms as the solution concentration is reduced. This is clearly a matter of considerable interest. These results obtained can be represented graphically by plotting either the Norrish double refraction, B, against the liquid film depth for solutions of known concentration, or the double refraction against the macroswelling, S. The observed macroswelling is, however, always greater than the swelling calculated from the Norrish data This fact indicates that the layers responsible for intercrystalline swelling are much thicker than those involved in intracrystalline swelling. This fact must clearly be taken into account if calculated and observed values of the double refraction are to be compared with a view to determining the water layer anisotropy. On the other hand, it can be proven that the two curves, one showing double refraction as a function of macroswelling, S, and the other double refraction as a function of the Norrish microswelling, tend to merge as the swelling increases. This indicates that macroswellings in excess of ten-fold give the same results as Norrish microswellings.

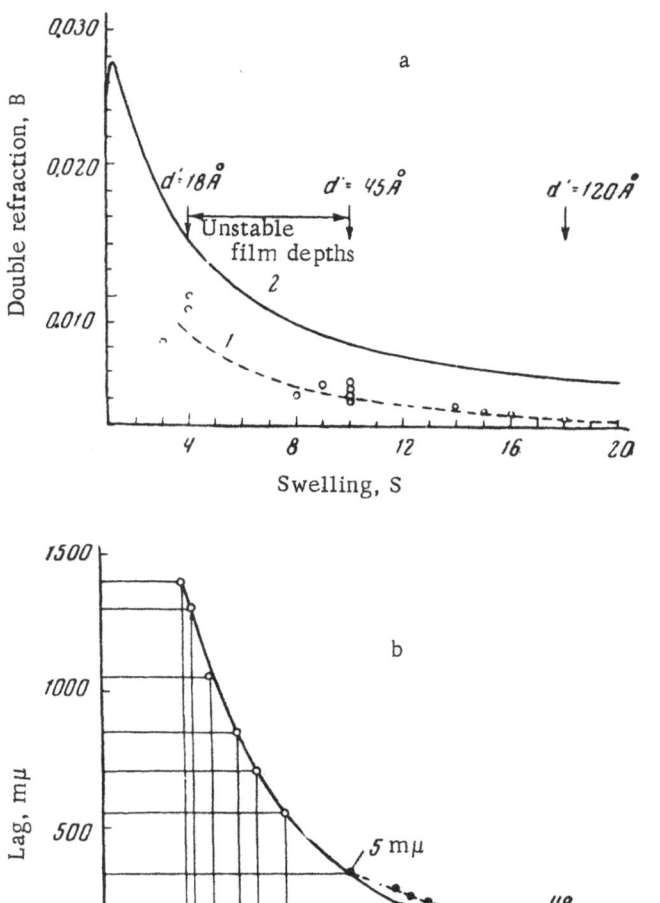

Fig. 2. Aggregate double refraction (a), and optical path difference (b), plotted against the degree of swelling, S: 1) experimental data; 2) calculated from the Wiener equation. Values of the depth, d', taken from the work of Norrish.

Figure 2a shows: a) the double refraction calculated from the Weiner equation under the assumption that $\delta_1 = 1/S$, and b) the directly measured values of the double refraction in NaCl solutions of various concentrations (open circles), plotted against the macroswelling. The spread of values is to be explained by the fact that each point represents a measurement on a new aggregate which had been brought into equilibrium with a NaCl solution of predetermined concentration.

The relative error associated with depth measurements on the nonuniform montmorillonite sheets increases here.

It will be shown below that the point spread is considerably reduced in measurements on a single sample; the accuracy of determination of the path difference, Δ, falls at the same time, however, since the degree of swelling varies from point to point in the sample if equilibrium is not established. Figure 2a shows that the experimental points fall below the Wiener equation curve at all values of the degree of swelling. The indication is either that the silicate sheet double refraction is different from the accepted value for the bulk phase, or that the water layers contribute to the optical anisotropy of the system. The Wiener equation curve is a hyperbolic at the higher S values, and cannot, therefore, be brought into coincidence with the experimental curve by any choice of intrinsic double refraction of the silicate sheets. Explanation of the results obtained here then requires the assumption that liquid layers 100-200 A deep show double refraction.

Additional information is available here. Thus it was observed that the double refraction fell off slowly as the swelling increased in experiments where the swelling was in excess of twenty-fold. The double refraction rapidly fell off to an immeasurably small value without a simultaneous increase in the swelling when there was a forty-fold swelling, the swollen sample showing reduced cohesion and a tendency to disintegrate. Spontaneous disorientation clearly set in at this point.

The effect of swelling on the double refraction was followed by two different methods in order to show that swelling independent of disorientation does not occur when the degree of swelling is low. In the first of these, the double refraction was measured after equilibrium had been established at each new value of the salt concentration; in the second, the measurements were made over a long period of time, working at fixed salt concentration in the region where the swelling was altering but slowly. The two methods give essentially the same results for a fixed degree of swelling (Fig. 2b). The two curves of the figure, the one passing through experimental points (open circles) corresponding to rapid establishment of apparent equilibrium with variation in swelling due to alteration of salt concentration, and the other through experimental points (filled circles) showing the degree of swelling and difference in optical path as functions of time for extended holding at a single concentration in the swollen state (48 hours), are almost completely identical. Both curves were developed with the same sample. The ordinate scales were chosen so that points corresponding to the same state (S = 18, time of swelling, 4 minutes) would coincide. Spontaneous disorientation occurred when S = 40.

It can be concluded from this that disorientation does not essentially alter the double refraction, so long as there is less than thirty-fold swelling.

The effect of swelling on the disorientation is still unknown. Two arguments can, however, be advanced to support the belief that disorientation from this source must be of little significance. In the first place, the theory of the diffusion double layer indicates the potential energy of the system to be at a minimum with the silicate sheets arranged in parallel. In the second place observations show a preferential swelling along the c axis up to a thirty-fold increase in volume, with the optical absorption of the aggregate remaining constant and well defined. Thus microscopic examination of samples set with the crystal layers parallel to the object plate showed the swelling in the horizontal direction to be less than 10% of the original thickness. Variations of this order are within the limits of the experimental error, resulting from inhomogeneity in the swollen samples. Attempts were made to compress these samples with a view to detecting hysteresis on the curve showing double refraction as a function of swelling, but disintegration occurred before the measurement could be made.

A curve covering the difference, ΔB, between the observed and calculated B values shows that ΔB alters only from + 0.004 to + 0.003 over the interesting interval where the swelling, S, ranges from 10 to 18-fold. If the entire ΔB value is due to the intrinsic double refraction of the liquid layers, one must conclude that this diminishes by 25% as the concentration of the NaCl is reduced and the liquid layer thickness* calculated by Norrish increases from 45 to 120 A.

The curve based on the Wiener equation can fall still lower and still maintain hyperbolic form at the higher S's if the intrinsic double refraction of the silicate sheets is assumed to differ from the accepted value. The l o w e s t position for the Wiener curve and the l o w e s t value for ΔB are determined by the condition that ΔB be independent of the degree of swelling.

The value of ΔB corresponding to this quite reasonable limit proves to be + 0.002, but this implies an improbably low value for the double refraction of the silicate sheets. Thus, the minimum value of the intrinsic water layer double refraction consistent with the present data, is, according to [5], to be compared with the maximum double refraction of ice (1.3104-1.3090)=+0.0014.

These results indicate the existence of a special anisotropic structure in thin water films such as has been predicted by B. V. Deryagin [6] on the basis of other data. It should be noted that the sign of the double re-fraction of the montmorillonite aggregates is always positive for swelling in water, but changes when swelling is carried out in certain organic liquids (see below).

It is necessary to decide whether the observed anisotropy of the water films could be due to a Kerr effect from the strong electric field of the double layer established at the surface of the silicate sheets. It follows from the formulas for the Kerr effect that

$$\Delta B = \frac{\theta}{\lambda} \frac{1}{d} \int_0^d E^2 dx = \frac{\theta}{\lambda} \overline{E^2}. \tag{4}$$

Here, λ is the wave length of the light, θ is the associated value of the Kerr constant, $4.03 \cdot 10^{-7}$, d is the half-depth of the water layer, and E is the field strength, a function of the distance from the silicate sheet.

From this it follows that a minimum value of $\Delta B = 0.002$ would require that $E^2 = 8.5 \cdot 10^7$.

The theoretical equations developed for the interaction of planar double ionic layers [9] are such that

$$\overline{E^2} = \frac{8\pi^2 \sigma^2}{\varepsilon^2} \left[\frac{\coth \varkappa d}{\varkappa d} - \frac{1}{\sinh^2 \varkappa d} \right] \tag{5}$$

at low surface potentials, ε being the dielectric constant, $\varkappa$ the reciprocal depth of the ion atmosphere, and σ the charge density of the diffuse portion of the double layer (per 1 cm^2).

The m i n u m u m value of σ consistent with the calculated m i n i m u m value of E^2 is that correspond-ing to a maximum in the expression in square brackets; this maximum, in turn, is the value obtained in passage to the limit $\varkappa d \rightarrow 0$. Calculation for this limiting case shows that

$$\overline{E^2} = \frac{16}{3} \frac{\pi^2 \sigma^2}{\varepsilon^2}. \tag{6}$$

It can be proven that this equation is applicable for any value of the interfacial potential, regardless of how high this value may be.

The minimum CGS value of σ calculated on this basis is

$$\sigma = 10^5.$$

This result is many times larger than the physically permitted values of σ. Thus, Green−Kelly [7] has obtained a value of $3.8 \cdot 10^4$ for the charge density of the entire double layer. Since ΔB is proportional to σ^2, the m a x i m u m permitted value of σ could explain only about $\frac{1}{10}$ of the m i n i m u m water film double refrac-tion. Another explanation such as the existence of a specific anisotropic structure must be advanced here.

———————

*These data were confirmed by measurements on the montmorillonite sheets with which we worked.

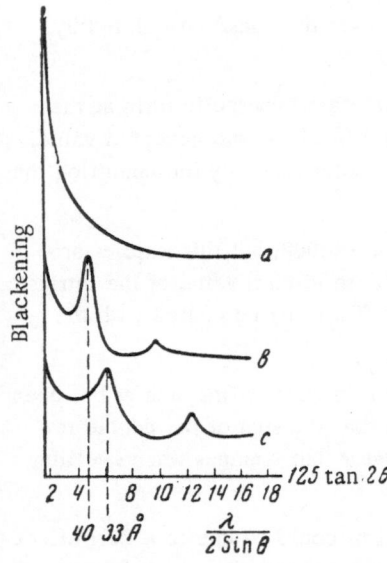

Fig. 3. Microphotograms of the diffraction patterns of dodecylammonia montmorillonite in: nitrobenzene (a), pyridene (b), and quinoline (c) solutions.

Jordan has shown [8] that montmorillonite saturated with long chain amide cations will swell in certain organic solvents. Lack of x-ray diffraction data makes it impossible to determine the intracrystalline swelling here, but measurements on the macroswelling indicate that it must be of considerable magnitude. For this reason x-ray diffraction and double refraction were studied together in these systems.

Dodecylammonia montmorillonite was prepared by washing sodium montmorillonite plates with an aqueous solution of the amine containing a slight excess of dilute hydrochloric acid. The excess salt was removed by washing with ethyl alcohol. The x-ray diffraction pattern of a sample of water-formed flocks showed a basic interplanar distance of 17.4 A. This was indication that the major portion of the sodium had been replaced by organic cations. The oriented aggregates were dried, cut into blocks, and measurement made of the optical path difference and swelling in several selected solvents. Certain of the blocks were put into capillaries and then treated with the same solvents. The swollen flocks were oriented under the polarizing microscope and then placed in the chamber of the x-ray camera for study of the low-angle diffraction of the filtered radiation from a copper anticathode operating at 30 kv. The chamber was so arranged that there was negligible background scattering in the absence of the sample.

Figure 3 shows microphotograms obtained from the various photographs. The Norrish method of one-dimensional Fourier analysis was used in interpreting the results. For the present purposes it is enough to point out that these diffraction patterns indicate the formation of polymolecular solvent films. These films are probably bound to the organic cation carbon chains. The accompanying table contains data on the double refraction, swelling, and diffraction maxima of the dodecylammonia montmorillonite aggregates in various solvents. It is seen that the double refraction of the pyridene and nitrobenzene dodecylammonia montmorillonites is quite different from that of the monolayer Na montmorillonite complexes. Thus pyridene gives zero double refraction in the monolayer complexes and a double refraction of + 0.002 in the polylayer type crystals. The corresponding values for nitrobenzene are −0.004 and + 0.002. This is clear-cut proof that the anisotropic structure of the polymolecular layers is different from that of the monolayers. It should be noted that there is an alteration in the sign of the optical path difference, the situation here being different from that met with the Na montmorillonites in water.

Solvent	Double refraction, B	Swelling (linear), S*	Diffraction maximum, $\dfrac{\lambda}{2 \sin \alpha}$, A	Gel volume,* ml
Water	−0.02	1.0	17.4	2.0
Decaline	−0.01(5)	1.0	−	−
Quinoline	+0.02	1.5	33.0	−
Pyridene	+ 0.002	2.5	40.0	28.0
Nitrobenzene	+ 0.002	4.0	Very minute angle	88.0

*Values of the gel volume were taken from Jordan [8].

The present study was carried out during the nine-month stay of Green—Kelly in the Soviet Union as part of the exchange between the Academy of Sciences of the USSR and the Royal Society of Great Britian. We would like to express our sincere thanks to V. A. Koved, Corresponding Member, Acad. Sci., USSR, for permission to carry out this work in the Department of Soil Science of Moscow University, and to the members of this Department, especially Associate Professor N. G. Zyrin, for aid and advice. We also wish to thank T. N. Voropaev, Assistant in the Laboratory of Surface Phenomena of the Academy of Sciences, for his help in carrying out this work.

Conclusions

1. Measurement of the optical path difference in montmorillonites swollen in various liquids makes it possible to calculate the double refraction of the polymolecular liquid layers formed between the silicate sheets of the crystals.

2. The double refraction of water layers was found to alter very little when the film depth was increased to 200 A by reducing the NaCl concentration, remaining definitely above 0.02.

3. The values obtained for the minimal double refraction proved to be one order higher than could be accounted for by a Kerr effect in the diffusion double layer at the maximum permissible charge density. Thus the results obtained can be explained only by assuming isotropic structure in the interfacial water layers.

4. Optical studies on dodecylammonia montmorillonites swollen in quinoline, pyridene, and nitrobenzene led to positive values of the double refraction, showing thus anisotropic liquid film structure.

The sign of the double refraction for swelling in nitrobenzene was opposite to that for the monolayer crystal, indicating a unique polymolecular layer structure.

Literature

1. N. N. Fedyakin, Dokl. Akad. Nauk SSSR 138:1389 (1961).
2. Wiener, Abhandl. Sächs. Ges. Wiss. Math.-Phys., Kl. 32:509 (1912).
3. Ambronn and Frey, Polarisations Mikroskop., Leipzig (1926).
4. Norrish, Discussion Faraday Soc., No. 18:120 (1954).
5. Dorsey, Properties of Ordinary Water-Substance, Am. Chem. Soc. Monograph, Reinhold Publishing Corp. (1940).
6. B. Deryagin, et al., Discussion Faraday Soc., No. 18:31 (1954).
7. Green—Kelly, Clay Minerals, Bull. 5:1 (1962).
8. Jordan, Mineral. Mag. 28:598 (1949).
9. B. Deryagin, Trans. Faraday Soc. 36:203 (1940).
   E. Verwey and J. Overbeek, Theory of the Stability of Lyophobic Colloids, New York (1948).

THE EFFECT OF A SOLID SURFACE ON THE PROPERTIES
OF CERTAIN LIQUIDS

N. N. Fedyakin

Institute of Physical Chemistry, Acad. Sci., USSR
Laboratory for Surface Phenomena
Kostroma Technological Institute

Experiment has shown that the pressures required for condensation in capillaries and cracks are lower than predicted by the Kelvin Equation relating vapor pressure at saturation and radius of curvature of the liquid surface [1-3]. The deviations in question cannot be accounted for by the increase in curvature resulting from an adsorption film, the transition curvature [4], or the variation of surface tension with curvature. The possibility that the reduction in saturation vapor pressure observed in glass capillaries [1, 2] might be due to salt dissolution from the capillary walls can be eliminated by using quartz capillaries [3], but there is still a marked deviation between the experimental results and the predictions of the Kelvin Equation. The explanation of these facts requires an understanding of the mechanism of the vapor—liquid transition in the capillary. The vapor can either condense on the newly formed liquid in the narrower sections of the capillary, or condense on the walls to form a metastable film which then passes over into a more highly stable liquid column, the capillary pressure of the meniscus at fixed surface tension being twice as large as that of the film surface in cylindrical capillaries. The first of these mechanisms limits condensation to capillaries closed at one end and requires that the liquid column rise from the bottom, gradually filling the entire capillary. The second mechanism also permits condensation in capillaries open at both ends, and gives the possibility of the simultaneous formation of several liquid columns in both open and closed capillaries. We have carried out various experiments to fix the condensation mechanism. Aqueous sulfuric acid solutions of various concentrations were put into carefully cleaned glass vessels. Capillaries, 1.5-2 mm in length, each attached to a special paraffin-covered base were introduced, the vessel sealed, and then put into a thermostat where their temperature could be controlled to 0.01°C. A liquid column eventually appeared in each capillary, the time required for column formation varying with the solution concentration. The length of each column was measured over 5-7 days, using a specially equipped microscope and viewing the system through the side wall of the vessel. The numerous experiments of B. V. Deryagin and his coworkers [5-8] have shown the properties of an interfacial film to be different from those of the bulk liquid. This difference traces back to the fact that the film molecules are in a surface force field and arranged in a definite structural pattern. The high absolute energy of molecular interaction assures that this be a more stable arrangement corresponding to a lower internal energy level in the liquid. It is natural to suppose that this structure would be maintained in passage over to the column of bulk liquid. The internal energy of the liquid column in the capillary should then be low and the pressure of the saturated vapor above it reduced. It is to be recalled that the studies of B. V. Deryagin and Z. M. Zorin [5, 8] on smooth surface vapor condensation have shown the properties of polar liquids to change discontinuously on passing from adsorption layer to bulk phase. Liquid droplets appear on an adsorption film in an atmosphere of supersaturated vapor, droplets and film being in equilibrium with one another. Droplet formation indicates enhanced chemical potential in the liquid film which is laid down on the surface, with passage from the metastable state to the more stable modification. The instability of the cylindrical film also contributes to column formation in capillaries. Passage from film to column can occur here even below saturation vapor pressure.

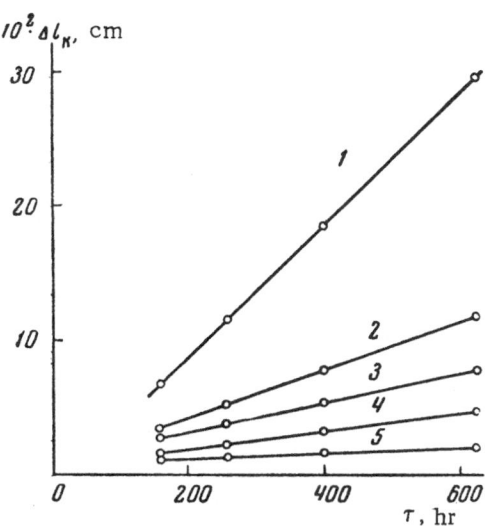

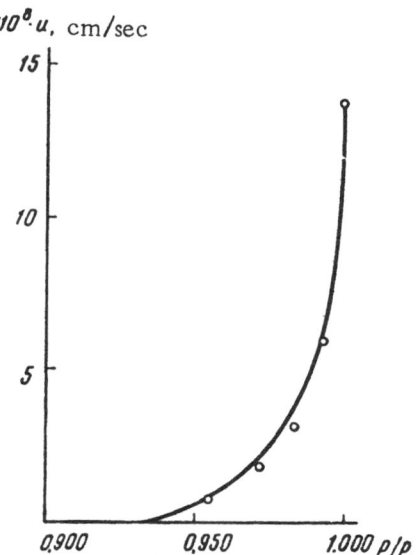

Fig. 1. Time dependence of the total length of condensed liquid column: 1) $p/p_s = 1$; 2) $p/p_s = 0.994$; 3) $p/p_s = 0.983$; 4) $p/p_s = 0.972$; 5) $p/p_s = 0.953$; r = 1.5 mm; t = 20°C.

Fig. 2. Relation between the rate of increase of length of liquid column and the relative vapor pressure.

Thus the liquid which passes from film to column has lower vapor pressure than the bulk liquid, just as does the film itself. The alteration of molecular energy resulting from passage to the more stable orientation can be calculated from the reduction in saturation vapor pressure which accompanies column formation from the film. Statistical physics gives the relation [9]

$$p_0 = \left(\frac{2\pi m}{\tau_0^2} \right)^{\frac{3}{2}} \frac{1}{\sqrt{kT}} e^{-\frac{|U_0|}{kT}},$$

(1)

p_0 being the saturation vapor pressure, m the mass of the molecule, τ_0 the mean vibration period of the surface molecules, k the Boltzmann constant, T the absolute temperature, and U_0 the mutual potential energy of the molecules.

Increase of the molecular interaction energy by ΔU entails a reduction of the saturation vapor pressure by Δp. Equation (1) gives

$$\Delta p = p_0 \left(1 - e^{-\frac{|\Delta U|}{kT}} \right).$$

(2)

An experimental determination of Δp makes it possible to calculate ΔU. Figure 1 shows the time dependence of column length in capillaries of uniform diameter for liquid condensation from vapors of various pressures in the vessel. The slope of each line represents the rate of increase of the total column length. Figure 2 shows the relation between the rate, u, of increase in column length and the relative vapor pressure in the capillary atmosphere, as obtained from the data of Fig. 1. The relative vapor pressure at equilibrium is determined by extrapolating this curve to intersection with the axis of abscissas, and proves to be equal to 0.935.

Equation (2) applied to 1 cm³ water gives

$$\Delta U = 4.35 \cdot 10^7 \text{ erg}.$$

The effect of meniscus curvature has been neglected in this work, the pressure differences calculated from the Kelvin Equation being much less than those determined experimentally.

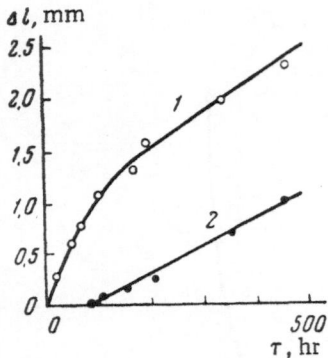

Fig. 3. Liquid transfer in the cylindrical capillary: 1) decrease in length of normally structured liquid column; 2) length of film-structured liquid column.

When this factor is taken into account, the expression for the saturation vapor pressure becomes

$$p = p_0 \exp\left(-\frac{\Delta U r + 2\sigma V \cos\theta}{rRT}\right).$$

(3)

Here, ΔU is the change in internal energy, per mole of liquid; r is the capillary radius; σ is the surface tension, and v the specific volume, of the liquid; θ is the contact angle; and R is the molar gas constant.

Freshly drawn capillaries were used in our work and it was therefore possible that the glass salts dissolved into the condensing liquid, thus leading to a reduction in saturation vapor pressure. Transfer of the liquid to sealed cylindrical capillaries showed that salt dissolution alone would not explain this effect.

It was found that liquid column formation would take place in the free space in a cylindrical capillary which had been partially filled with liquid, sealed at both ends, and then placed in the thermostat, the lengths of the newly formed columns increasing regularly while the length of the original column diminished. Figure 3 shows the relation between the column length and elapsed time, the latter measured from introduction into the cylindrical capillary. The upper curve shows the reduction in length of the original liquid column, and the lower, the length of the newly formed column, each plotted as a function of elapsed time. The apparent transition from ordinary liquid to special form which is reflected by this figure could not be explained in terms of salt dissolution since both liquid modifications were observed under the same conditions.

Transfer rates for passage from normal to "special" liquid columns separated by a distance L in a capillary of radius r are given below:

r, μ	1.3	1.3	1.3	0.8	1.1	1.5	2.0
L, mm	84.0	34.0	19.4	80.0	35.0	50.0	57.0
$\Delta l/\Delta\tau \cdot 10^3$, mm/hr	2.90	2.86	2.90	2.80	2.94	2.76	2.90

These data show the transfer rate to be independent of both the capillary radius and the distance between the two types of column. This is no longer true, however, when the column separation is of the order of several millimeters, or the capillaries quite narrow (microcapillaries). Assuming invariance of liquid properties in capillaries having radii of the order of several microns, radius determinations were made from the pressure of compressed air required for balancing the capillary pressure. The capillaries selected for this work were about 1 cm in length; these were sealed at one end and the other (open) end then brought into contact with the test liquid which rose in the capillary and compressed the entrapped air. Movement of the meniscus continued until the pressure of compressed air balanced the capillary pressure. Knowing the length of the capillary and the length of that portion of it which remained free, the capillary radius could be readily calculated from the equation

$$r = \frac{2\sigma}{p_0\left(\dfrac{l_0}{l_1} - 1\right)}.$$

(4)

Here, σ is the surface tension, p_0 is the air pressure in the capillary at the instant of contact between capillary and liquid (ordinary atmospheric pressure), l is the length of capillary bore, and l_1 is the length of the free portion of the capillary. Application of the Boyle—Mariotti Equation led to no appreciable error here, the air being compressed to only 2-3 atmos in these experiments. The time required for each experiment was of the order of several seconds, and the absorption of the compressed air therefore negligibly small. The experimental accuracy was improved by repeating the measurements with shorter and shorter lengths of the same capillary.

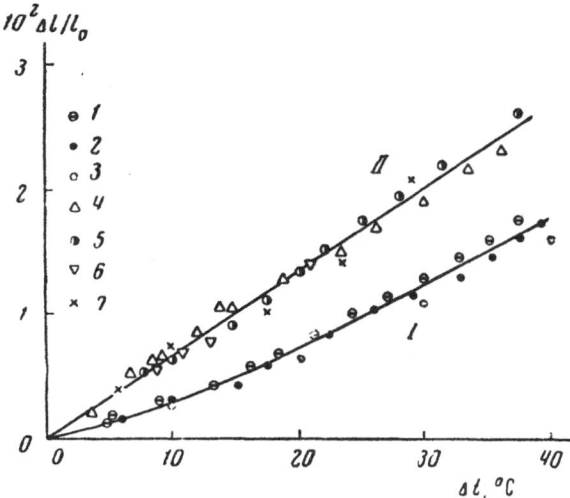

Fig. 4. Thermal coefficient of expansion of ordinary liquid and of the film-structured liquid: I) ordinary liquid, r = 1.69 μ; 1) heating; 2) cooling; 3) data of the literature; II) film-structured liquid, r = 1.69 and 3.16 μ; 4) heating; 5) cooling, r = 3.16 μ; 6) heating; 7) cooling.

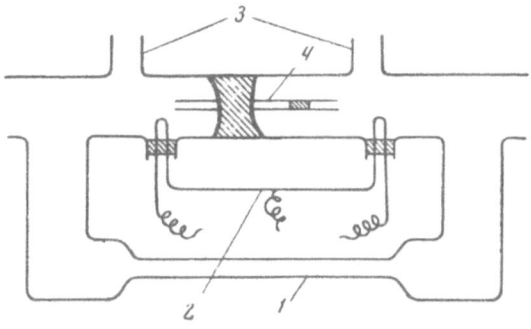

Fig. 5. Apparatus for viscosity measurements: 1) pressure differential tube; 2) thermocouple; 3) take-off for differential manometer; 4) capillary for the liquid viscosity measurement.

There was very little variation in the results obtained with any one well chosen capillary. The capillaries used in this work were broken from both ends and the radius measured in the manner described above in each of the broken pieces. Uniformity in the radius of the selected capillary was checked microscopically. The capillary channel scatters light when illuminated from the side, appearing as a bright streak under the microscope. The intensity of scattering varies with the diameter of the channel, so that any lack of uniformity can be detected as an alteration of the index reading when the capillary is moved over the objective of a microphotometer. The fact that the condensed water has a different structure from water in the bulk must entail differences in certain other physical properties, as well. Thus there must be differences in the coefficient of thermal expansion, viscosity, and in fact, in all properties which depend on the molecular packing and the energy of molecular interaction.

We also carried out measurements on the column elongation at various temperatures. One of the capillaries contained the original (ordinary) liquid and liquid which had formed from the film. This capillary was heated in a water bath and the relative elongation of the column determined as a function of temperature. The results obtained are shown in Fig. 4. Curve I shows the temperature dependence of the relative elongation of the column of ordinary liquid. The figure makes it clear that the points fall somewhat above the accepted values, this being the result of either experimental error or of salt dissolution from the glass. Curve II shows the temperature dependence of the relative elongation of the column of film-form liquid. The coefficient of volume expansion of this structure clearly remains constant over the indicated temperature interval. The effect of the liquid transfer from the one column to the other can be neglected since the measurements were usually completed in 5-6 minutes.

Viscosity measurements were carried out by inserting the capillary containing the special liquid column [11] in the apparatus shown schematically in Fig. 5. A pressure differential could be established at the two ends of the working capillary by varying the rate of flow of air through tube 1, and this differential measured with a differential manometer. The differential thermocouple 2 was used so as to assure the absence of temperature gradients; a third junction immersed in a Dewar flask could be switched into the thermocouple circuit, thus permitting measurement of the absolute temperature at any time. The entire apparatus was attached to the microscope stage in such manner that the column of test liquid was continually held in the viewing field. The rate of movement of the liquid column was observed with various pressure differentials at the ends of the capillary and the viscosity of the liquid calculated from the results obtained.

TABLE 1

r, μ	Δl, mm	η, 10^2 g·cm·sec	r, μ	Δl, mm	η, 10^2 g·cm·sec
1.06	0.39	12.5	2.17	1.57	11.0
1.03	0.39	12.0	2.13	1.33	10.0
1.78	0.63	15.0	1.37	0.69	12.5
0.94	0.38	11.0	1.85	0.61	12.5
2.13	1.5	18.4	2.54	1.08	6.5
2.64	0.98	8.0			

Movement of a liquid column of length Δl in the capillary bore is equivalent to liquid flow through a capillary of length Δl. The Poiseuille Equation can be written as

$$8\,\frac{v\,\Delta l}{r^2} = \frac{1}{\eta}\,\Delta p,$$

(5)

v being the rate of movement of the liquid along the capillary bore, Δl the length of the liquid column, r the capillary radius, and Δp the pressure differential, and η the viscosity.

Effects from forces other than those of internal friction (e.g., meniscus distortion under motion, capillary taper, etc.) must be eliminated in determining the viscosity from the liquid column movement in the capillary.

The specially structured liquid may also prove to be Non-Newtonian, showing limiting shear stress, or slip along the capillary wall. For this reason, the relation between rate of meniscus movement and pressure differential must be known before the viscosity can be calculated. Equation (5) is valid only for newtonian liquids and must be altered before application to cases involving a limiting shear stress (f) [10]. When

$$f < \frac{r}{2\Delta R}\,\Delta p,$$

one has

$$v = \frac{r^2}{8r\,\Delta l}\left(1 - \frac{8\Delta\,lf}{3\Delta\,pr}\right)\Delta p.$$

(6)

Thus it follows that the straight line showing the relation between $8\,v\Delta l/r^2$, Δp will intersect the axis of abscissas at

$$a = \frac{8\Delta\,lf}{3\Delta\,pr},$$

so that

$$f = \frac{3\,ar}{8\Delta\,l}.$$

(7)

The presence of such an intercept in a liquid already known to be Newtonian is indication that the movement is either retarded or accelerated by other forces.

Our own experiments studied the movement of liquid columns approximately 1 mm long in capillaries whose radii were of the order of one micron, the ratio of length to radius being here of the order of 10^3. Contact effects or the possibility of passage to laminar flow could therefore be neglected.

Measurements were made 1-2 seconds after pressure alteration, presumably in the stationary state, since the time required for establishing this state was negligibly small [12]. The experiments lasted only a few minutes and there was therefore no need of thermostating. The capillary radius was established by determining the air pressure required for balancing the capillary pressure [4].

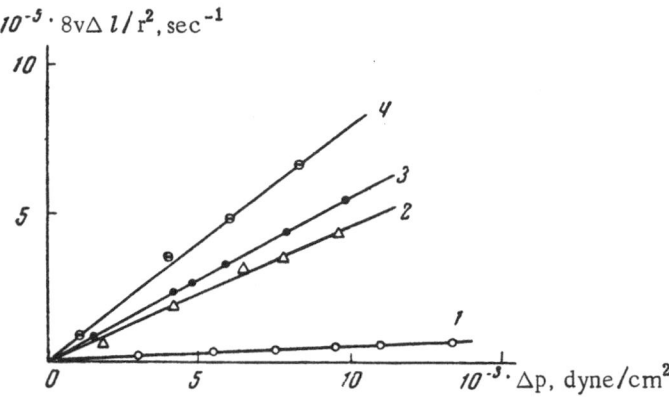

Fig. 6. The relation between the rate of movement of the specially structured liquid column and the pressure differential, for movement in one direction and over the same section of the capillary: 1, 2, 3, 4) first, second, third, and fourth movements; $r = 2.23\ \mu$; $\Delta l = 1.5$ mm; $t = 19°C$.

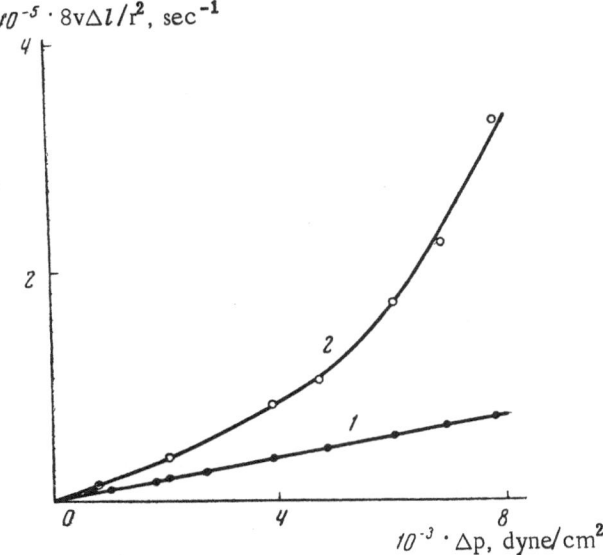

Fig. 7. Relation between the rate of movement and the pressure differential over a single section of the capillary: 1) first movement; 2) second movement; $r = 2.13\ \mu$; $\Delta l = 1.33$ mm; $t = 20°C$.

Viscosity measurements on liquids newly formed in the capillary, and on liquids which had been held in the capillary for days, or even months, gave results which agreed with the reported values, at least to within the limits of experimental error. This error was principally due to the uncertainty in the measurement of the capillary radius and was of the order of $\pm 2\%$.

Departures of the viscosity of the daughter liquid columns from the values of Table 1 could not, therefore, have been due to long holding in the capillary.

The results of one of these experiments are shown in Fig. 6. Here measurement was made of the rate of movement of a daughter column of water, 1.5 mm in length, over a 2-3 mm section of capillary. Curve 1 was obtained in first movement measurements, curve 2 in measurements over the same section and in the same direction, after the column had been returned to its original position. Curves 3 and 4 were obtained in the same manner for subsequent movements. Movements over a single section and in the same direction are desirable to assure uniformity of conditions and thus avoid the possibility that velocity differences could be due to some side effect such as capillary taper.

Figure 7 shows the relation between the pressure and the rates of forward and reverse movement of the daughter column.

Curve 1 was obtained from the first movement, the pressure being gradually increased over a definite period of time so that there was no pause between measurements. Curve 2 was obtained for the reverse movement, with pause between each change. It is seen from the figure that a greater and greater force was required to reestablish movement, although the velocity at fixed pressure was greater in the second case than the first. The viscosity fell off sharply as the velocity at the beginning of movement increased, rapidly approaching

TABLE 2

r, μ	Δl, mm	η_1^*, 10^2 g/(cm·sec)	η_2, 10^2 g/(cm·sec)	η_T, 10^2 g/(cm·sec)
1.4	4.2	0.52	0.34	0.33
1.45	1.64	0.58	0.32	—
1.85	6.58	0.55	0.31	—

* η_1 is the viscosity at first movement, and η_2 the viscosity at
second movement, over the same section and in the same direction; η_T is the accepted value of the viscosity.

the normal value at velocities of the order of 1 mm/sec. The situation in regard to the first movement was analogous. A limiting shear stress appeared when movement of the column was stopped after several points had been obtained and then once more resumed at the minimum velocity, and this despite the fact that the viscosity was lower than before. The value of this limiting shear calculated from Eq. (6) was approximately 2 dyne/cm^2. Table 1 gives the results of viscosity measurements on the daughter water column at the initiation of the first movement.

Results of viscosity measurements on daughter columns of acetone are shown in Table 2.

No alteration could be observed in the viscosity of newly formed methyl alcohol columns, and though there was a limiting shear stress, this disappeared after the column had been displaced about one millimeter.

It is significant that movement of the daughter column led to reestablishment of the original high (but lower than normal) value of the viscosity after a marked preliminary reduction.

An increase in the viscosity became apparent after 6-7 hours, and the initial value for the daughter column was once more reached in the course of the day.

Discussion of Results

The results outlined here indicate that two distinctly different modifications of water exist in the columns observed in glass capillaries at fixed temperature. Indication of unmistakable structural differences in the two modifications is found in differences in the coefficient of thermal expansion, viscosity, and saturation vapor pressure. Establishment of a new structure in the "anomolous" column clearly traces back to the fact that this column originates as a thickening of the adsorbed wetting film on the inner capillary wall. Contact with the meniscus of a "normal" column, or with nearly saturated vapor, can increase the depth of the adsorption layer to the point where the instability of the cylindrical liquid surface cannot be compensated by the action of a splitting pressure, $\Pi(h)$ which varies with $\partial p/\partial h$. This effect is well known from capillary theory.

Loss of stability causes the liquid film to acquire undulating form and results in the appearance of secondary columns. Conclusion of this process with the formation and growth of columns in the neighborhood of the original, is indication of property differences (above all, vapor pressure differences) between the original and daughter columns. The formation of more than two secondary columns can indicate that either all columns have the same vapor pressure as the parent so that new column formation is limited by the length of the capillary alone, or that all secondary columns are formed simultaneously. This last situation could be favored by rapid equalization of the depth of the adsorbed layer prior to liquid compaction and column growth, this layer being originally thicker in the neighborhood of the original column, or capillary constriction. The fact that the second of these two cases is the one commonly met is indicated by the observation that the number of secondary columns increases when passage is made to wider capillaries where diffusion is more important and the time of column formation extended.

In view of this proposed mechanism, the appearance of a "peculiar" structure in the second liquid column can be supposed to result from retention of the boundary layer structure while adsorbed film is thickening to a depth in excess of that corresponding to equilibrium. This peculiar structure in the boundary layer (and the anomolous column) can, in turn, be looked on as originating in hydroxyl groups on the glass surface, these groups serving as focal points for chains of liquid molecules joined by hydrogen bonds. This is consistent with the fact that such structures are met in water and acetone which show hydrogen bonding. Thus the anomolous column of liquid is polymerized, as it were, by the glass surface, through a special long-range topochemical action. It follows that the number of hydrogen bonds per unit volume must be greater in an anomolous column than in a column of normal water; this increases the density of molecular packing thereby lowering the vapor pressure and raising the viscosity. The superfluous hydrogen bonds can clearly break down in flow, and the column thus acquire the structure and properties of bulk water. The accompanying increase in vapor pressure, chemical potential, and free energy, would be at the expense of work expended by external forces over and above the thermal energy dissipation in column flow.

The breakdown of superfluous bonds can proceed so far as to give rise to a potential barrier that makes it impossible to reestablish the original state, even though this would lead to a reduction in the chemical potential. Experiment shows that the original anomolous structure can be completely reestablished when there is only an insignificant breakdown of bonds. This is clear indication that the region of structural alteration is limited and the potential barrier low.

Literature

1. I. L. Shereshefsky. J. Am. Chem. Soc. 50:2966 (1928); 72:3682 (1950).
2. K. V. Chmutov. Zh. Fiz. Khim. 9:357 (1937).
3. K. V. Chmutov, Kolloidn. Zh. 11:44 (1949).
4. B. V. Deryagin. Zh. Fiz. Khim. 14(2):137 (1940).
5. B. V. Deryagin, V. V. Karasev, and Z. M. Zorin. The Structure and Physical Properties of Matter in the Liquid State, Material from the Kiev Meeting of May 28-30, 1958. Izd. Kievsk Univ. (1954).
6. B. V. Deryagin, V. V. Karasev, N. N. Zakhavaeva, and V. P. Lazarev. Zh. Teor. Fiz. 27(5):1077 (1957).
7. V. V. Karasev and B. V. Deryagin, Zh. Fiz. Khim. 33(5):100 (1959).
8. B. V. Deryagin and Z. M. Zorin. Dokl. Akad. Nauk SSSR 98(1):98 (1954).
9. Ya. I. Frenkel'. Statistical Physics, Moscow, Izd. Akad. Nauk SSSR (1948), p. 391.
10. E. Gatchek. Liquid Viscosities, Moscow—Leningrad, GONTI (1935).
11. B. V. Deryagin and N. N. Fedyakin. Dokl. Akad. Nauk SSSR 147(2):403 (1962).
12. I. S. Gromeka, Collected Works, Moscow, Izd. Akad. Nauk SSSR (1952), p. 149.

A MICROPOLARIZATION STUDY OF BENZENE FILMS ON MERCURY

Z. M. Zorin

Institute of Physical Chemistry, Acad. Sci., USSR
Laboratory for Surface Phenomena
Tula Pedagogical Institute

The thermodynamic properties of supported liquid films can be studied by two different methods:

1. A thin film is formed on the supporting surface, either by adsorption from nearly saturated vapors or by condensation from saturated vapors. Study is made of the relation between the depth, h, of the adsorbed polymolecular film and the relative pressure, p/p_S, of the parent vapors.

This method has been used by B. V. Deryagin and the present author to study the adsorption of polar and nonpolar liquids on smooth glass surfaces [1].

2. The thin film is produced on the supporting surface by bulk phase thinning. Here one studies the relation between splitting pressure and film depth. This method was developed by A. Sheludko and D. Platikanov [2] for the study of benzene films on mercury.

The results obtained by the two methods can be compared [3] by passing from the adsorption isotherm to the splitting pressure isotherm through the relation

$$\Pi(h) = -\frac{RT}{V_\mu} \ln p/p_S.$$

Here, $\Pi(h)$ is the splitting pressure in the film of depth h, R is the molar gas constant, T is the absolute temperature, V_μ is the molar volume of the liquid, and p/p_S is the relative pressure at which the adsorbed film has depth h.

Comparison of the results obtained by the two methods of studying liquid film thermodynamics is meaningful only for experiments carried out on the same system.

We have chosen to work with benzene films on mercury, trustworthy data having been obtained for this system by A. Sheludko and D. Platikanov through the second of the above methods [2]. Final measurements were made on benzene and mercury kindly supplied by Professor Sheludko, the aim being to avoid uncertainties due to differences in the working materials.

Experimental Methods

The apparatus used for studying the adsorption of benzene on mercury is shown schematically in Fig. 1. Benzene vapors were admitted into the system from the ampule 3 which was held in a thermostat at temperature T_1. The mercury on which adsorption and condensation were to be studied was introduced into the apparatus through the capillary 2. A water jacket 4 and a plexiglas water thermostat 5 were used to maintain the entire apparatus at temperature T_2, close to room temperature.

The pressure of benzene vapors was varied through adjustment of the temperature difference $(T_2 - T_1)$ between mercury surface and benzene ampule. The film depth was calculated from the parameters of the elliptically polarized light reflected from the mercury surface carrying the film. The light rays were directed onto the mercury surface through narrow slits in the thermostat 5 and an optical glass hemisphere, 1, which was carefully ground to the apparatus. The apparatus was evacuated to 10^{-3} mm Hg and then flushed out with benzene vapors before the measurements were made.

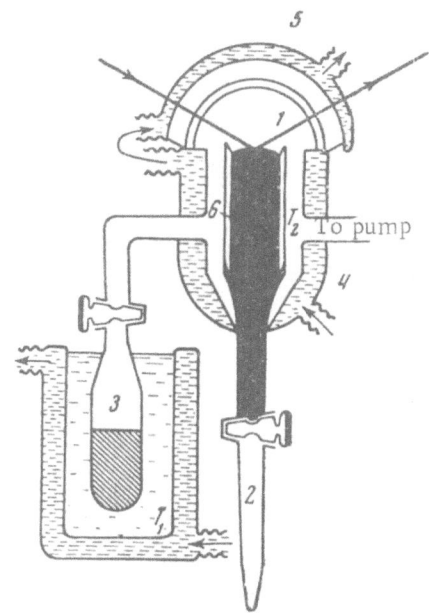

Fig. 1. Apparatus No. 1 for studying the adsorption of benzene on mercury: 1) hemisphere; 2) capillary; 3) ampule; 4) water jacket; 5) thermostat; 6) tube.

Constancy of temperature was maintained by two TS-15 thermostats. The temperature control was made more exact by feeding these thermostats with stabilized current, by assuring steady flow of cooling water, and by a three-fold increase in the length of the condenser. Temperature pulses were smoothed out in this manner and the temperature held constant to 0.02°C.

The temperature in the thermostat containing the ampule and the temperature of the water in the thermostating jacket were compared with Beckmann thermometers, the accuracy of comparison being 0.01°.

This apparatus permitted development of the adsorption isotherm, and determination of the film depth, at saturation vapor pressure, but could not be used to obtain steady condensation from supersaturated vapors. Condensation occurred in a ring-shaped space some two millimeters wide between the mercury tube 6 and the thermostating jacket 4 when the temperature of the benzene was much higher than that of the mercury ($T_1 > T_2$). Vapor saturation was therefore established under these conditions and condensation could not be observed.

When condensation was the process of interest, measurements were carried out in the apparatus of Fig. 2. Here the temperature control was less precise. This lack of control was an inconsequental detail in condensation studies since the degree of supersaturation over the mercury surface was dependent on both the temperature difference, $T_1 - T_2$ and the rate of vapor condensation in the lines, and could not be calculated from the temperature difference alone. The mercury was introduced into the sphere 1, of ordinary glass, through capillary 2. The vapor pressure was once more determined by the temperature difference between mercury and benzene. The temperature of the surroundings was held constant to 0.1° during the experiments and measured with an accuracy of 0.05°. The ampule containing the benzene was placed in a thermostat. The temperature of the latter was also measured to 0.05°.

Measurements of adsorbed film depths were carried out with the micropolarimeter designed by the Laboratory for Surface Phenomena of the Institute of Physical Chemistry, Acad. Sci., USSR, for determining the thickness of oxide films on metals. The polaroids of this apparatus was replaced by polarizing prisms to increase the accuracy of measurement. The optical system of this apparatus is shown in Fig. 3.

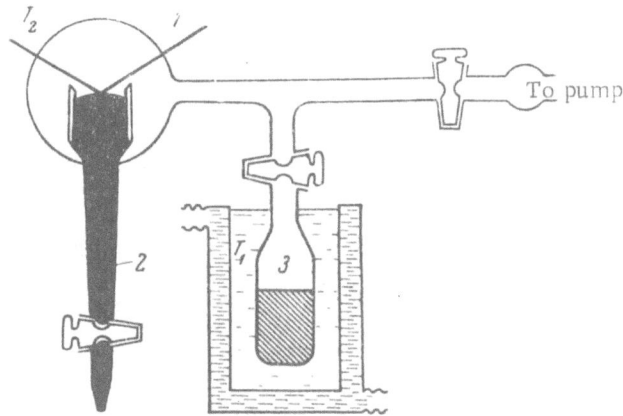

Fig. 2. Apparatus No. 2 for measuring the adsorption of benzene on mercury: 1) sphere; 2) capillary; 3) ampule with benzene.

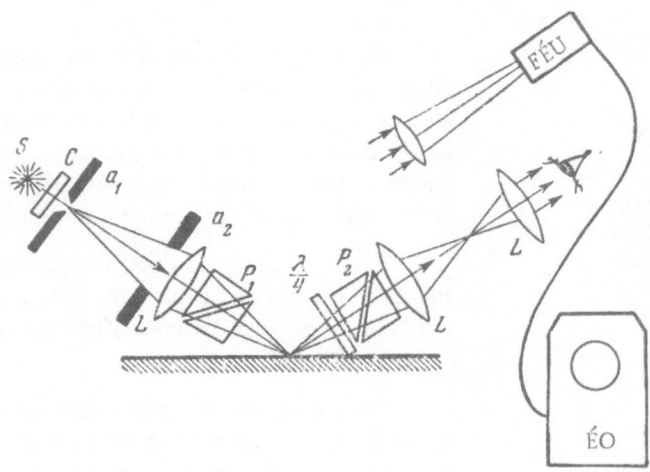

Fig. 3. Optical system for measuring film thicknesses.

A PRK-4 mercury lamp, S, served as the source of illumination. The filter C was used to separate out the yellow mercury lines. The light passed through the working slit, a_1, the aperture a_2, and the lens L, being focused by the latter to form an image of slit a_1 on the test surface. A Franck—Ritter prism, P, was used as the polarizer. The reflected light from the mercury surface was passed through a quarter-wave plate ($\lambda/4$) set with its principal axes at 45° and 135° to the plane of incidence. This plate served as a compensator. The light then passed through a second polarizing prism, P_2, which served as the analyzer. The image of slit a_1 on the test surface was viewed through a microscope equipped with a long-focus 1.5-power objective and an eye-piece ($\times 15$).

Measurements were carried out by setting the polarizer and analyzer to extinction. This setting was made visually when it was a matter of determining whether the film had come to equilibrium; when, however, the film had come to equilibrium and it was a matter of determining its depth with high accuracy, the setting was made with the aid of a FÉU-19 photomultiplier. Here the FÉU-19 output was fed to the vertical axis of an ÉO oscillograph.

Vibrations of the mercury surface made for considerable difficulty in the measurements. For this reason, the micromanipulator and the apparatus were separated from the other parts of the system and attached to angle irons in the wall.

Calculations and Determination of the Accuracy of Measurement

The pressure of benzene vapors in the system could be calculated (in mm Hg) by the usual integrated saturation vapor pressure equation

$$\log p = b - \frac{0.05223a}{T_1}$$

when the temperature of the benzene in the ampulte (T_1) was lower than the system temperature (T_2) and there was no condensation.

Over the interval from 0 to 42°C, the values of the constants for benzene are: a = 34172, b = 7.9622.

The pressure of the saturated vapor, p_S, over the mercury surface at temperature T_2 is given by

$$\log p_S = b - \frac{0.05223a}{T_2}.$$

The relative vapor pressure could be obtained by dividing these two equations, the one by the other,

$$\log p/p_S \cong 0.0522a \left(\frac{1}{T_1} - \frac{1}{T_2} \right) \approx 0.522a \frac{\Delta T}{T_1 T_2}.$$

136

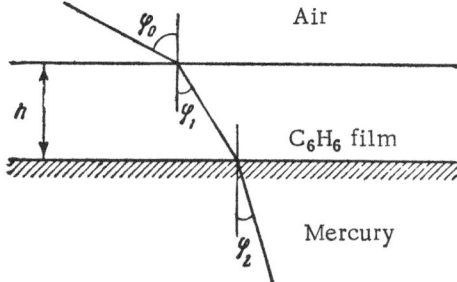

Fig. 4. Wave path in the film.

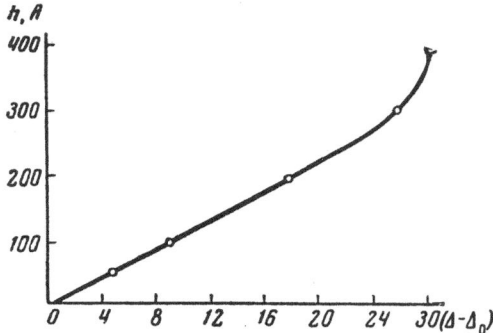

Fig. 5. The relation between film depth and change in phase difference, $(\Delta - \Delta_0)$, between parallel and perpendicular electric vector components in the reflected wave.

The accuracy of determination of the temperature difference in the apparatus of Fig. 1 was $\approx 0.04°C$. The derived above equation shows that an error of this magnitude led to less than a 0.002 uncertainty in the relative pressure, p/p_S. The maximum error in the temperature determination could have been as high as 0.2°C when working with the air thermostated apparatus of Fig. 2. The corresponding error in the determination of the relative pressure was then 0.009.

The depth of the benzene film on the mercury was calculated from the phase difference between the parallel ($\parallel$) and perpendicular ($\perp$) electric vector components of the reflected wave.

The angle of incidence, φ_0, onto the film was 60° in every case. The polarizer and analyzer were rotated to extinction of the light reflected from the mercury surface carrying the film, and readings then made of ψ, the angle between the plane of incidence and the plane of the vibrations transmitted by the polarizer (angle for reestablishment of polarization), and γ, the angle between the plane of the vibrations transmitted by the analyzer and one of the principal directions of the quarter-wave plate. The phase difference, Δ, between the parallel and perpendicular components of the electric vector of the reflected wave was then equal to 2γ. The optical constants of the mercury, that is to say, the coefficient of refraction, n, and the absorption coefficient, k, for 60° incidence were first determined, the aim being to develop a relation between the phase difference, Δ, and the film depth, h. These determinations were made by measuring the angles ψ and γ for the reflection from a clean mercury surface.

The values obtained at $\lambda = 5780$ A were $\psi_0 = 40°17'$ and $\gamma = 73°52'$. These results are in good agreement with the data of the literature [4].

The values of n and k for mercury were then calculated through the Vasicek equations [5]

$$n^2 = \sin \varphi_0 = \left[1 + \frac{\operatorname{tg}^2 \varphi_0 \cos^2 2\psi_0}{(1 + \sin 2\psi_0 \cos \Delta_0)^2} \right],$$

$$k = \frac{\operatorname{tg} \varphi_0 \sin 2\psi_0 \sin \Delta_0}{\sqrt{(1 + \sin 2\psi_0 \cos \Delta_0)^2 + \operatorname{tg}^2 \varphi_0 \cos^2 2\psi}}.$$

Here ψ_0 is the angle for reestablishment of polarization, and Δ_0 the phase difference, for reflection from the clean mercury surface.

These data led to the values n = 1.614 and k = 2.828 for mercury.

Knowing the index of refraction of benzene to be n = 1.501, it was then possible to obtain the values of the Fresnel coefficients, $r'_{\parallel}$ and $r'_{\perp}$ for reflection at the air−film interface, and $r''_{\parallel}$ and $r''_{\perp}$ for reflection at the film−mercury interface, and the phases of the parallel and perpendicular electric vector components, $\delta''_{\parallel}$ and $\delta''_{\perp}$, for reflection at the film−mercury interface (Fig. 4):

$$r'_{\parallel} = \frac{n_1 \cos \varphi_0 - n_0 \cos \varphi_1}{n_1 \cos \varphi_0 + n_0 \cos \varphi_1},$$

$$r'_\perp = \frac{n_0 \cos \varphi_0 - n_1 \cos \varphi_1}{n_0 \cos \varphi_0 + n_1 \cos \varphi_1},$$

$$r''^2_\parallel = \frac{(A \cos \varphi_1 - n_1 p)^2 + (B \cos \varphi_1 - n_1 q)^2}{(A \cos \varphi_1 + n_1 p)^2 + (B \cos \varphi_1 + n_1 q)^2},$$

$$r''^2_\perp = \frac{(n_1 \cos \varphi_1 - p)^2 + q^2}{(n_1 \cos \varphi_1 + p)^2 + q^2},$$

$$\operatorname{tg} \delta''_\parallel = \frac{2 n_1 \cos \varphi_1 (Aq - Bp)}{(A^2 + B^2) \cos^2 \varphi_1 - n_1^2 (p^2 + q^2)},$$

$$\operatorname{tg} \delta''_\perp = \frac{2 n_1 q \cos \varphi_1}{n_1^2 \cos^2 \varphi_1 - (p^2 + q^2)}.$$

In these equations

$$A = n^2 (1 - k^2), \qquad B = 2 n^2 k \cos \varphi_2,$$

$$q = nk, \qquad p = n \cos \varphi_2, \qquad \cos \varphi_2 = \frac{\sqrt{n^2 - \sin^2 \varphi_0}}{n}.$$

The phases, $\delta_\parallel$ and $\delta_\perp$, of the parallel and perpendicular electric vector components for reflection from the film covered mercury surface were calculated from the equations

$$\operatorname{tg} \delta_\parallel = \frac{L_\parallel \sin (x - \delta''_\parallel)}{M_\parallel + N_\parallel \cos (x - \delta''_\parallel)},$$

$$\operatorname{tg} \delta_\perp = \frac{L_\perp \sin (x - \delta''_\perp)}{M_\perp + N_\perp \cos (x - \delta''_\perp)},$$

where

$$L_\parallel = - r''_\parallel (1 - r'^2_\parallel), \qquad L_\perp = - r''_\perp (1 - r'^2_\perp), \qquad M_\parallel = r'_\parallel (1 + r''^2_\parallel),$$

$$M_\perp = r'_\perp (1 + r''^2_\perp), \qquad N_\parallel = r''_\parallel (1 + r'^2_p), \qquad N_\perp = r''_\perp (1 + r'^2_\perp),$$

and x, the phase difference resulting from single reflection in the film, is given by

$$x = 2 n_1 \cos \varphi_1 h \frac{2\pi}{\lambda} = 4 \pi n_1 \cos \varphi_1 \frac{h}{\lambda}.$$

It is obvious that $\delta_\parallel$ and $\delta_\perp$ are periodic functions of the film depth, with values repeating when $x = 2\pi$, that is to say, when the alteration of film depth, Δh, amounts to 2360 A.

Calculated values of the phase differences are plotted against the film depth in Fig. 5.

Here the film depth, h, (in A) has been plotted along the axis of ordinates and the alteration of phase difference, $(\Delta - \Delta_0)$, for reflection, first from a clean mercury surface then from a surface carrying a film, along the axis of abscissas.

The same results are also presented below:

h, A	50	100	200	300	400
$\Delta - \Delta_0$	$-4°46'$	$-9°18'$	$-17°51'$	$-25°52'$	$-30°34'$

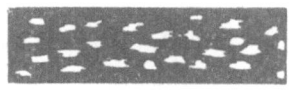

a

b

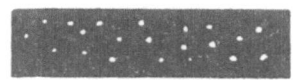

c

Fig. 6. Film as seen through the microscope: a) view of the rippled surface resulting from gentle heating; b) field of view after thin film extinction; c) light patches on the dark background of the thin film.

The relation between h and $(\Delta - \Delta_0)$ is almost linear up to 200 A. An alteration of 5' in the phase difference (or an alteration of the measured angle, γ_0, by 2.5') corresponds to an increase in the film depth by 1 A. The mean-square error in the determination of the angle γ was less than 6', except in special cases which will be discussed separately. This corresponds to an error of ±3 A in the determination of the film depth, h.

It is to be noted that it is $(\Delta - \Delta_0)$ and not the phase difference which has been plotted on the axis of abscissas. Passage of the light through the glass sphere gives rise to a supplementary phase difference of the order of 1° as a result of double refraction in the glass itself. This additive factor is, however without influence on the accuracy of the measurements since it appears with the clean mercury surface and with film-covered surface as well.

Experimental Results

1. Preliminary measurements were carried out with the apparatus sketched in Fig. 2. A film of thickness h ≅ 70 A appeared on the mercury surface when the ampule with the benzene was connected to the system. "Rippling" of the surface was observed when the ampule was gently heated, say by being held in the hand for 3-4 seconds, the appearance of the surface being that shown in Fig. 6a. The field of view could be darkened and the ripples illuminated, or the ripples could all be simultaneously darkened and the field illuminated, by rotating the analyzer and polarizer.

This experiment makes it clear that the ripples are film sectors of a fixed depth, which is different from that of the film itself.

The ripples disappeared when the heating of the benzene ampule was discontinued, and a uniform film, ≈70-80 A deep, was established.

Highly flattened lenses appeared if the heating of the benzene ampule was continued after the ripples had been formed. Each lens proved to be a system of 5-8 concentric light and dark rings, when observed under the microscope. Since repetition of the polarization state of the reflected light corresponded to a 2360 A alteration, Δh, in the thickness, the height of the lens was 1-2 μ and its diameter 0.1-0.5 mm. The portion of the film between these lenses was covered with a system of interference fringes which could be displaced by heating the ampule.

Cooling the ampule caused the benzene to vaporize from the surface, the interference fringes becoming wider and wider as the film thinned and finally became uniform. The polarizer and analyzer settings were read at the instant the film became homogeneous. The phase difference between the parallel and perpendicular components of the electric vector of the reflected light then corresponded to a film depth of h=(200 ± 20) A.

TABLE 1

$(\Delta t + 0.04)$, °C	$p/p_S \pm 0.002$	$(\Delta - \Delta_0)\pm 12'$	$(h\pm3)$, A
1.52	0.930	0°24′	4
1.21	0.945	1°06′	11.5
0.54	0.975	2°02′	21
0.45	0.979	3°42′	39
0.40	0.982	4°08′	43
0.09	0.996	6°02′	64
0.00	1.000	6°50′	73
0.00	1.00	18°24′±20′	207±4

TABLE 2

$(\Delta t \pm 0.04)$, °C	$p/p_S \pm 0.002$	$(\Delta - \Delta_0) \pm 15'$	$(h \pm 3)$, Å
1.60	0.929	0°56'	10
1.24	0.944	1°42'	18
0.60	0.972	3°26'	36
0.24	0.989	3°40'	39
0.13	0.994	5°16'	55
0.03	0.999	6°22'	67
0.00	1.000	6°22'	67

2. Data applying to adsorption isotherms developed with the aid of the apparatus of Fig. 1 are presented in Tables 1 and 2 in curve I of Fig. 7, where the circles indicate results from the first series of experiments and the crosses, results from the second series. This figure makes it clear that the results were satisfactorily consistent.

A nonuniform film was formed on the mercury surface in the first experiment which was carried out at saturation. Figure 6b shows the field of view at extinction of reflection from the thinner portion of the film. Here the film depths are 73 ± 3 Å for the darkened sections and 210 ± 5 Å for the illuminated sections.

3. Since we were interested in condensation processes, experiments were also carried out in the apparatus of Fig. 2 which was designed for preliminary work.

Satisfactory depth measurements could be carried out here, since vibration of the mercury surface had been eliminated. The double refraction of the glass altered the phase difference by about 1°, an effect which has been considered above. The fact that the apparatus was not exactly spherical in form affected the angle required for reestablishment of the polarization, but not the phase difference at reflection. The relative pressure could be measured only roughly. It was shown that the experimental error, $\Delta(p/p_S)$, could be of the order of 0.009.

The data obtained are shown in Table 3 and curve I of Fig. 7.

The films obtained here were thicker than those observed under the same conditions in the first case. Saturation led to the appearance of a uniform film, 210 ± 10 Å thick.

This film remained uniform on passage to supersaturation, and its depth continued to increase. Light spots eventually appeared on the dark background (see Fig. 6c), these being quite obviously benzene microdrops of high curvature on the film.

Phase difference readings were made at the instant of appearance of these microdroplets, and a calculation made of the film depth. The value obtained was 360 ± 20 Å. Further increase in the supersaturation led

TABLE 3

$(\Delta t \pm 0.2)$, °C	$p/p_S \pm 0.009$	$(\Delta - \Delta_0) \pm 15'$	$(h \pm 3)$, Å
1.3	0.939	1°44'	18
0.9	0.958	3°32'	37
0.7	0.967	4°04'	43
0.3	0.986	7°16'	76
0.0	1.00	19°58' + 40'	226 ± 10
0.0	1.00	18°10' ± 40'	204 ± 10
Supersaturated	1	29°00' ± 1°	360 ± 20

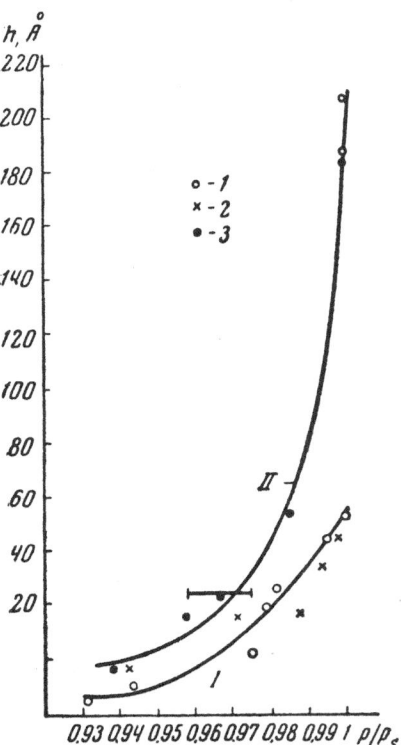

Fig. 7. Adsorption isotherms for benzene on mercury.

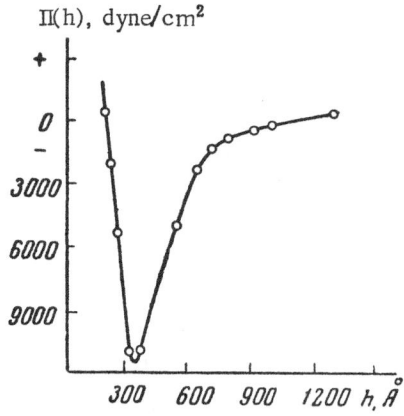

Fig. 8. Splitting pressure isotherm for benzene films on mercury.

to the formation of a nonuniform film carrying compressed benzene lenses, just as in the previous experiments. These lenses disappeared, and the film became uniform, when the ampule was cooled. A situation similar to that of Fig. 6c could not be reproduced by vaporizing the film. Account must be taken of the fact that the relative pressure was only crudely measured in this experiment, so that the smooth course of the curve must be viewed critically, especially since there is only one point in the region of high values.

Discussion of Results

The data indicate that two films of different depth, one uniform and one about 210 A thick, can exist on the mercury surface at saturation, or slight supersaturation.

These films are in equilibrium with the saturated vapor ($p/p_S = 1$), and with one another. A situation of this kind can arise only if the films structures are different. The possibility that polar liquid films adsorbed on glass show polymorphism over certain intervals of relative pressures (with $p/p_S > 0$) was shown by B. V. Deryagin and L. M. Shcherbakov [6] from an analysis of data from experimental studies on adsorption and condensation [1].

Of the two films in equilibrium with the saturated vapor, the one 70 A thick will be designated as the α modification, and the other, 210 A thick, designated as the β modification.

It is interesting to note that the experimental results of A. Sheludko and D. Platikanov [2] (Fig. 8) indicate passage from negative to positive splitting pressures ($\Pi(h) = 0$) at a film depth of $h = (240 \pm 30)$ A. This is in good agreement with out own determination of the depth of the β film in equilibrium with the saturated vapor.

The depth of homogeneous film obtained by vaporizing condensed benzene (results of the preliminary experiments) was also 200 ± 20 A.

The experimentally developed isotherm, curve II of Fig. 7, also shows the formation of a homogeneous film, 210 ± 10 A deep, in adsorption from saturated vapors, the indication being that the entire mercury surface carries a β modification benzene film.

The film remains homogeneous on passage to a slight supersaturation, and its depth increases to 360 ± 20 A. Small droplets of benzene appear on the homogeneous film at this point. A depth of 340 ± 10 A corresponds to the minimum on the splitting pressure curve obtained in the experiments of A. Sheludko and D. Platikanov.

We will now consider the splitting pressure isotherm developed for film formation by liquid phase thinning (see Fig. 8). The left branch of this curve running from positive to negative values shows a reduction in pressure (or an increase in the pressure of vapor over the film) with increasing film depth:

$$\frac{d\Pi}{dh} < 0, \frac{dP}{dh} > 0 .$$

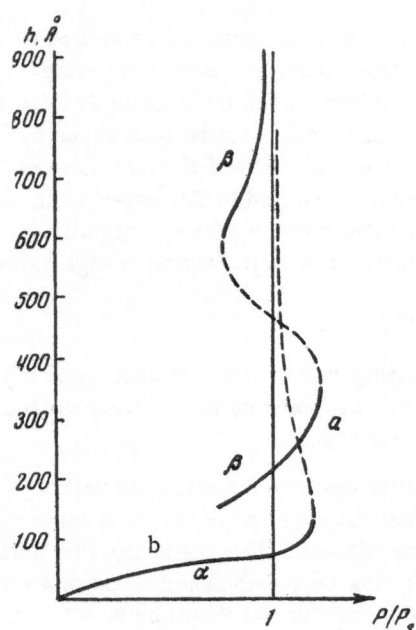

Fig. 9. Adsorption isotherms for the α and β modifications.

States in which $d\Pi/dh > 0$ are either stable or metastable; they can, in principle, be realized in films formed either by vapor adsorption or condensation.

The right-hand branch of the isotherm reflects the state of affairs under which film thickening leads to an increase in the splitting pressure $(d\Pi/dh > 0)$; here the pressure of vapor above the film must diminish as the film depth increases $(d\Pi/dh < 0)$. Such states are unstable and cannot be realized through vapor condensation. Therefore thickening under supersaturation of a β film which had been formed by condensation cannot proceed beyond the minimum on the splitting pressure curve; here the film itself becomes unstable and droplets form on it.

Thus our own results are in good agreement with the experimental data of A. Sheludko and D. Platikanov. There is reason to believe that our adsorption isotherm for the β film and the splitting pressure isotherm for liquid phase thinning [2] are identical.

Possibilities for the α and β phase adsorption isotherms are shown in Fig. 9.

The thick liquid film corresponds to that section of curve a of this figure which asymptotically approaches the $p/p_S = 1$ axis. Metastable states are covered by the dashed portions of the curves. There is an alteration in the sign of dp/dh at 360 A, and the section of the curve ranging up to p/p_S corresponds to metastable states.

Curve b of the figure corresponds to the α modification. The section of the curve for $p/p_S > 1$ is shown in full and reflects metastable states, while the region of unstable labile states is covered by the dotted portion of the curve. The vapor pressure for an α film is higher than that of a β film of the same thickness. The assumption of a functional dependence of vapor pressure on structure was advanced in a paper which we have published with B. V. Deryagin [1].

Comparison of the experimentally developed α and β isotherms (see Fig. 7) shows the possibility of polymorphism even below saturation pressure ($p/p_S < 1$). Here it is natural to assume the existence of structural differences in the adsorbed films. Thus the preferential orientation of the benzene rings is parallel to the mercury surface in the α modification and perpendicular to this surface in the β modification.

The benzene molecule does not have a permanent dipole moment. Thus the polymorphism of a benzene film on mercury is something different from the polymorphism of such polar liquids as water and alcohol. For this reason, another type of polymerization [6] and adsorption isotherm has also been proposed here. The assumption that the oriented structure is invariably more stable than the ordinary liquid in polymerization at the orienting interface (at the same film depth) has proven to be poorly founded.

There are two ways of looking on the transition from oriented film to ordinary liquid in benzene adsorption and condensation on mercury:

1. An α benzene phase is first formed on the mercury surface and reaches a depth of 70 A saturation. Condensation leads to lens formation on the oriented film. The film is an instance of a transition phase, passage to the ordinary liquid taking place discontinuously. Reorientation of the benzene molecules can occur over certain sectors in other cases. A terraced film is then established at saturation, the depth being 70 A over sectors corresponds to the α, and 210 A over sectors corresponding to the β, modification; condensation is possible only after this point has been passed. It is likely that all of the molecules will be phase oriented when the benzene layer is thick and covers the entire mercury surface.

2. A film forms spontaneously over all sections in adsorption. A u n i f o r m film, 210 A deep, is established at saturation. Passage to slight supersaturation causes the film depth to rise to 360 A, where condensation

begins. It is a question as to whether there is a sharply demarcated interface between the oriented adsorption film and the ordinary liquid, or a gradual passage from the one to the other as one moves away from the wall and the surface effect diminishes.

The theory of the van der Waals forces would require that the splitting pressure, Π, fall off in inverse proportion to the third power of the film depth, h, when electromagnetic retardation is neglected. The film depth increased in our experiments to 360 A, but the splitting pressure diminished much more slowly than required by the theory, Π being approximately proportional to h^{-1}. It is therefore natural to suppose that the structure continually alters as one moves away from the wall, with passage to the ordinary liquid occurring gradually. I.E. Dzyaloshinskii, E. M. Lifshits, and L. P. Pitaevskii [7] have advanced an explanation for the appearance of a region of instability in a metal-supported dielectric layer which is based on the theory of van der Waals forces.

Conclusions

1. An optical micropolarization method has been used to study the adsorption and condensation of benzene on mercury.

2. The results obtained are in quantitative agreement with those obtained by A. Sheludko and D. Platikanov in investigating the splitting pressure isotherm for thin benzene films on mercury.

3. The assumption is advanced that adsorption can lead to the formation of two film modifications, the α and the β, these differing in the benzene molecule orientation on the mercury.

We would like to conclude by thanking B. V. Deryagin, Corresponding Member, Academy of Sciences, USSR, for consultations and for his interest in this work, L. M. Shcherbakov for discussion, and G. Chusov and Yu. Dubovenko for active participation in the development of the apparatus and the measurement.

Literature

1.   B. V. Deryagin and Z. M. Zorin. Zh. Fiz. Khim. 29(6):1010 (1955); 29(10):1755 (1955).
2.   A. Sheludko and D. Platikanov, Dokl. Akad. Nauk SSSR 138(2) (1961).
3.   B. V. Deryagin and M. M. Kusakov. Izv. Akad. Nauk SSSR, Ser. Khim. 5:741 (1936); 5:1119 (1937).
4.   A. Vašiček, Sb. Česke vysoke školy v Brne 6:26 (1931).
5.   A. Vašiček. Optica tenkých vrstev, Prague (1956), pp. 288-304.
6.   B. V. Deryagin and L. M. Shcherbakov, Kolloidn. Zh. 23(1):40 (1961).
7.   I. E. Dzyaloshinskii, E. M. Lifshits, and L. P. Pitaevskii. Usp. Fiz. Nauk 73:3 (1961).

THE INDEPENDENCE OF DEPTH AND DIAMETER IN EQUILIBRATED FREE FILMS

D. Ekserova, Iv. Ivanov, and A. Sheludko

*Institute of Physical Chemistry,
Bulgarian Academy of Sciences (Sofia)*

Duyvis has recently shown [1] that the equilibrium depth of a free film of aqueous electrolyte alters markedly with the film diameter. This "diameter effect" leads to a film depth maximum at a diameter of some $2 \cdot 10^{-2}$ cm, the data of Duyvis agreeing at this point with our own values for such solutions. The effect is usually more marked when the electrolyte concentration is low and the equilibrium depth high. Thus the equilibrium depth of a film of containing $5 \cdot 10^{-4}\%$ saponin and 10^{-3} mole KCl per liter falls to approximately 300 A when the diameter is increased to 10^{-1} cm, while a similar alteration in the diameter of a film of a solution containing 10^{-3} mole sodium oleate and 10^{-3} mole KCl per liter reduces the depth to 200 A.

Such an unexpected effect casts doubt on the validity of the widely accepted Deryagin—Verwey—Overbeek equilibrium splitting pressure theory of thin liquid films [2-5]. Duyvis' discussion [1] of equilibrated films of $2 \cdot 10^{-2}$ cm diameter is unacceptable since his choice of parameters is arbitrary.

This situation has led us to a careful test of this "diameter effect."

The equilibrium depth of a microscopic film formed in the capillary of a vertically oriented tube by pumping away the liquid can be diminished by increasing the meniscus curvature, thereby reducing the capillary pressure which balances the splitting pressure and holds the system in equilibrium. This trivial effect was allowed for in the Duyvis experiments by measuring the capillary pressure directly and proves to be quite small in comparison with the observed alteration in the equilibrium depth.

In our own experiments, the capillary pressure of the meniscus was calculated from the curvature and the solution surface tension. Here allowance had to be made for the effect of the film diameter on the capillary pressure. This was also necessary in order to show that the equation previously used ($p_0 = 2\sigma/R$, p_0 is the pressure, σ the surface tension, and R the radius of the tube) is correct, even though it takes no account of a dependence of film diameter on curvature, thus eliminating the possibility that the Duyvis effect be due to these trivial factors.

The problem reduces to a determination of the principal radii of curvature, R_1 and R_2, in the Laplace Equation: $p_0 = \sigma(1/R_1 + 1/R_2)$. Since the meniscus is a surface of revolution (Fig. 1), its meridians and parallels are lines of curvature [6], and R_1 and R_2 can be obtained from the equation

$$\frac{1}{R_i} = \pm \frac{\cos \varphi_i}{\rho_i}.$$

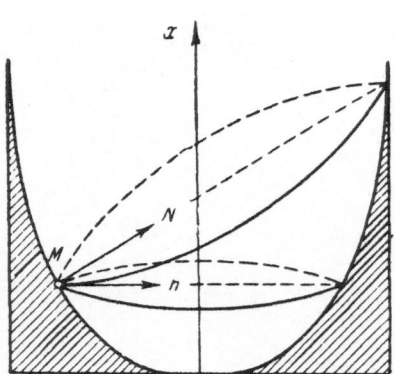

Fig. 1. Schematic representation of the measurements.

Here, the ρ_i are the radii of curvature of meridians and parallels at the point M and the φ_i are the angles between the normals to the

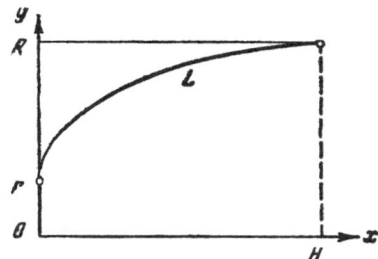

Fig. 2. Independence of depth and diameter in the free equilibrated film.

surface (N) and the normals to the corresponding lines of principal curvature at this point. When the system of coordinates is oriented so that the axis of abscissas coincides with, and the axis of ordinates is perpendicular to, the axis of the meniscus, the equations for parallels and meridians become

$$\left.\begin{array}{c} \cos \varphi_1 = \dfrac{1}{\sqrt{1 + y'^2}}, \\[2mm] \bar{\rho}_1 = y, \qquad \dfrac{1}{R_1} = \dfrac{1}{y\sqrt{1 + y^2}}, \\[2mm] \cos \varphi_2 = 1, \dfrac{1}{R_2} = \dfrac{1}{\rho_2} = -\dfrac{y''}{(1 + y'^2)^{2/3}}. \end{array}\right\} \quad (1)$$

To obtain $1/R_1 + 1/R_2$ from (1), it is necessary to find the equation, $y = y(x)$, of the curve L (Fig. 2) which generates the meniscus surface by rotation around the axis of abscissas. A complete solution requires determination of the free energy minimum for fixed volume V.

This problem will be simplified by reduction to a determination of the mean curvature of a surface of rotation limited by a cylinder of radius R and the surface $x = 0$, and having volume

$$V = \pi \int_0^H (R^2 - y^2)\, dx$$

and surface area

$$S = 2\pi \int_0^H y\sqrt{1 + y'^2}\, dx,$$

with the curve L subject to the boundary conditions

$$\left.\begin{array}{c} y = r, \quad y' = \infty, \\ y = R, \quad y' = 0. \end{array}\right\} \quad (2)$$

The conditions of (2) reflect the assumption that tube and film are completely wetted.

A determination of minimum S at fixed V requires solution of the Euler Equation

$$\frac{\partial F}{\partial y} - \frac{d}{dx}\left(\frac{\partial F}{\partial y'}\right) = 0,$$

with

$$F = y\sqrt{1 + y'^2} + \lambda(R^2 - y^2), \quad (3)$$

λ being an arbitrary constant.

Since F is independent of x, the first integral of the Euler Equation takes the form

$$F - y'\frac{\partial F}{dy'} = C,$$

C being a constant.

In addition to (3), one can also write

$$\sqrt{1 + y'^2} = \frac{y}{C - \lambda(R^2 - y^2)}. \quad (4)$$

145

Eliminating y' from (3) and y" after differentiation of (3), one obtains

$$\frac{1}{R_1} = \frac{C - \lambda(R^2 - y^2)}{y^2},$$

$$\frac{1}{R_2} = -\frac{C - \lambda(R^2 + y^2)}{y^2},$$

$$\frac{1}{R_1} + \frac{1}{R_2} = 2\lambda.$$

The boundary condition of (2) can be combined with (4) to give

$$\lambda = \frac{R}{R^2 - r^2}$$

or, finally,

$$\frac{1}{R_1} + \frac{1}{R_2} = \frac{2R}{R^2 - r^2}.$$

Thus the expression for the capillary pressure is obtained in the form

$$p_0 = 2\sigma \frac{R}{R^2 - r^2};$$

(5)

this passes into the equation previously used for low values of the r/R ratio, namely

$$p_0 = \frac{2\sigma}{R}$$

(5')

when $r \to 0$.

The trivial part of the effect of the diameter on the equilibrium depth, h_0, can be easily handled in the case of thick equilibrated films of dilute electrolytes since the van der Waals component of the splitting pressure is then negligibly small in comparison with the electrostatic component.

By drawing on the approximation equation of the Deryagin—Landau theory for the electrostatic component, Π_{el}, of the splitting pressure, the equilibrium condition can be formulated as

$$\Pi_{el} = \frac{2\sigma R}{R^2 - Z^2} = 64\, cKT\gamma^2 e^{-\varkappa h_p},$$

c being the electrolyte concentration, and $1/\varkappa = \sqrt{\varepsilon KT/8\pi z^2 e_0^2 C}$ the depth of the diffuse electric layer.

For the rather small film of depth h_p^0, one has

$$\frac{2\sigma}{R} = 64 cKT\gamma^2 e^{-\varkappa h_p^0}.$$

It follows from the last two equations that the expression for the trivial part of the diameter effect is

$$h_p^0 - h_p = \frac{1}{\varkappa} \ln \frac{1}{1 - \left(\frac{r}{R}\right)^2}.$$

(6)

Calculations based on (6) readily show that the alteration, $h_p^0 - h_p$, of the equilibrium depth with r/R must have been negligibly small in comparison with h_p^0, even at the very lowest of the electrolyte concentrations, c, used by Duyvis [1].

146

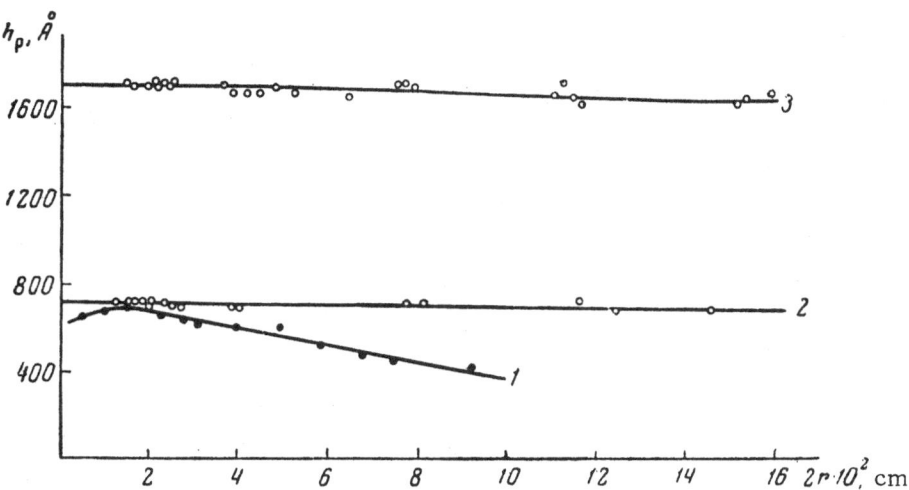

Fig. 3. Relation between the equilibrium depth, h_p, and the diameter in free films of a $5 \cdot 10^{-4}\%$ saponin (Schering) solution: 1) experimental data of Duyvis for a 10^{-3} molar saponin solution in KCl; 2) the same, but from our own experiments; 3) the same, but for a $6.3 \cdot 10^{-5}$ molar solution.

Our own measurements were carried out on homogeneous solutions containing $5 \cdot 10^{-4}\%$ saponin and 10^{-3} mole/liter of sodium oleate, which had been prepared in the same manner as in [1]. The KCl concentrations in these two solutions were 10^{-3} and $1.44 \cdot 10^{-3}$ mole/liter, respectively, these values being close to the lowest concentrations at which Duyvis observed a maximum diameter effect. Measurements were also carried out on saponin solutions containing KCl at a lower concentration ($6.3 \cdot 10^{-5}$ mole/liter) where the observations of Duyvis would lead one to expect a still larger diameter effect and where Eq. (6) would indicate the trivial part of the effect to be of appreciable magnitude. The electrolyte concentration values cited above were corrected for the electrical conductivity by the method described in [3]. In addition the working range of r/R values for the study was almost twice that of [1].

Measurements were made in the apparatus described in [3, 5]. Since the rate of thinning falls off markedly at the higher diameters, thus opening the possibility of error through variations of the intensity of illumination, the mercury lamp was replaced by an incandescent projection lamp fed from a storage battery, and monochromatization of the light achieved with an interference, rather than an adsorption, filter. The resulting decrease in the intensity of illumination necessitated improvement in the sensitivity of the photometeric section of the system. This was done by introducing an electrometric bridge-circuit balanced amplifier between the photomultiplier and the galvanometer and feeding the multiplier from an adjustable voltage divider. The desired sensitivity could be obtained at high stability and low background noise by a Zeiss photometric photomultiplier operating at 1000 V.

The results of the measurements are shown by the open circled points in Figs. 3 and 4. The curves passing through these points were developed through Eq. (6). The excellence of the agreement here is indication that the film diameter only weakly affects the equilibrium thickness, the effect being limited to the trivial interaction of film radius and meniscus curvature. The results also indicate that the previously used formula, namely, $p_0 = 2\sigma/R$, is sufficiently exact even for films of larger diameter than those studied so far.

For comparison, Duyvis' results have been plotted in Figs. 3 and 4 and the plotted points joined by smooth curves. Either true equilibrium was not established in these measurements, or there was some sort of systematic error in the depth determinations. Since our measurements disclosed no nontrivial effect of diameter on equilibrium depth, the problem of Duyvis' results must be sought in work on Duyvis' system; it is to be hoped that this will be done.

It should be pointed out that there is less possibility of a variation in the errors of measurement with the film radius in our system than in that of Duyvis where dependence was placed on visual photometry. The in-

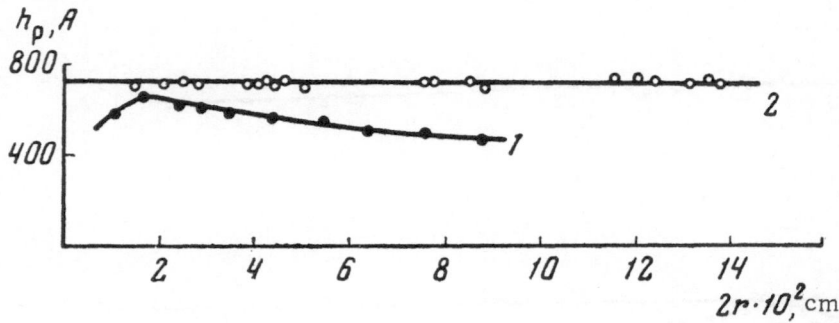

Fig. 4. Relation between the equilibrium depth, h_p, and the diameter in free films of various solutions of sodium oleate in KCl: 1) experimental data of Duyvis for a 10^{-3} molar sodium oleate solution in KCl; 2) our own experimental data for a $1.44 \cdot 10^{-3}$ molar sodium oleate solution in KCl (curve developed from Eq. (6)).

tensity of the light reflected from the film was recorded automatically on a moving light-sensitive tape in our own case. Here interference maximum and minimum were obtained in the course of each measurement, these values being used in the calculation of the film depth to eliminate the effect of alteration of the optical "situation" in the course of the diameter measurements. Continuous registration of the intensity of the light reflected from the film also made it possible to very accurately determine the instant of establishing the equilibrium depth.

The validity of our results showing the independence of the film dimensions and equilibrium diameter have been confirmed by certain unpublished results of Maisel's. Here working with similar compositions, values of the equilibrium depth quite close to our own were obtained in large vertically oriented soap films of the order of several square centimeters.

The values of the equilibrium depths of the saponin films agree exactly with the values for the same electrolyte concentrations reported earlier in [3]. The equilibrium depths obtained for the sodium oleate solutions differ both from those which we reported in [7] and from those obtained by Duyvis at a diameter of $2 \cdot 10^{-2}$ cm [1]. It should be noted in this connection that sodium oleate is quite unsuitable for testing the Π_{el} theory, and has never been used by us for this purpose. Solutions of this material do not give stable reproducible values of σ, and it is impossible to accurately determine their content of electrolyte. Not only does the compound itself always contain a considerable, but indefinite, amount of electrolyte, but it adsorbs a part of the KCl which is introduced into solution. The measured values of the conductivity are always quite different from the calculated, and the corrections necessary for recalculation to KCl are large and untrustworthy. Values of the φ_0-potential from equilibrium film depths in 10^{-3} molar sodium oleate solutions (70 mV in [7] and 100 mV in [1]) should therefore be considered as rough approximations.

Literature

1. Duyvis. Doctoral dissertation, The Hague (1962).
2. B. Deryagin and A. Tityevskaya. Proceedings of the Second International Congress of Surface Activities, Vol. 1 (1957), p. 211.
3. A. Sheludko and D. Ekserova. Kolloidn. Zh. 165:148 (1959).
4. A. Sheludko. Dokl. Akad. Nauk SSSR, 123:1074 (1958).
5. A. Sheludko and D. Ekserova. Kolloidn. Zh. 168:24 (1960).
6. V. I. Smirnov. Higher Mathematics, Moscow, Gosfizmatizdat (1952).
7. A. Sheludko. Proc. Koninkl. Nedl. Akad. Wetenschap., Ser. B 65(1):97 (1962).

AN EXPERIMENTAL STUDY OF THE THERMODYNAMIC FUNCTIONS OF THE THIN LIQUID LAYER

Yu. M. Popovskii

Institute of Physical Chemistry, Acad. Sci., USSR
Laboratory for Surface Phenomena
Odessa Higher Institute for Marine Engineers

The kinetic and thermodynamic properties of the thin liquid layer are frequently quite different from those of the bulk phase. The thermodynamic properties of thin liquid layers largely determine the characteristics of dispersed systems, and the values of these parameters must be drawn on in studies of polymolecular adsorption, flotation theory, and so on. A theory developed by B. V. Deryagin makes it possible to evaluate the splitting pressure, the most important thermodynamic characteristic of the thin layer of liquid electrolyte.

We have been engaged for a number of years in the study of the properties of thin films of organic liquids bounded by uniform solid surfaces. There is no generally accepted theory to account for the behavior of this type of liquid film, and all the more need for experimental studies.

Our first work was carried out in collaboration with E. N. Ovchinnikova and involved a study of the heat capacity of polar nitrobenzene (dipole moment, μ, 3.94 D) mixed with various dispersing media to form dispersed two-component systems. The mean heat capacity was measured with an isothermal mercury calorimeter, the results obtained being accurate to 1%. The temperature of the calorimeter was determined with a Beckmann thermometer. The heat capacity measurements were made over the interval from 20 to 45°C. The sequence of the studies was such that the heat capacities of the pure liquid and solid phase (powder) were first determined in the calorimeter, and the heat capacity of the dispersed system then measured over the same temperature interval. The heat capacity of the liquid was calculated as the difference between the heat capacities of dispersed system and solid phase, allowance being made for the amount of the latter phase present in the system. The results of these measurements are presented in Table 1.

The only departures from additivity observed here were in the nitrobenzene−amber and nitrobenzene−glass systems. In the first, the dispersed amber powder swelled and partially dissolved in the nitrobenzene, so that an additive equation could not be used to obtain the heat capacity of the dispersing medium. Study of the thermodynamic properties of dispersing liquids is best carried out by working with systems in which the

TABLE 1

System	Solid phase content, % by volume	C_p, nitro- benzene, cal/(g · deg)
Nitrobenzene	—	0.340 ± 0.010
Nitrobenzene−glass	53.8	0.313 ± 0.007
Nitrobenzene−quartz	56.2	0.346 ± 0.004
Nitrobenzene−graphite	26.2	0.341 ± 0.005
Nitrobenzene−amber	49.0	0.419 ± 0.003

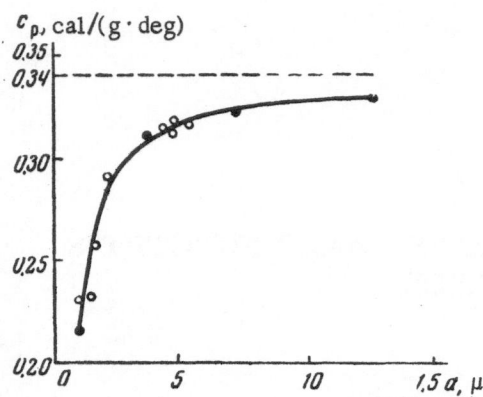

Fig. 1. The relation between the layer depth (d) and the heat capacity (C_p) of nitrobenzene in the dispersed system nitrobenzene—glass. O) Dispersed system in which the solid phase had a surface area of 824 cm^2/g; ●) the same, but with a surface area of 1113 cm^2/g.

properties of the solid phase remain unaltered on wetting, and departures from additivity are observed in the properties of the system itself. The nitrobenzene—glass system which was studied here is of this type.

We have studied the heat capacity of nitrobenzene in the dispersed nitrobenzene—glass system, and its dependence on the liquid—solid phase volume ratio. Specific surface areas of the glass powders employed in this work were determined from specially developed microscopic particle size distribution curves.

The results of these measurements are shown in the graph (Fig. 1). Here the heat capacity, calculated as described above, has been plotted on the axis of ordinates, and the mean depth of nitrobenzene film on the axis of abscissas, this depth being calculated on the assumption that the liquid was uniformly distributed between the solid particles. It is clear that the heat capacity of the nitrobenzene in the dispersed system diminished as the layer depth was reduced. The fall-off in the heat capacity was most marked when the volume of nitrobenzene was below the limit of natural wetting, the greater part of the liquid in the system being then concentrated in the thin films which separate the individual grains of glass powder. It should be pointed out that the experimental error increases markedly at the lower nitrobenzene concentrations, so that quantitative calculations cannot be carried out with this technique.

It is of interest to attack the glass with various physical and chemical agents so as to alter its surface and then investigate the effect of this structure on the heat capacity of adhering thin nitrobenzene surface films. Glass powders which had been subjected to the following treatments were used in this work.

1. Chromic acid treatment. Clean portions of glass tubing were ground in a mortar, and the powder obtained allowed to stand in chromic acid for 24 hours at 20°C. The powders were then carefully washed with distilled water, shaken with water for eight hours, once more washed with water, and then dried.

2. Hydrofluoric acid treatment. Glass powders prepared in the manner described above were immersed in dilute (0.3%) HF, allowed to stand in contact with the latter for 15 minutes, and then washed as in the preceding experiment.

3. Electric discharge treatment. The glass powders were inserted in a discharge tube, dried, and the tube evacuated to a pressure of approximately 1 mm Hg. A glow discharge was then set up in the tube and maintained for 5-7 minutes. The attempt was also made to treat the powders in the flame, but it was found that the more highly dispersed powder fractions expanded markedly, thus reducing the specific surface area to such a degree that it was no longer possible to observe a positive effect. The nitrobenzene used in this work had been more carefully purified than that employed earlier and its heat capacity was different for this reason. The order of purification was such that the nitrobenzene was triply distilled in vacuum, and ions removed from the product by the application of a fixed potential of ~200 V directly before measurement. This last step led to a 20-fold increase in the specific resistance of the nitrobenzene in the course of several minutes, the value rising to $\sim 2 \cdot 10^8$ Ω cm.

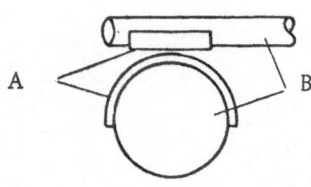

Fig. 2. Schematic representation of the mutual placement of the steel cylinders carrying the glass films.

The results of these experiments are presented in Table 2. Here the nitrobenzene—glass powder volume ratio had been held constant.

Departures from regularity were observed only in the system containing the freshly ground powder where the heat capacity of the nitrobenzene

TABLE 2

Agent	C_p, nitro-benzene, cal/(g·deg)	Volume of nitrobenzene in dispersed system, %
Chromic acid	0.335 ± 0.009	47.0
Hydrofluoric acid	0.378 ± 0.017	49.0
Electrical discharge	0.301 ± 0.015	45.3
"Fresh surface"	0.384 ± 0.011	46.2
Pure nitrobenzene	0.369 ± 0.002	—

was higher than for the bulk liquid. This effect can be tentatively ascribed to the fact that the freshly formed glass surface is electrically charged and nitrobenzene chemisorbs on it. The deactivation of glass surfaces by hydrofluoric acid has been observed by others [1, 2].

Theoretical analysis shows that a study of the temperature dependence of the heat capacity of the dispersed nitrobenzene should lead to a determination of the stability interval for the thin "anomolous" nitrobenzene film, and to the laws applying to its breakdown and return of bulk liquid properties. Measurements of this kind are being carried out at the present time.

Study was also made of the splitting pressures of thin nitrobenzene films held between mutually perpendicular cylindrical glass surfaces (Fig. 2).

Glass films, A, approximately 1 μ in depth, were affixed to the outer surfaces of two magnetized steel bars, B. The cylindrical diameters were 4 and 0.4 mm. All measurements were carried out on surfaces which had been subjected to glow discharge. No effect was obtained in control experiments on glass surfaces which had been deactivated by treatment with HF. The cylinder of 0.4 mm diameter was attached to a spring balance. The cylinders were allowed to approach under the action of the magnetostatic force and the one of larger diameter then slowly drawn away by a motor. The force required for detachment of the cylinders, one from the other, was read off from the spring balance, or determined in arbitrary units as the time of movement of the cylinders prior to separation in air, this followed by the measurement in the liquids, and the work concluded by the control experiments in air. A series of experiments covering the dependence of the magnetic force on the distance of separation showed a gradual

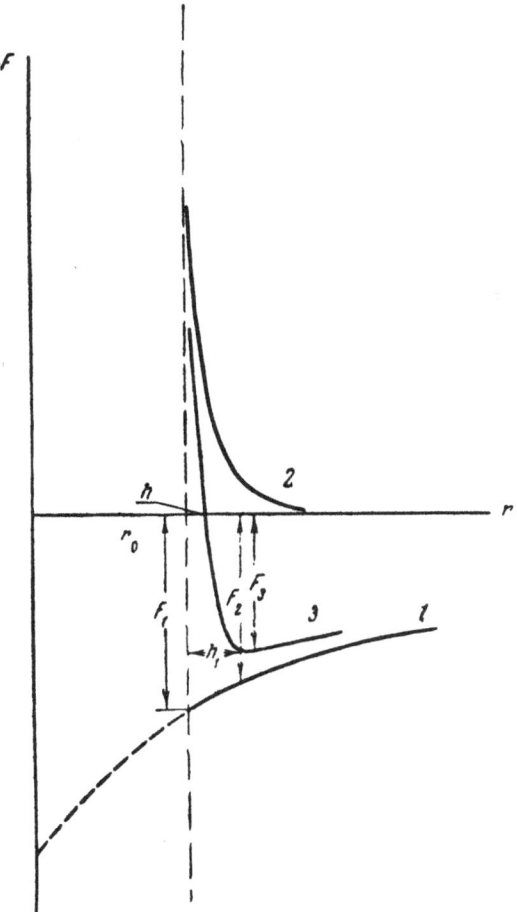

Fig. 3. Relation between the distance of separation and the force of interaction between the cylinders.

fall-off according to an approximate $r^{-1/4}$ law. The dependence of the forces of attraction and repulsion on the distance of separation of the cylinders is shown in Fig. 3. The variation of the force of magnetic attraction with the distance of separation is represented by curve 1, the splitting pressure by curve 2, and the resultant force of interaction between the cylinders, by curve 3.

Separation of the cylinders in air occurs when the elastic force becomes equal to the magnetostatic force, F_1; the positive splitting pressure established in the thin liquid film becomes equal to the magnetostatic force of attraction at a certain film depth, $\bar{h}$, when the cylinders are allowed to approach in the liquid. The resultant force of compression falls off at breakaway of the cylinders because of the elastic force of deformation of the suspension, and the film depth changes discontinuously to the value $\bar{h}_1$ corresponding to the minimum on curve 3. The force of magnetic attraction decreases so slowly with the distance that its value at the instant of break can be considered as $F_2 \approx F_1$. The splitting pressure, N, at depth $\bar{h}_1$ can then be set equal to $F_1 - F_3$. It should be noted that the slow fall-off of the force of magnetostatic attraction with the distance of separation tends to flatten the minimum on curve 3 and thus leads to a pronounced scatter in the experimental points. The results obtained in the measurements of the force required for separating the cylindrical glass surfaces are presented below in arbitrary units, the order being the same as that in which the experiments were carried out:

Medium	Air	Benzene	Nitrobenzene	Benzene	Air
Experiment 47	35.5	34.8	—	—	34.9
Experiment 19	20.4	—	16.25	16.1	19.8
Experiment 80	29.3	29.5	27.0	27.8	29.0

Differences in the absolute value of the separating force in these experiments trace back to the lack of uniformity in the magnetization of the steel cylinders, while the variation in the value of the splitting pressure (Expt. No. 19, nitrobenzene, $N = F_1 - F_3 = 4.15$, arbitrary units, and Expt. No. 80, $N = 2.3$, arbitrary units) results from the lack of complete homogeneity of the glass surfaces obtained in glow discharge.

The data presented here make it clear that the force required for separating the cylinders in the nonpolar liquid benzene is the same as in air, while the force required in the polar nitrobenzene is less because of the positive splitting pressure which acts in the thin liquid film. The layer of nitrobenzene which remains on the glass after washing with benzene is so thin that it does not affect the force for separation in air, but it chemically modifies the surface to the extent that immersion in the nonpolar benzene (Expt. No. 19) yields a positive splitting pressure of the same order of magnitude as that met in nitrobenzene.

We have used the equation of B. V. Deryagin [3] relating the free energy of the thin liquid film and the splitting pressure

$$N = - \frac{2\pi}{\sqrt{\varepsilon \, \varepsilon'}} \, f(h_0)$$

to determine the free energy of thin nitrobenzene films. Here, N is the splitting pressure between convex surfaces, ε and ε' are factors characterizing the surface geometry in the neighborhood of the point of contact, and $f(h_0)$ is the free energy per 1 cm^2 of liquid film.

For the present case of mutually perpendicular cylinders, this equation takes the form

$$N = - 2\pi \sqrt{r_1 r_2} \, f(h_0).$$

The values of the free energy of the thin nitrobenzene film obtained from this equation fall within the interval 4-6 erg/cm^2.

We have shown earlier [4] that calculation of the thermodynamic functions of the thin liquid film requires the measurement of the equilibrium film depth at various values of the acting force and temperature. For this purpose we have constructed an apparatus [5] which permits the film depth to be determined to within 3-5 mμ, and the acting force to be measured to within 0.3-0.5 dyne. Splitting pressure isotherms have been developed on this apparatus for nitrobenzene and CCl$_4$. The results obtained here will be presented in a later communication.

Literature

1. G. L. Michnevitch and E. N. Ovchinnikova. Acta Phys. Chim. 11:603 (1939).
2. C. Jech. Nature, No. 4546:1343 (1956).
3. B. V. Deryagin. Zh. Fiz. Khim. 6:1306 (1935).
4. Yu. M. Popovskii. Kolloidn. Zh. 24:117 (1962).
5. Yu. M. Popovskii. Summaries of Reports, Fifth All-Union Conference on Colloidal Chemistry, Odessa (1962).

THERMOOSMOTIC FLOW IN HIGH POLYMER FILMS

V. V. Karasev, B. V. Deryagin, and E. N. Khromova (Efremova)

Institute of Physical Chemistry, Acad. Sci., USSR
Laboratory for Surface Phenomena

It has been shown in an earlier publication [1] that two types of flow occur when a temperature gradient is established in a solid-supported wetting film, thermoosmotic creep and thermocapillary movement.

1. Thermoosmotic creep results from an alteration of the specific enthalpy of the thin film. Theory indicates that the velocity gradient is here localized in an interfacial layer at the solid wall. The thermoosmotic forces vanish outside this layer, so that the rest of the liquid undergoes displacement as a whole. The liquid flux is here expressed through the equation

$$Q_1 = \varkappa \, \frac{dT}{dl} \, h,$$

$\varkappa$ being the thermoosmotic coefficient, dT/dl the tangential temperature gradient, and h the film depth.

2. Thermocapillary flow results from surface tension differences which trace back to the presence of the temperature gradient. Here the equilibrating force acts on the outer film surface. The tangential tension is transmitted through the wetting film to the supporting surface and is there compensated by the latter's resistance. The flux under the action of the thermocapillary forces is expressed through the equation

$$Q = \frac{1}{2\eta} \cdot \frac{d\sigma}{dT} \cdot \frac{dT}{dl} \cdot h^2,$$

η being the liquid film viscosity and σ the surface tension.

This theory has been experimentally tested and confirmed in studies on flow in water films which have been carried out by B. V. Deryagin and M. K. Mel'nikova [2]. It is quite natural to attempt the application of this theory to other liquids as well.

The present work is a report of preliminary data on flow under the action of a temperature gradient in wetting films of oils and liquid high polymers.

These studies were carried out by the following method (Fig. 1).

A film of the test liquid was deposited on the steel gauge plate, 1, one of the larger faces of this plate having been given an additional polishing to obtain a Class 14 surface. This plate was clamped between a heater and a cooler, these being "boxes" through which steam and tap water, respectively, were passed. A temperature gradient was set up along the plate and at right angles to the plane of incidence of the light rays. The plate was illuminated with monochromatic light from a water-cooled quartz lamp which was set in a housing, 2, and equipped with an optical filter, 3, to separate out a definite portion of the spectrum. A frosted glass plate, 4, was introduced beyond the filter for better viewing of the interference fringes through the MIN-1 polarizing microscope. A screen, 5, was inserted in the eye-piece through an adjustable microscope tube and focused on the film of a Zenith-S camera.

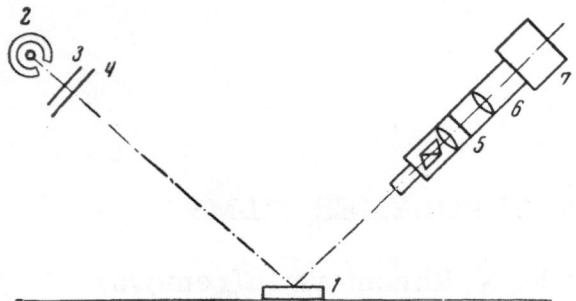

Fig. 1. Schematic representation of apparatus for studying thermoosmosis in high polymer films. 1) Polished steel plate; 2) quartz lamp; 3) optical filter; 4) frosted glass; 5) screen; 6) focusing tube for eye-piece; 7) Zenith-S camera.

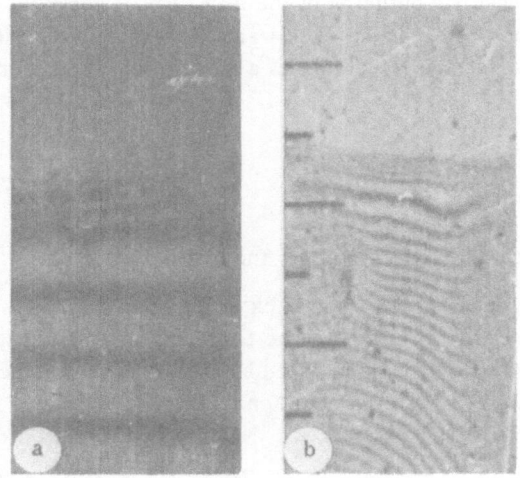

Fig. 2. Thermoosmotic flow in films of GKZh-94 (a) and benzontron (b).

The liquid film proved to be wedge-shaped at the beginning of the experiment but became flatter and flatter with the passage of time, the distance between the interference fringes increasing. The fringe movement was continually followed and the pattern photographed when the fringe separation had altered appreciably.

The fringe separation was uniform and constant during the entire duration of experiments on turbine oil. This is indication that the film profile remained linear, flow occurring under the action of thermocapillary forces alone. It has been shown that such forces act at the film surface and are analogous to the tangential component of the air current used in determining the surface viscosity by the blowing method. The profile observed here was therefore similar to that established by blowing a boundary film in which the surface viscosity and bulk viscosity are identical. Thus it appears that there is no detectable thermoosmotic effect in this case.

Experiments with a silane polymer GKZh-94 (Fig. 2a) showed narrow, almost regularly spaced, interference fringes around the wetting boundary, the remaining fringes being equally, but much more widely, separated. From this it can be concluded that the angle of inclination of the wedge was greatest near the wetting edge.

Study of the surface viscosity of the GKZh-94 by the blowing method showed the interference fringe separation on the film to remain uniform and constant with the passage of time. Thus the film had the form of a linear wedge. The same observations would have been made for flow under the action of thermocapillary forces alone. Thus fringe compression here is the result of the thermoosmotic forces predicted by theory.

From this it is seen that the thermoosmotic forces give rise to liquid creep along the wall in the direction of the hot end of the plate. Flow in the thicker portions of the film is toward the cold end of the plate, and results from the forces of thermocapillary action.

A pronounced tendency for fringe compression in the neighborhood of the wetting boundary was also observed in the case of incompletely hydrogenated benzontron (Fig. 2b), the indication being that thermocapillary forces come into play here, also.

Blowing studies on incompletely hydrogenated benzontron [3] have shown that the surface viscosity even falls off somewhat in the neighborhood of the wall, this fall-off propagating smoothly over a short distance. Fringe concentration under a temperature gradient is indication of the existence of thermoosmotic flow (just as in the case of the silane polymer), which masks the effect of a slight reduction of the viscosity along the wall. Thus the thermoosmotic flow which was experimentally detected earlier in water has now been detected in organic liquid films.

Conclusions

Thermoosmotic flow resulting from alteration of the specific enthalpy has been detected in thin films of high polymers, thus confirming the prediction of B. V. Deryagin.

Literature

1. B. V. Deryagin and G. P. Sidorenkov, Dokl. Akad. Nauk SSSR 32:622 (1941).
2. B. V. Deryagin and M. K. Mel'nikova, In collection: Papers in Honor of Academician A. F. Ioffe on His Seventieth Birthday; Moscow, Izd. Akad. Nauk SSSR (1950).

   B. V. Deryagin and M. K. Mel'nikova. VIe Congrès International de la JOL (1956), pp. 305-314.
3. B. V. Deryagin and V. V. Karasev, Dokl. Akad. Nauk SSSR 101(2):289 (1955).

   B. V. Deryagin et al., Proceedings of the Second International on Surface Activities Butterworths, London (1957).

THE EFFECT OF MOLECULAR WEIGHT AND STRUCTURE ON THE VISCOSITY AND STABILITY OF THE THIN LIQUID FILM

N. N. Zakhavaeva, B. V. Deryagin, and A. M. Khomutov, and S. V. Andreev

Institute of Physical Chemistry, Acad. Sci., USSR
Laboratory for Surface Phenomena

The properties of a thin liquid film supported on a solid surface are different from those of the bulk liquid. It has been repeatedly pointed out that the surface viscosity is not the same as the bulk viscosity. The passage from surface to bulk viscosity occurs discontinuously in normal liquids. With chain molecule liquids, however, the viscosity can alter either continuously or discontinuously, or a transition region can be established. The effect of surface forces extends out for distances in excess of 10 μ. A method for determining the viscosity at various distances from a solid wall has already been described in the literature [1-8]. The principle in question here is that of blowing the liquid film from a plane surface by an air balst from a slit-form capillary burette. The viscosity, η, at any point in the thin wedge-shaped film which remains on the surface after blowing for a time, τ, with a tangential force, F, is related to the coordinates of this point by an equation of the type

$$ \eta = F \tau \frac{dy}{dx} , $$

y being the distance of the point from the solid surface (film depth at the given point), x the distance from the wetting edge (Fig. 1), and dy/dx the wedge slope.

The layer profile y = y(x) can be obtained from the interference fringe pattern which arises when the wedge-shaped film is illuminated with monochromatic light; from this profile equation one can evaluate the derivative dy/dx at any point and thus determine the viscosity [2]. Increase in the viscosity at the solid interface is associated with close fringe spacing. This spacing increases on moving away from the wall, thus indicating a reduction in the viscosity (Fig. 2a). The opposite situation prevails when the surface viscosity is reduced at the wall (Fig. 2b.). The apparatus for the viscosity measurements must be carefully thermostated since the viscosity is known to be temperature dependent. We have investigated the viscosities of a number of liquids which are similar in structure but differ in chain length and in the nature of the final groups.

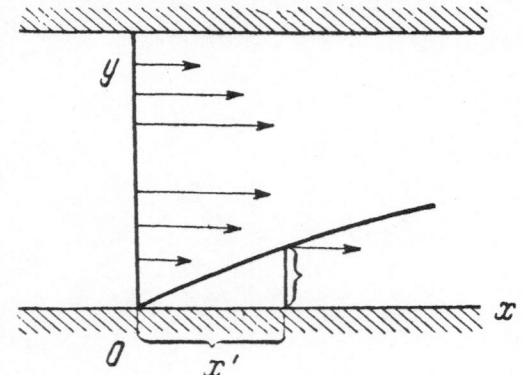

Fig. 1. Wedge-shaped film profile optained by tangential blowing of the liquid from a solid surface.

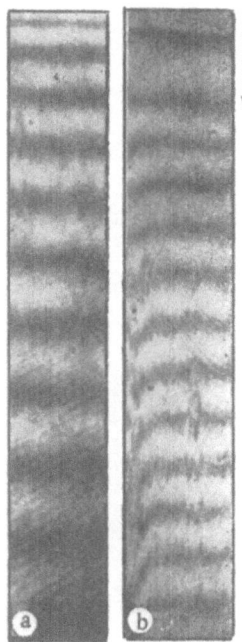

Fig. 2. Interference pattern obtained in the blowing of films in which the surface viscosity is higher than (a), and lower than (b), the bulk viscosity.

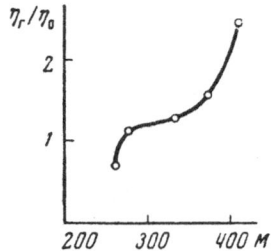

Fig. 3. The relation between the relative surface viscosity (η_r/η_0) and the molecular weight (M) of the liquid.

Tetraethoxyhexane (mol. wt. 262.4)

$$CH_3 - CH - CH_2 - CH - CH_2 - CH - OC_2H_5$$
$$\quad\quad\; | \quad\quad\quad\quad\quad | \quad\quad\quad\quad | $$
$$\quad\quad OC_2H_5 \quad\quad\; OC_2H_5 \quad\quad OC_2H_5$$

Here the surface viscosity is lower than the viscosity in bulk. The ratio of surface viscosity to bulk viscosity is $\eta/\eta_0 = 0.8$.

Tributoxybutane (mol. wt. 274)

$$CH_3 - CH - CH_2 - CH - OC_2H_5$$
$$\quad\quad\; | \quad\quad\quad\quad\quad | $$
$$\quad\quad OC_2H_5 \quad\quad\; OC_2H_5$$

The surface viscosity is somewhat higher than the viscosity in bulk: $\eta/\eta_0 = 1.12$.

Pentethoxyoctane (mol. wt. 334.4)

$$CH_3 - CH - CH_2 - CH - CH_2 - CH - CH_2 - CH - OC_2H_5$$
$$\quad\quad | \quad\quad\quad\quad | \quad\quad\quad\quad | \quad\quad\quad\quad | $$
$$\quad OC_2H_5 \quad\; OC_2H_5 \quad\; OC_2H_5 \quad\; OC_2H_5$$

The surface viscosity is somehwat higher than the viscosity in bulk: $\eta/\eta_0 = 1.24$.

Tetrabutoxyhexane (mol. wt. 374.6)

$$CH_3 - CH - CH_2 - CH - CH_2 - CH - OC_4H_5$$
$$\quad\quad | \quad\quad\quad\quad | \quad\quad\quad\quad | $$
$$\quad OC_4H_5 \quad\; OC_4H_5 \quad\; OC_4H_5$$

The surface viscosity is appreciably higher than the viscosity in bulk: $\eta/\eta_0 = 1.45$.

Hexethoxydecane (mol. wt. 409.5)

$$CH_3 - CH - CH_2 - CH - CH_2 - CH - CH_2 - CH - CH_2 - CH - OC_2H_5$$
$$\quad\quad | \quad\quad\quad\quad | \quad\quad\quad\quad | \quad\quad\quad\quad | \quad\quad\quad\quad | $$
$$\quad OC_2H_5 \quad\; OC_2H_5 \quad\; OC_2H_5 \quad\; OC_2H_5 \quad\; OC_2H_5$$

The surface viscosity is considerably higher than the viscosity in bulk: $\eta/\eta_0 = 2.45$.

The present work involved only the determination of the relative viscosity of the liquid in the thin surface layer, comparison being made with the bulk viscosity. The experiments show (Fig. 3) that the surface viscosity alters with the molecular weight and structure. The orientation of the liquid molecule at the solid interface may be of significance here. The orientation can be such as to direct the molecules upward from the solid surface and perpendicular to it, this position being maintained through various types of interactions. It is also possible to have the molecules oriented in a "prone" position. Unstable, easily perturbed, orientations can also occur. Cases arise in which the film is highly stable and maintains its profile long after blowing has ceased. In other cases, the film is entirely unstable and slips off the supporting surface under the action of the tangential force. Here the surface viscosity naturally exerts an effect on the position and state of the film. The results of experiments on the stability of the thin liquid film on a steel supporting surface are presented in Fig. 4. This figure shows the patterns obtained in blowing vaseline oil vinylbutylene ester solutions. There was a clear-cut creep at the wetting edge, which became more pronounced as blowing progressed. The film gradually broke up into parts connected solely by strands of oriented molecules extended in the direction of blowing. These strands stretched under the action of the force, and finally they too, ruptured. Knowing the air pressure in the jet, the tangential force required for break-up could be calculated, and a determination of stability thus obtained.

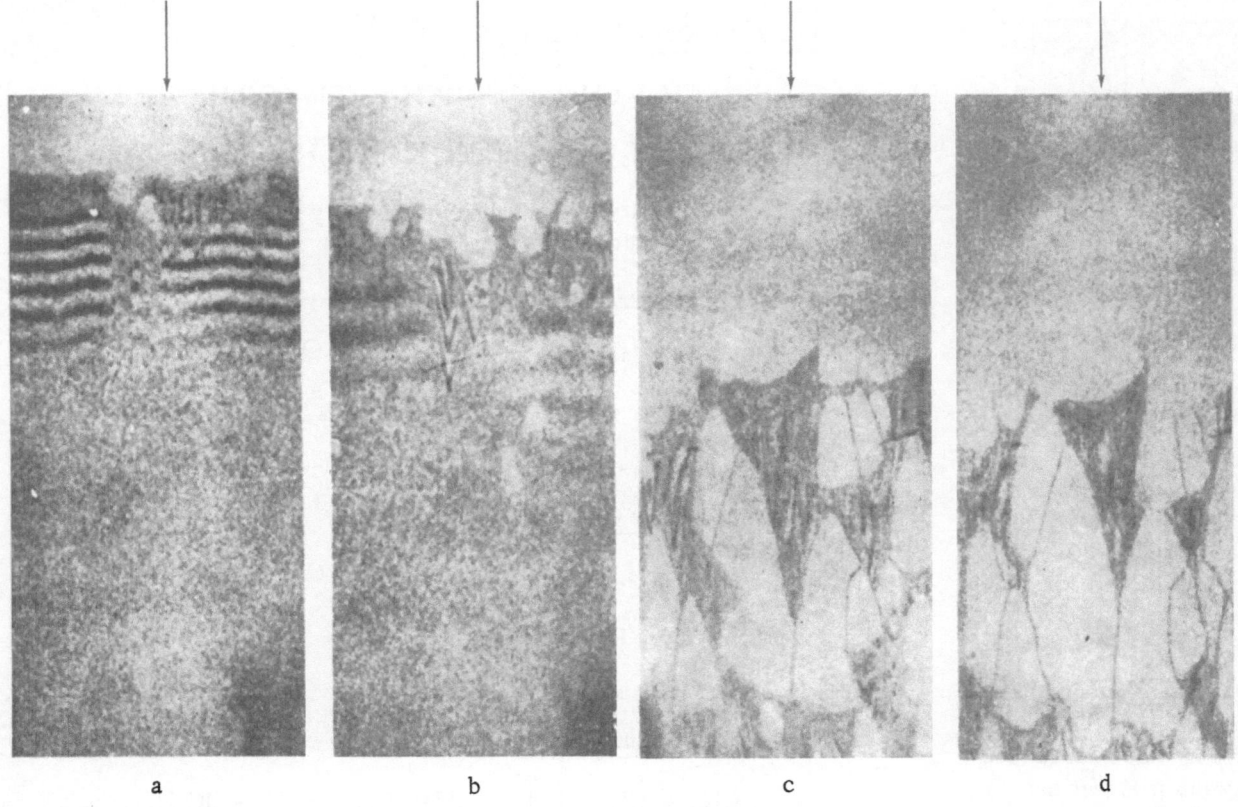

Fig. 4. Patterns obtained by blowing of vaseline oil solutions of a vinylbutylene ester. a) Initial stage of film creep (interference fringes visible); b, c, d) the same, but at later times (strands of oriented molecules visible in c and d).

Conclusions

Experiment has shown that the molecular weight and structure have great influence on the wall viscosity of the liquid. The orientation of the molecule on the solid surface is a factor of significance here. This orientation determines the surface viscosity and the stability of the wetting, wedge-shaped film which remains on the supporting surface after blowing. A study of the factors influencing the stability of the surface film is significant for work on the friction of lubricated surfaces.

Literature

1. B. V. Deryagin, L. M. Strakhovskii, and D. S. Malysheva. Zh. Éksperim. i Teor. Fiz. 16:171, 179 (1946).
2. B. V. Deryagin and N. N. Zakhavaeva. Transactions of the Sixth Conference on High Molecular Compounds, Moscow, Izd. Akad. Nauk SSSR (1949), p. 233.
3. N. M. Kusakov. Dokl. Akad. Nauk SSSR 54:145 (1946).
4. B. V. Deryagin and V. V. Karasev. Dokl. Akad. Nauk SSSR 62:761 (1948); 101:289 (1955); Kolloidn. Zh. 15:365 (1953).
5. B. V. Deryagin, V. V. Karasev, N. N. Zakhavaeva, and V. P. Lazarev. Zh. Teor. Fiz. 27(5):1076 (1957).
6. B. V. Deryagin, N. N. Zakhavaeva, S. V. Andreev, and A. A. Milovidov. In collection: Research in Surface Forces, Moscow, Izd. Akad. Nauk SSSR (1961), p. 139. [English translation: Consultants Bureau, New York (1963).]
7. B. V. Deryagin, N. N. Zakhavaeva, S. V. Andreev, and A. M. Khomutov. Kolloidn. Zh. 24(3):289 (1962).
8. B. V. Deryagin, N. N. Zakhavaeva, and S. V. Andreev. Inzh.-Fiz. Zh. 5(5):92 (1962).

THE POLYMOLECULAR COMPONENT
OF THE LUBRICATING BOUNDARY LAYER

G. I. Fuks

The Scientific-Research Institute for the Watchmaking Industry
Section on Friction and Lubricating Materials

The Components of Static Friction Under Boundary Lubrication

Distinction is generally made between three types of lubrication and lubricating action, liquid, semi-liquid, and boundary. The hydrodynamic theory of lubrication [1, 2] gives an adequate description of the first two types of friction and permits a determination of the liquid friction [2-4] from the coefficient of viscosity alone. The value of the coefficient of viscosity also determines the conditions required for passage from semi-liquid to liquid or boundary lubrication.

The relation between boundary friction and the composition and properties of the lubricating liquid is much more complex. Not only is there no generally accepted theory of boundary friction, but there is not even agreement on the physics of this process, especially regarding the question as to whether shear is localized in the liquid boundary layer or in the surface layer of the solid. Experiments on the effect of surface active substances on boundary friction also fail to give consistent results. Hardy [5] has shown the coefficient of friction of the solid body lubricated with fatty acid, alcohol, or aliphatic hydrocarbon to fall off linearly with the molecular weight of the liquid. Bowden [6] confirmed Hardy's results in the lower members of these series, but detected no relation between boundary friction and molecular weight in the higher homologues. Explanation of these and other inconsistencies requires a more detailed physical analysis of boundary friction [7].

Studies on the molecular-mechanical properties of thin (boundary) liquid films on solid surfaces are of great significance here. B. V. Deryagin and his coworkers discovered the existence of a splitting pressure in the thin boundary layer [8, 9] and showed that the properties of this layer must differ from those of the bulk liquid [10]. These same authors have developed the theory of the bulk aspects of surface phenomena [11]. A. S. Akhmatov [12] has treated the boundary layer as a quasi-solid body. Electron diffraction [14] and x-ray analyses [13] have shown that the fatty acids are oriented almost at right angles to the solid surface in layered structures. It has been proposed that the lubricating effect of an oil be evaluated in terms of boundary layer resistance to thinning and shear [7, 15] and indices have been established to show the variation of these properties with the distance from the solid surface [16, 17].

The boundary layer depth usually falls in the interval from $1 \cdot 10^{-5}$ to $1 \cdot 10^{-4}$ cm* [16, 17] and is therefore of the same order of magnitude as the micro-nonuniformities which remain after polishing to a Class 11 surface.† It is a well-known fact that surface roughness causes the actual area of physical contact to fall to a small fraction of the nominal geometrical area [20]. Contact stresses are high on the surface microrelief

---

*The measurements in question here were carried out by the method of approach of plane parallel disks [18, 19]. Passage to the boundary layer is marked by the onset of departures from the Stefan—Reynolds thinning law in the space between the two oil-immersed disks. This method of marking out the boundary layer obviously gives no information about the molecular nature of the wall effect.

†Only on the specularly ground and polished surface (Class 13-14) is the height of the roughness projections essentially less than the boundary layer depth.

TABLE 1

Oil	Contact pressure, kg/cm²	h_{min}, μ	Shear stress, g/cm²	h_{min}, μ
Turbine oil, "L"	0.6	0.12	81	0.08
The same, plus 0.1% palmitic acid	0.6	0.19	38	0.12
	1.75	0.12	90	0.09
The same, plus 0.1% behenic acid	0.85	0.19	32	0.13

projections and the projection peaks can enter into direct adhesional binding even at moderate contact pressures, the adhesional forces over the remainder of the surface being screened by monomolecular (more correctly, bimolecular) surface active layers. The surface areas of the contacting bodies are at the same time largely separated by fissures of the order of the boundary layer depth. The general expression for the force, F, of static friction is

$$F = \alpha S \tau_m + \beta S \tau_M + \gamma S \tau_n.$$

Here S is the nominal contact area, α, β, and γ are the fractions of this area involved in shear or breakage of the roughness projections, mutual shear of surfaces separated by monomolecular surface active layers, and shear in the polymolecular boundary layer, and τ_m, τ_M, and τ_n are the shear resistances on these various portions of the surface S.

Although there are few satisfactory measures of these three resistances, there is no doubt that $\tau_m \gg \tau_M \gg \tau_n$. Bailey and Courtney—Pratt [21] have shown that the mutual shear resistance of freshly cleaved mica surfaces (10 kg/mm²) can be decreased to one-fortieth of its original value by a monomolecular layer of potassium stearate. The shear resistance in a polymolecular boundary layer is 1.5-2.5 orders lower still [17, 19]. Although the coefficient of friction depends on the contact area as well as the resistance to the tangential forces, a comparison of values for dry friction (0.5-1.0, and higher [22]), monomolecular lubrication (~0.1-0.2 [6, 23]), and polymolecular boundary lubrication (< 0.01-0.02, see Fig. 1) confirms the validity of this ordering of the shear resistance forces in the contact zone.

It is obvious that the various contributions to the integral force of friction, F, also depend on the values of α, β, and γ. The values of these parameters vary with the surface roughness, the contact pressure, the resistance of profile protuberances to deformation, and the thinning resistance of the boundary layer. The energy of bonding of the monomolecular layer of fatty acid or surface active substance to the metal or mineral [24, 32] is so high that the film can withstand pressures of the same order of magnitude as the hardness of the supporting surface. Such being the case, the value of α will differ from zero only if there is deformation of the irregularities under loading or shear, with the formation of new surface sections on which an adsorption layer cannot form. The firmness of bonding of the monomolecular layer to the solid body is such that the α to β ratio depends on both the mechanical properties of the irregularities and the angle of their inclination to the shear direction (according to the friction theory of Ernst and Merchant [25], see also [26]). It follows from the nature of the surface microrelief that $\alpha + \beta \ll \gamma$ at moderate contact pressures.

It has been shown in [16, 17, 19] that the polymolecular boundary layer of lubricating liquid differs from an essentially nonlubricating film in possessing intrinsic resistance to both shear and thinning, and this fact leads to a further increase in the value of γ.

Tangential stress carries such film into a state of complex strain. An analysis of the mechanics of frictional contact by S. B. Ainbinder [26, 27] indicates that tangential stress diminishes the normal stress required for the onset of deformation. Shear in a boundary layer of oil held at constant pressure between two plane parallel steel disks results in a decrease in the layer depth (Table 1). The measurements in question here were carried out at the limit of sensitivity of the experimental method and the results obtained cannot serve as a

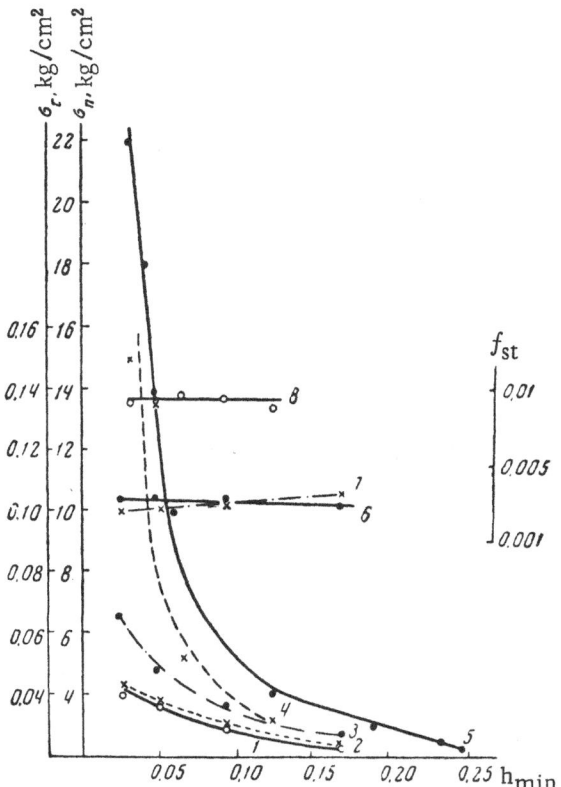

Fig. 1. Residual layer depth and properties of boundary layers of fatty acid solutions. 1) σ_n, 0.05% $C_{17}H_{35}COOH$ in isooctane; 2) σ_T, 0.05% $C_{17}H_{35}COOH$ in isooctane; 3) σ_T, 0.05% $C_5H_{11}COOH$ in isooctane; 4) σ_T, 0.01% $C_{15}H_{31}COOH$ in PPF (petroleum—paraffin fraction); 5) σ_n, 0.01% $C_{15}H_{31}COOH$ in PPF; 6) f_{st}, 0.05% $C_5H_{11}COOH$ in isooctane; 7) f_{st}, 0.05% $C_{17}H_{35}COOH$ in isooctane; 8) f_{st}, 0.01% $C_{15}H_{31}COOH$ in PPF.

quantitative test of the validity of the theory [26, 27]. But since boundary layer thinning increases shear resistance, these experiments do indicate that there is a preliminary displacement effect for boundary friction similar to that for dry static friction discovered by A. S. Verkhovskii [28]. The value of γ can diminish over this period if the contact pressure is sufficiently high. It is interesting to note that boundary layer thinning depends on film composition, just as was to be expected (see Table 1). The table shows the results obtained in a study of the relation between the residual (stationary) oil film depth (h_{min}) and the total normal and tangential stresses. The technique of measurement was the same as that used in [18, 19].

The mechanism of lubricating action depends on certain factors associated with boundary friction. The lubricating liquid may alter the mechanical properties of the rubbing surfaces (Rebinder effect [29, 30]) or lead to their chemical modification (for example, by oxidation [31]). The friction will be determined by the molecular-mechanical properties of the liquid if shear is localized in the boundary layer of lubricant. From this it follows that there is a specific relation between lubricating properties and lubricant composition for each component of boundary friction. The present work represents an attempt at tracing out these relations for "pure" polymolecular boundary friction.

The Composition of the Lubricating Liquid and Static Polymolecular Boundary Friction*

Study was made of boundary layers of mineral oils, and solutions of fatty acids and certain surface active substances in liquid hydrocarbons, between plane parallel steel disks, following the method of [18, 19]. The experiments were carried out at 20 ± 1°C (exceptions are noted) on residual stationary layers under fixed pressure, the film depth, h_{min}, remaining constant over a period of one hour. Measurement was made of the thinning resistance, σ_n, and the shear resistance, σ_T, and the coefficient of static resistance, f_{st}, then calculated. A narrow petroleum paraffin fraction of MS-20 oil separated by chromatography will be designated as PPF in what follows, and a highly purified paraffin-base S-100 cable oil as CO.

Resistance to thinning and shear increased rapidly as the normal load rose and the residual layer depth diminished (Fig. 1). σ_n and σ_T either increased in parallel or the first rose slightly more rapidly than the second. The value of σ_n was greater than that of σ_T by 1.5-3.0 orders for a fixed value of h_{min}. The corresponding values of the coefficient of friction for polymolecular boundary friction fell in the interval from 0.001 to 0.05 and were essentially constant for the layer depths covered by the experiments (from 0.02 to 0.2 μ).

Interest attaches to the relation between the values of σ_n, σ_T, and f_{st}, and the composition of the lubricating liquid. The values of these parameters are higher for solutions of surface active substances in oils than

*The experimental data of G. S. Bratova was used here.

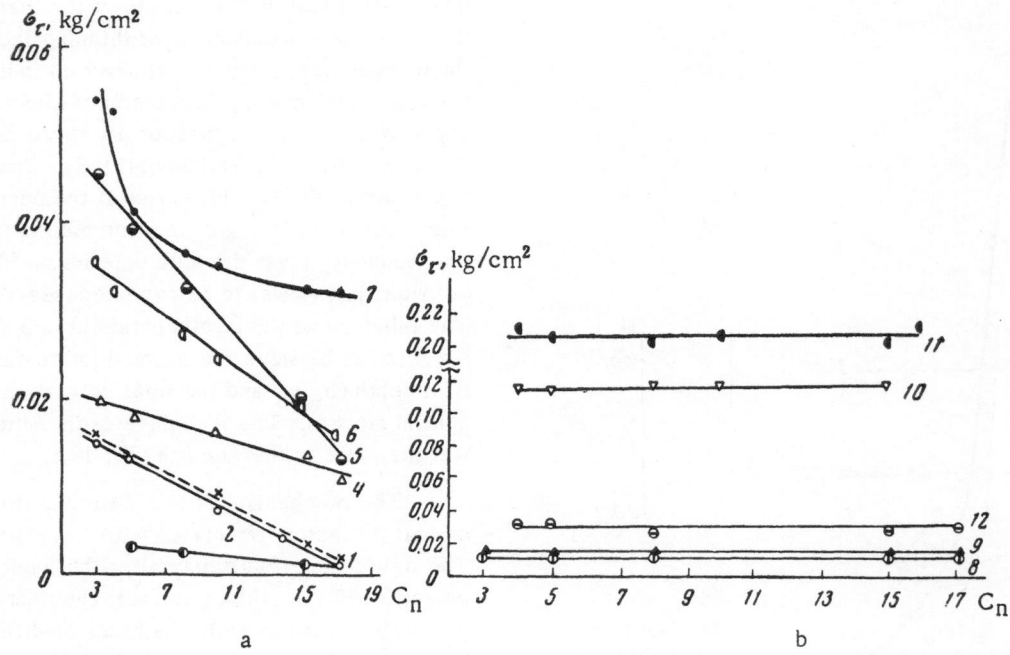

Fig. 2. a) The relation between boundary layer shear resistance in fatty acid solutions (%) and the length of the acid carbon chain (C_n). 1) $\sigma_n = 2$ kg/cm², 0.05% fatty acid in benzene; 2) $\sigma_n = 2$ kg/cm², 0.1% fatty acid in isooctane; 3) $\sigma_n = 2$ kg/cm², 0.1% fatty acid in PPF; 4) $\sigma_n = 2$ kg/cm², 0.1% fatty acid in CO (cable oil); 5) $\sigma_n = 4$ kg/cm², 0.1% fatty acid in isooctane; 6) $\sigma_n = 4$ kg/cm², 0.1% fatty acid in PPF; 7) $\sigma_n = 6$ kg/cm², 0.1% fatty acid in PPF. b) Depth of film, h_{min}, constant. 8) $h_{min} = 0.10$ μ, 0.05% fatty acid in benzene; 9) $h_{min} = 0.10$ μ, 0.1% fatty acid in isooctane; 10) $h_{min} = 0.13$ μ, 0.1% in PPF; 11) $h_{min} = 0.10$ μ, 0.1% in PPF; 12) $h_{min} = 0.06$ μ, 0.1%, in PPF.

they are for solutions of these same substances in low molecular hydrocarbons. It is generally true that these solutions will not form residual films without the addition of some surface active substances [19]. Both σ_n and σ_T depend on the molecular weight, or, more exactly, on chain length, of the fatty acid component of the solution.

It has been suggested [33, etc.] that values of the coefficient of friction less than 0.05 are realized only in fluid lubrication, and are therefore directly proportional to the viscosity, and inversely proportional to the depth, of the lubricating layer. Our experiments did indeed show a correlation between boundary layer shear resistance and bulk viscosity (Figs. 1 and 2a) but they also disclosed certain peculiarities in the effect of the lubricating liquid composition on the friction. Additions of fatty acids to benzene, isooctane, hexane, and heptane increase the boundary layer thinning resistance to the point where it is equal to, or even exceeds, the value for the pure oil. Shear localization in the boundary layer leads to a more gradual increase of coefficient of friction with liquid viscosity than would be anticipated from hydrodynamic laws, but the increase of shear resistance with diminishing film depth is more rapid than predicted. Oils containing surface active substances may show no correlation between bulk viscosity and the coefficient of boundary friction, this being the case for temperatures below the "melting point" of the film itself. This point is clearly illustrated in Table 2 which presents our data on the effect of narrow temperature variations on the coefficient of static friction under polymolecular lubrication at a boundary layer depth of 0.12 ± 0.03 μ. The existence of a relation between static boundary friction and viscosity clearly traces back to the fact that the viscosity of the lubricating liquid, and the boundary layer structure are determined by the same factors, viz., the volume and the elongated shape of the molecules, and the various molecular interactions (see [36] for details). It is to be noted that constancy of the coefficient of friction under a one-order alteration of the boundary layer depth does not justify the conclusion that polymolecular boundary lubrication is a combination of monomolecular boundary layer lubrication and liquid lubrication.

162

TABLE 2

Lubricant composition	Coefficient of static friction, f_{st}		
	0°C	20°C	40°C
PPF	0.026	0.020	0.017
PPF +0.05% $C_7H_{15}COOH$ . . .	0.013	0.013	0.016
PPF +0.05% $C_{15}H_{31}COOH$. . .	0.009	0.009	0.010
KM +0.05% $C_{15}H_{31}COOH$. . .	0.011	0.012	0.015

It has already been shown that an equation of the form [17, 19]

$$h_{min} = K_1 + K_2 C_n$$

applied to liquid hydrocarbon solutions of the fatty acids, C_n being the number of carbon atoms in the chain, and K_1 and K_2 constants, the first representing h_{min} for the solvent, and the second depending on the solvent's molecular structure. Under moderate contact pressures, the boundary layer shear resistance in these solutions is also a linear function of the length of the carbon chain in the surface active molecule (Fig. 2a):

$$\sigma_\tau = K_3 - K_4 C_n.$$

K_3 increases with the contact pressure, and is inversely proportional to the h_{min} value for the solvent at fixed pressure; while K_4 varies with the composition of the solvent, being lower for aromatic and petroleum hydrocarbons than for the paraffins.

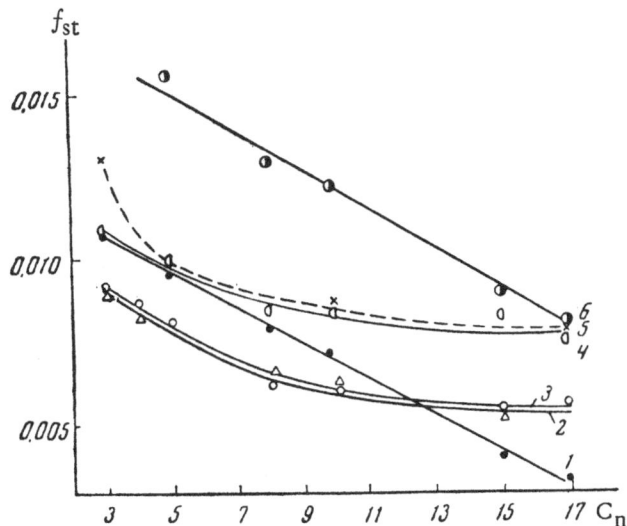

Fig. 3. The relation between the coefficient of static friction, f_{st}, of the contacted surfaces, and the length of the fatty acid hydrocarbon chain (C_n). 1) $\sigma_n = 4$ kg/cm², 0.1% fatty acid in isooctane; 2) $\sigma_n = 6$ kg/cm², 0.1% fatty acid in PPF; 3) $\sigma_n = 4$ kg/cm², 0.1% fatty acid in PPF; 4) $\sigma_n = 14$ kg/cm², 0.1% fatty acid in PPF; 5) $\sigma_n = 22$ kg/cm², 0.1% fatty acid in PPF; 6) $\sigma_n = 22$ kg/cm², 0.5% fatty acid in benzene.

TABLE 3

Acid (0.025% benzene solution)	h_{min}, μ σ_n, kg/cm²		Acid (0.025% benzene solution)	h_{min}, μ σ_n, kg/cm²	
	0.2	2.0		0.2	2.0
Stearic	0.11	0.08	Enanthic	0.08	0.06
Oleic	0.10	0.07	2C1-4F - Enanthic	0.06	0.05
Phenylstearic	0.15	0.13	Benzoic	0.07	0.04
Sebacic	0.05	0.03	Thioenanthic	0.04	0.03

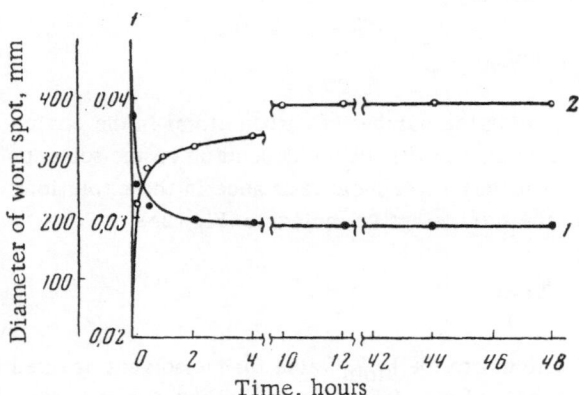

Fig. 4. Coefficient of friction (curve 1), and abrasion (diameter of worn spot, curve 2), under polymolecular boundary lubrication. Contact pressure, 145 kg/mm².

The linear relation between σ_τ and C_n breaks down at nominal contact pressures in excess of 6-10 kg/cm² (curve 7, Fig. 2a). Under these loads, the boundary layer depth in solutions of the lower fatty acids falls to 0.03-0.05 μ. Since the maximum height of the roughness projection is 0.02 ± 0.01 μ, and tangential stress thins the boundary layer still further (see above), it is quite possible that solutions of this kind cannot assure polymolecular boundary layer lubrication of the entire surface. This fact leads to a breakdown of the linear relation. The divergence between the results obtained by Hardy and Bowden in studying the lubricating effect of fatty acids is explained by the fact that the experiments of the first were carried out at low contact pressures (plane slide) and those of the second at high contact pressures (spherical slide).

The relation between the lubricant composition and the coefficient of friction, f_{st}, for polymolecular boundary lubrication is the same for lubrication with individual acids and for lubrication with these acids in solution (curves 1 and 6, Fig. 3). The fact that Hardy's rule applies here is indication of layer structure similarities. The effect of the solvent is limited to the slope of the plot. Hardy's rule breaks down when the molecular weight (more exactly, the molecular volume) of the solvent exceeds the molecular weight of the surface active material, as is the case with oil solutions of the fatty acids (see Fig. 3). This breakdown is due to the fact that both the solvent and the surface active substance affect h_{min}, so that there is a clear-cut alteration in the structure of the boundary layer. This conclusion is confirmed by the following data which illustrate the effect of solvent composition on the properties of residual boundary layers of a 0.05% stearic acid solution at a fixed layer depth of 0.06 μ:

Solvent	Benzene	Hexane	Isooctane	Dodecane	PPF
σ_n, kg/cm²	3.0	3.8	4.0	5.2	22.0
σ_τ, kg/cm²	0.021	—	0.032	0.052	0.217

The effect of fatty acids on σ_τ and h_{min} was determined separately by measuring the shear resistance in the boundary layer with the value of h constant at h < h_{min} by adjusting the external pressure. The results proved (Fig. 2b) that the shear resistance was independent of the length of the fatty acid molecule, being fixed, by the value of h, and then, by the composition of the liquid hydrocarbon which was used as solvent for the acid.

From this it follows that separation of solid surfaces by a film of fatty acid or fatty acid solution weakens the interaction not only by reducing the boundary layer friction, but also by forming a thinning-resistant layer of linear molecules, the layer depth and the surface separation being proportional to the chain length of the acid molecule.

164

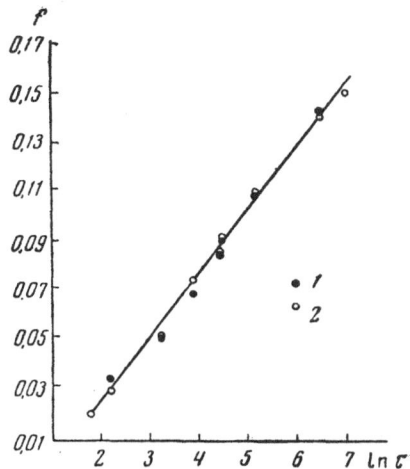

Fig. 5. Relation between the coefficient of friction, f, and metal hardness, for polymolecular lubrication. Working materials: metal-ruby pairs. Hardness measured in kg/mm² on a PMP-3 machine. Lubricating liquids based on: 1) polysilane; 2) SAS.

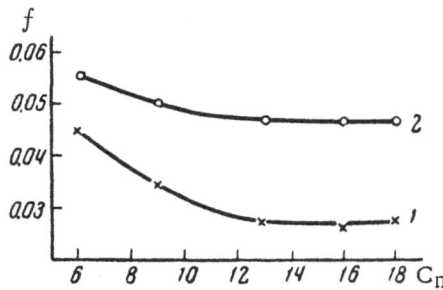

Fig. 6. Relation between the metal-ruby coefficient of friction, f, under polymolecular boundary lubrication and the length of the hydrocarbon radical in the fatty acid added to the lubricating liquid.

The data of Fig. 2b are consistent with the familiar molecular mechanism of boundary lubrication [3] which assumes boundary layer shear to be localized in planes passing through the terminal hydrocarbon radical of the oriented surface active molecules. The length of the molecule determines the residual layer depth and thereby fixes the distance between the slip plane (or planes) and the solid surface. The chemical activity of the molecule is of no significance here. This conclusion follows both from Fig. 2b and from a comparison of the thinning resistances of solutions of the fatty acids and their more active chloro-derivatives (Table 3). The intermolecular interaction is a factor of importance. The fatty acids dimerize through hydrogen bonding and thereby producing boundary layers that are more "stable" than those formed by the closely related thioacids which are free of hydrogen bonding (see Table 3). The significance of molecular elongated shape and molecule—solvent interaction (solvation) has been pointed out [19].

The Polymolecular Boundary Layer at High Contact Pressures*

Localization of frictional shear in the polymolecular boundary layer requires satisfaction of the condition $P/S \leq \gamma S \sigma_n$, P being the normal load, and σ_n the thinning resistance at $h \approx h_{min}$. The methods of [17-19] assured the satisfaction of this condition by working with a mirror surface at low values of the P/S ratio. The contact pressures are considerably higher at the separations actually met in frictional processes. The method proposed in [35] can be used to increase γS and σ_n and thus raise the value of the right-hand member of this inequality by several orders. This makes it possible to reach conditions such that $\gamma \approx 1$ or $\gamma \gg (\alpha + \beta)$ at contact pressures of the order of 100 kg/mm², thereby obtaining a 1:3-10 reduction of the coefficient of boundary friction.

The friction was measured in a small-scale four-ball tribometer [34] at nominal contact pressures of 50-500 kg/mm², working at a shear rate 0.5 cm/sec and a temperature of $20 \pm 1°$C. Study was made of metal-ruby and metal-leucosapphire friction pairs.

The frictional regime was established over a brief period in which abrasion developed (wear spot diameter increased), the contact surface increased, and the coefficient of friction fell. Abrasion ceased, and the coefficient of friction became stabilized, in the steady state (Fig. 4).

The variability of the hardness of the working materials led to alteration in the contact area and the shear resistance under both dry friction and monomolecular boundary lubrication. The force of friction generally fell as the hardness increased, but there were also cases in which it increased. The hardness determines the contact area under polymolecular lubrication, but has no direct effect on the shear resistance. At the same time, a reduction in the hardness increases S and h_{min}, all other conditions being fixed. The relation between metal hardness and the coefficient of friction under polymolecular boundary lubrication is expressed by the simple equation (Fig. 5)

$$f = k_1' + k_2' \lg \tau,$$

*Use was made here of the experimental data of M. M. Blekherov and I. V. Rozen.

TABLE 4

Liquid	Surface active additive			
	Without SAS	Fatty acid	Lard	Decyl alcohol
Cetane	0.24	0.05	0.12	0.11
Decane	0.14	0.08	0.12	0.13
PPF	0.10	0.04	0.06	0.08
Polysilane	0.32	0.01	0.03	0.06

k_2' being a constant which depends on the composition of the lubricating liquid. This relation proved to be applicable to our own experiments in which the metal hardness was varied from 3 to 500-800 kg/mm^2.

The data of Table 4 illustrate the effect of the composition of the lubricating liquid on the steel-ruby coefficient of friction under polymolecular boundary lubrication, lubrication being by the method of [35] with all other materials fixed and the contact pressure held constant at 145 kg/mm^2. It is seen that polymolecular lubrication at high pressures is quite sensitive to the composition of the lubricating liquid, just as was to be anticipated. This component of the friction does not acquire predominance without increasing σ_n through the addition of a surface active substance (SAS), even though there is a considerable increase in the values of the γS term and a marked effect on the mechanical properties of the metal. The coefficient of friction does not fall below 0.1 for this same reason.

The relation between the coefficient of friction and the length of the hydrocarbon radical of the surface active additive (Fig. 6) is the same as that observed at low contact pressures and fixed h_{min} value (see Fig. 3). The lubrication is essentially the same in these two cases, although the coefficient of friction does not fall below 0.01 at the higher contact pressures. There is no reason to suppose that $\gamma < 1$, since abrasion did not develop with the passage of time (see Fig. 4). It is more likely that the above described independence of f and h_{min} no longer applies when h_{min} is less than 0.02 μ.

Conclusions

1. Boundary friction is determined by three factors: the adhesional interaction of the solid surfaces, the screening effect of monomolecular and bimolecular adsorption films, and the shear resistance of the polymolecular boundary film. The contribution of each of these factors to the integral force of boundary friction is determined by the roughness and mechanical properties of the solid, and the thinning resistance of the polymolecular boundary layer.

2. The polymolecular component alone determines the force of friction when the contact pressure is less than the derivative of the thinning resistance of the boundary layer with respect to the actual contact area. This fact makes it possible to obtain coefficients of friction in the interval from 0.001 to 0.05. Polymolecular boundary lubrication can be realized at high contact pressures.

3. Resistance to thinning and shear increase rapidly as the depth of the boundary layer diminishes, but the coefficient of friction remains constant if shear is localized in the layer. It is possible that the situation changes when the film depth falls below 0.02-0.03 μ.

4. The polymolecular boundary layer is formed by the surface active substance but involves molecules of the hydrocarbon solvent, especially when the molecular weight of the latter is considerably higher than that of the surface active material. The effectiveness of the surface substance is determined by the length and elongated shape of the molecule, and by the molecular interactions.

5. Resistance to shear and thinning in boundary layers of fatty acid solutions in liquid hydrocarbons vary linearly with the length of the acid hydrocarbon radical. The Hardy law for acids is found to be applicable to solutions of acids in low molecular liquids as well.

Literature

1. N. P. Petrov. Selected Works (Inzhenernyi Zh., p. 1883), Moscow, Izd. Akad. Nauk SSSR (1948).
    O. Reynolds. Phil. Trans. Roy. Soc. London, Ser. A 177:173 (1886).
2. A. Cameron, The Theory of Lubrication in Engineering [Russian translation], Moscow, Mashgiz (1962).
    M. Hersay. Theory of Lubrication, John Wiley and Sons, Inc., New York (1938).
3. L. Aribyut and P. Dilei. Friction, Lubrication, Lubricating Materials [Russian translation], second edition, Moscow, Gostoptekhizdat (1940).
4. M. V. Korovchinskii. The Applied Theory of Liquid Friction Bearings, Moscow, Mashgiz (1954).
5. W. Hardy. Collected Works, Cambridge University Press (1936).
6. F. Bowden and D. Tabor. The Friction and Lubrication of Solids, Cambridge University Press (1954).
7. G. I. Fuks. "The lubricating properties of instrument oils," In collection: Timing Mechanisms, No. 1, Moscow, Mashgiz (1955).
8. B. V. Deryagin and E. V. Obukkov. Kolloidn. Zh. 1:385 (1935).
    B. V Deryagin and M. M. Kusakov. Izd. Akad. Nauk SSSR, Otd. Khim. Nauk No. 5:471 (1936).
9. B. V. Deryagin and M. M. Kusakov. Zh. Fiz. Khim. 26:586 (1952).
    B. V. Deryagin. Kolloidn. Zh. 17:207 (1955).
10. B. V. Deryagin and E. F. Pichugin. Dokl. Akad. Nauk SSSR 63:53 (1948).
    B. V. Deryagin and V. V. Karasev. Dokl. Akad. Nauk SSSR 101:289 (1955).
11. B. V. Deryagin. "Two and three dimensional aspects of surface phenomena," this volume, p. 3.
12. A. S. Akhmatov. Transactions, Second All-Union Conference on Friction and Wear in Machines, Vol. 3, Moscow, Izd. Akad. Nauk SSSR (1939).
13. J. Trillat. J. Phys. Radium 10:32 (1929); Usp. Fiz. Nauk 2:493 (1931).
14. G. Clark, R. Sterret, and B. Lincoln. Ind. Eng. Chem. 28:1318 (1936).
15. G. I. Fuks. In collection: Additives for Greases and Fuels, Moscow, Gostoptekhizdat (1961), p. 228.
16. G. I. Fuks. Kolloidn. Zh. 20:748 (1958); Dokl. Akad. Nauk SSSR 103:635 (1957).
17. G. I. Fuks and N. I. Kaverina. Transactions, Third All-Union Conference on Friction and Wear in Machines, Vol. 3, Moscow, Izd. Akad. Nauk SSSR (1960), p. 397.
18. G. I. Fuks. Zavodsk Lab., No. 12:1455 (1955); Transactions, Third All-Union Conference on Friction and Wear in Machines, Vol. 3, Moscow, Izd. Akad. Nauk SSSR (1956), p. 301.
19. G. I. Fuks. In collection: Research in Surface Forces, Moscow, Izd. Akad. Nauk SSSR (1961), p. 99. [English translation: Consultants Bureau, New York (1963).]
20. I. V. Kragel'skii. Friction and Wear, Moscow, Mashgiz (1962); Chapters I and II.
    E. M. Shvetsova. In collection: Friction and Wear in Machine Parts, No. VII, Moscow, Izd. Akad. Nauk SSSR (1953).
21. A. Bailey and Courtney–Pratt. Proc. Roy. Soc., A, Vol. 227 (1955).
22. I. V. Kragel'skii and I. E. Vinogradova. Coefficients of Friction, second edition, Moscow, Mashgiz (1962), Chapter VI.
23. B. V. Deryagin and M. M. Samygin. Transactions, First All-Union Conference on Friction and Wear in Machines, Vol. 2, Moscow, Izd. Akad. Nauk SSSR (1940).
    B. V. Deryagin and V. P. Lazarev. Transactions, Second All-Union Conference on Friction and Wear in Machines, Vol. 3, Moscow, Izd. Akad. Nauk SSSR (1949).
24. G. I. Fuks and A. S. Mikhailyuk. Priborostroenie, No. 9:18 (1957); Kolloidn. Zh. 22(6) (1960).
    G. I. Fuks, 22(2) (1960).
25. G. Ernst and M. Merchant. Surface Treatment of Metals, Soc. of Metals (1941).
26. S. V. Ainbinder. Transactions of the Seminar: The Quality of Machines Part Surfaces, Collection 5, Moscow, Izd. Akad. Nauk SSSR (1961), Isv. Akad. Nauk SSSR Otd. Tekhn. Nauk, No. 6:21 (1962).
27. J. McFarlane and D. Tabor. Proc. Roy. Soc., (A)202:2, 44 (1950).
28. A. S. Verkhovskii. Zh. Prikl. Fiz. 3:311 (1926); Tr. Tomsk. Politekhn. Inst. 61:1 (1947).
29. P. A. Rebinder. Z. Phys. 72:191 (1931); Izd. Akad. Nauk SSSR Otd. Khim. Nauk, No. 11:1284 (1957).
30. S. Ya. Veiler and V. I. Likhtman. The Action of Lubricants in the Pressure Treatment of Metals, Moscow, Izd. Akad. Nauk SSSR (1960).
    V. I. Likhtman, E. D. Shchukin, and P. A. Rebinder. Physical, Chemical, Mechanics of Metals, Moscow, Izd. Akad. Nauk SSSR (1962), Chapter 2.

31. G. V. Vinogradov. In collection: Additives for Greases and Fuels, Moscow, Gostoptekhizdat (1961), p. 197.
32. G. S. Khodakov and P. A. Rebinder. Kolloidn. Zh. 23:482 (1961).
   A. S. Aslanova and P. A. Rebinder. Dokl. Akad. Nauk SSSR 96:299 (1954).
33. A. Perry. Metallurgia 48:285 (1953).
34. G. I. Fuks. Zavodsk Lab., No. 5:594 (1956).
35. G. I. Fuks, M. M. Blekherov, and N. A. Smagunova, Patent No. 152525 (1961).
36. G. I. Fuks. Viscosity and Plasticity of Heat Resistant Materials, Moscow, Gostoptekhizdat (1951).

THE ANOMALOUS THERMAL CONDUCTIVITY OF FILM WATER
IN MICA CRYSTALS

M. S. Metsik and O. S. Aidanova

Irkutsk State University
Department of General Physics

It is a well-known fact that surface effects tracing back to molecular forces are significant in fixing the properties of thin films. It has been shown that thin liquid films show enhanced elasticity, these anomalies being especially marked in water films [1-19, 31, 32].

Experiment has also shown that the humidity has a pronounced effect on the thermal conductivity of hydroscopic and porous bodies. From this it can be assumed that film water plays an important role in thermal transfer.

We have developed a special experimental technique for testing this hypothesis.

Experimental Methods

It was first necessary to obtain a body containing a reasonably large number of thin water films of controlled depth. This requirement is satisfied by a pile of thin mica sheets separated by water films. Such piles were built up from mica sheets 20-30 μ thick and 4×5 cm^2 in area. Optical methods were used to assure uniformity in the thickness of each mica sheet [20], departures from uniformity being less than 1 μ. The pile was compressed between steel plates with a hydraulic press, or vise, and the compression measured and fixed by the positions of the bolts which held the plates together (Fig. 1). A copper-constantan thermocouple constructed from wires 0.18 mm in diameter was inserted between the interior sheets of the pile prior to compression.

The pile with its enclosed thermocouple was set in an empty vessel under a vacuum jar. This jar was evacuated to a pressure of several millimeters of mercury, the vessel filled with water, and the pile allowed to saturate for about an hour. It was assumed that this method of adding moisture would assure the presence of water in all of the interstices between the sheets.

The thermal conductivity of the pile was investigated by the regular regime method [21-22].

Application of the stationary current method would have been unsatisfactory here since:

1) Experimental difficulties of considerable magnitude arise in attempting to keep the humidity of the pile in its holder constant over several hours while a stationary current is being set up, and

2) Temperature differences of the order of ten degrees are set up between the solid faces. This makes the method unacceptable when the thermal properties under study are markedly temperature dependent, as is to be anticipated in the case of film water. Mass transfer also comes into play to further complicate the thermal transfer.

The regular regime method is one of the simpler nonstationary procedures for the measurement of thermal conductivity, and one which is free of the defects just mentioned. Another favorable aspect of the method is the fact that the sample is studied in water so that special measures are not required to assure fixed depth of the water films between the sheets. The temperature difference between the center of the sample and the outer surface was 3-6° under measurement, and cooling or heating required approximately 10 minutes.

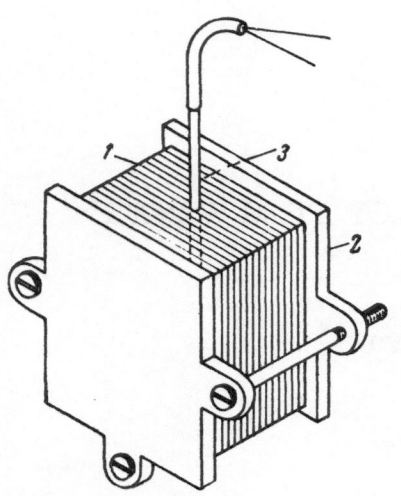

Fig. 1. Pile of mica sheets in clamp.
1) Pile of mica sheets; 2) clamp for
compressing the pile; 3) thermocouple.

The water film depth is determined by the compression and was evaluated from the amount of water taken up and the geometrical measurements of the pile. The pile was given a preliminary compression at 3500 kg/cm^2, since the thermal conductivity of the mica sheets might alter as a result of compaction under pressure. The pile experienced a certain small irreversible compression as a result of this treatment, and its coefficient of elasticity increased sharply.

Method of Calculation

The second task which arose here was that of separating the film component from the total pile conductivity.

The system in question here was a set of anisotropic sheets, each of thickness h_1, the thermal conductivity being $\lambda_1^{\parallel}$ along the sheet and $\lambda_1^{\perp}$ normal to the sheet. These sheets were separated by water films of depth h_2; these too could, in general, be anisotropic with coefficients $\lambda_2^{\parallel}$ and $\lambda_2^{\perp}$.

Such system can be considered as an anisotropic body in which the mean axial coefficients of thermal conductivity are given by

$$\lambda_{av}^{\perp} = \frac{\lambda_1^{\perp}\left(1 + \frac{h_2}{h_1}\right)}{1 + \frac{\lambda_1^{\perp}}{\lambda_2^{\perp}}\frac{h_2}{h_1}}, \qquad \lambda_{av}^{\parallel} = \frac{\lambda_1^{\parallel} + \lambda_2^{\parallel}\frac{h_2}{h_1}}{1 + \frac{h_2}{h_1}}.$$

(1)

The regular regime method employed in the present work was developed earlier by M. S. Metsik for anisotropic bodies [22].

The theory of this method shows the coefficients $\lambda^{\parallel}$ and $\lambda^{\perp}$ for an anisotropic parallelepiped with its axes parallel to the axes of anisotropy to be related by the equation [22]

$$\lambda_{av}^{\parallel} = \frac{m_{\infty}c\rho}{\pi^2\left(\frac{1}{X^2} + \frac{1}{Y^2}\right)} - \frac{\lambda_{av}^{\perp}}{\left(\frac{1}{X^2} + \frac{1}{Y^2}\right)Z^2}.$$

(2)

Here X, Y, and Z are the pile dimensions along the x, y, and z axes (see Fig. 1), Z being the measurement at right angles to the plane of the sheets, ρ and c are, respectively, the bulk density and specific heat capacity of the pile with allowance for the uptake of water, and m_{∞} is a coefficient which characterizes the rate of cooling of the pile in a regular regime when $\alpha \to \infty$, α being the coefficient of thermal transfer of the sample.

The coefficient m_{∞} was determined from the rate of cooling, or heating, of the pile in a thermostat in which the water was vigorously agitated [21].

Let the filling of the space between the sheets with water increase the pile height, Z, by dZ and thereby alter the rate of cooling, m, by dm, and change $\lambda^{\parallel}$ by $d\lambda^{\parallel}$. It follows from (2) that

$$d\lambda^{\parallel} = \frac{c\rho\,dm}{\pi^2\left(\frac{1}{X^2} + \frac{1}{Y^2}\right)} + \frac{2\lambda_{\perp}\,dZ}{\left(\frac{1}{X^2} + \frac{1}{Y^2}\right)Z^3}.$$

(3)

170

This change of thermal conductivity is compounded of an increase, $\lambda_2^{\parallel} \, dZ/Z$, resulting from the presence of the water and a decrease, $\lambda_1^{\parallel} \, dZ/Z$, resulting from a partial displacement of mica by water, i.e.,

$$d\lambda^{\parallel} = (\lambda_2^{\parallel} - \lambda_1^{\parallel}) \frac{dZ}{Z}.$$

(4)

The following equation can then be easily obtained by combining (3) and (4):

$$\frac{dm}{dZ} = \frac{\lambda_2^{\parallel} - \lambda_1^{\parallel}}{Z} \frac{\left(\dfrac{1}{X^2} + \dfrac{1}{Y^2} \right) \pi^2}{c\rho} - \frac{2\lambda_{\perp}\pi^2}{c\rho} \frac{1}{Z^3},$$

(5)

and integrated to give

$$\lambda_2^{\parallel} = \frac{1}{\pi^2 \left(\dfrac{1}{X^2} + \dfrac{1}{Y^2} \right)} \frac{m_2 - m_1}{\ln \dfrac{Z_2}{Z_1}} + \frac{2\lambda_1^{\perp}}{\pi^2 \left(\dfrac{1}{X^2} + \dfrac{1}{Y^2} \right)} \left(\frac{1}{Z_2^2} - \frac{1}{Z_1^2} \right).$$

(6)

The character of the alteration of m_{∞} resulting from the increase in Z caused by the film water is readily seen from (6). Such analysis was used experimentally to obtain an estimate of $\lambda_2^{\parallel}$.

1. Let $m_2 = m_1$ with $Z_2 > Z_1$, or

$$\frac{dm}{dZ} = 0.$$

(7)

Equation (6) then gives

$$\lambda_2^{\parallel} = \lambda_1^{\parallel} + \frac{2\lambda_1^{\perp}}{\pi^2 \left(\dfrac{1}{X^2} + \dfrac{1}{Y^2} \right)} \left(\frac{1}{Z_2^2} - \frac{1}{Z_1^2} \right).$$

(8)

The second member of (8) is considerably less than $\lambda_1^{\parallel}$ for piles of the dimensions of those used in these experiments, and it therefore follows that

$$\lambda_2^{\parallel} \lesssim \lambda_1^{\parallel}.$$

(9)

2. The thermal conductivity of the film is higher than $\lambda_1^{\parallel}$ when $m_2 > m_1$, and a good approximation is given by

$$\lambda_2^{\parallel} = \frac{1}{\pi^2 \left(\dfrac{1}{X^2} + \dfrac{1}{Y^2} \right)} \frac{m_2 - m_1}{\ln \dfrac{Z^2}{Z_1}} + \lambda_1^{\parallel}.$$

(10)

3. Finally, the thermal conductivity diminishes with increasing film depth when $m_2 < m_1$, gradually tending to the value for the thermal conductivity of bulk water,

$$\lambda_1^{\parallel} > \lambda_2^{\parallel} \gg \lambda_{wa}.$$

(11)

Thus the coefficient of thermal conductivity of the water films responsible for the observed increase in the pile can be obtained from the experimental data on the variation of m_{∞} with the pile height, Z, and this without the necessity of calculating a value for $\lambda_2^{\parallel}$.

Since $\lambda_1^{\parallel}$ is now less than $9 \cdot 10^{-3} \, \text{cal} \cdot \text{cm}^{-1} \cdot \text{sec}^{-1} \cdot \text{deg}^{-1}$ [22], an increase in cooling rate (dm > 0) with pile depth (dZ > 0) in a pile of our dimensions requires that the film thermal conductivity be in excess of $10^{-2} \, \text{cal} \cdot \text{cm}^{-1} \cdot \text{sec}^{-1} \cdot \text{deg}^{-1}$. This is a quite definite criterion and one which was used to obtain a first estimate of the results of the measurements. It was assumed in the calculations that: $\lambda_1^{\perp} = 1.4 \cdot 10^{-3}$ [23]; and

TABLE 1

Pile dimensions, cm^2	No. of sheets per 1 cm of pile	Mean thickness of crystal in pile	Depth of water film, 2h	Pile density, ρ, g/cm^3	Heat capacity of pile, c, cal/(g·deg)	Cooling rate, m_∞	$\lambda_2^{\parallel}$, mean thermal conductivity of a water film of depth, 2h cal/(cm·deg·sec)·10^2
4.0×5.0×3.560	267	34.6	1.0	2.65	0.215	1.68	0.04
4.0×5.0×3.496	272	34.6	0.3	2.68	0.210	1.73	1.6
4.0×5.0×3.473	274	34.6	0.1	2.69	0.208	1.76	14.0
4.0×5.0×3.467	274	34.6	∼0.0	2.69	0.208	1.72	—
4.0×5.0×4.235	236	41.7	0.7	2.65	0.214	1.60	0.9
4.0×5.0×4.194	238	41.7	0.4	2.67	0.212	1.61	1.7
4.0×5.0×4.172	239	41.7	0.06	2.68	0.210	1.62	9.0
4.0×5.0×4.166	240	41.7	∼0.0	2.70	0.208	1.60	—

$\lambda_2^{\perp} = 1.5 \cdot 10^{-2}$ cal·cm^{-1}·sec^{-1}·deg^{-1} [24]; the heat capacity of the crystals, c, was taken as 0.208 cal·g^{-1} [23]. It is seen from (1) that the contribution of $\lambda_2^{\perp}$ to $\lambda_{av}^{\perp}$ diminishes as the value of $\lambda_2^{\perp}$ increases and amounts to no more than 1-2%. Such a correction would lie within the limits of experimental error and no account was taken of it in this work.

Experimental Results and Discussion

Measurements were carried out on seven piles consisting of 1000, or more, mica sheets, working in both the compressed and decompressed states. Table 1 presents certain of the more characteristic data obtained in a study of two previously saturated piles in compression. Errors arising from incomplete water filling of the space between the mica sheets could be avoided by working in compression, and compression experiments were preferred for this reason.

The last column of the table shows that the pile cooling rate, m_∞, increased at first, and then diminished as the depth of the water films between the pile sheets rose, the curve passing through a maximum corresponding to a film conductivity $\approx \lambda_1^{\parallel}$. The existence of this maximum on the m curve was indication that compaction of the mica crystals under compression had little effect on the thermal conductivity, this compaction, in itself, leading to a monotonic alteration in m_∞.

The thermal conductivity of intercrystalline water films 0.1 μ deep proved to be more than an order higher than the thermal conductivity of bulk water (approximately 0.1 instead of $1.5 \cdot 10^{-3}$ cal·cm^{-1}·sec^{-1}·deg^{-1}).

This is an entirely new result. Before passing to its discussion, we will analyze the possible experimental errors.

A study of the working Eq. (6), makes it clear that the error in the calculated value of $\lambda_2^{\parallel}$ is largely determined by the accuracy of measurement of the cooling rate, m_∞, and the pile depth, Z,

$$\frac{\Delta\lambda_2^{\parallel}}{\lambda_2^{\parallel}} \approx \frac{2\delta m}{\Delta m} + \frac{2\delta Z}{\Delta Z}.$$

(12)

The value of m_∞ at each film depth was obtained from six experiments. The errors in question here could have been as large as: $\delta m = 0.01 \cdot 10^{-2} \cdot$ sec^{-1} (at $\Delta m = 1.5 \cdot 10^{-2} \cdot$ sec^{-1}); $\delta Z = 0.005$ cm (at $\Delta Z = 3 \cdot 10^{-2}$ cm).

This gives a statistical error of 50% for thin films. This error could have been reduced to 20-30% by using films one micron or more in depth in the measurement of mean thermal conductivity. Although the results obtained here must be considered as provisional in view of this possibility of a considerable experimental error, they do establish the fact that water films in mica crystals have an anomalously high thermal conductivity which varies with the film depth.

TABLE 2

Particle (lamella) coarseness, mica, mm	Bulk humidity, %	Density, g/cm³	Specific heat, c, cal/(g · deg)	Thermal conductivity, 10^3, cal/(cm · sec · deg)
3—5	0.0	0.574	0.208	0.63
	2.0	0.594	0.234	0.71
	3.6	0.611	0.255	0.79
	4.8	0.623	0.270	0.80
	9.2	0.667	0.317	0.81
0.2—0.3	0.0	0.600	0.208	0.44
	6.4	0.665	0.285	0.76
	7.9	0.682	0.300	0.84

On this basis it can be concluded that the molecular ordering is higher in water films between mica crystals than in bulk water. This ordering is such that the film structure approximates the structure of the solid. Such structures are generally characteristic of thin polar liquid films on solid substrate [6, 25, 12]. The boundary monolayer ordering in these liquids propagates to depths which are large in comparison with the depth of the monolayer itself. The properties of liquid boundary films have been studied in great detail by B. V. Deryagin and his coworkers [1-9]. Here it has been shown that the thin water film possesses dynamic shear elasticity [1, 7], while thin liquid film boundary layers have enhanced viscosity [3, 4, 9]. The existence of this dynamic shear elasticity has been shown by radio-techniques, and by other methods [10, 13].

It has been experimentally proven that adsorbed water has enhanced density [12] and reduced solubility [15, 18, 26, 28], that its phase transition temperatures are reduced [12, 17, 19], and that it shows other anomalies as well [16].

Adsorbed water films of considerable depth can be obtained on mica crystals [14, 25, 28].

It is clear that such structurally-order films transmit thermal vibrations along the film with considerable ease, and this fact accounts for their anomalously high thermal conductivity.

The thermal conductivity of the crystal is higher than that of the substance in the amorphous state. Thus the room temperature thermal conductivity of crystalline quartz is an order higher than the thermal conductivity of molten quartz under the same conditions ($5 \cdot 10^{-2}$ and $5 \cdot 10^{-3}$ cal $\cdot$ cm$^{-1} \cdot$ sec$^{-1} \cdot$ deg^{-1}, respectively) [30]. Despite its lower density, the thermal conductivity of ice is greater than the thermal conductivity of water ($5 \cdot 10^{-3}$ and $1.4 \cdot 10^{-3}$ cal $\cdot$ cm$^{-1} \cdot$ sec$^{-1} \cdot$ deg^{-1}, respectively) [24].

The structural ordering of water films on mica crystals can be inferred from the structure of the mica itself [25, 29]. The origin of this ordering is found in the intense dipole field which exists near the crystal surface.

Anomalously high film conductivity is obviously to be expected in other solid bodies as well as mica. This gives an understanding of the fact that the thermal conductivity of porous and layered materials increases rapidly with the humidity. This increase is due not only to the joining of various body elements by water bridges, but also to the conduction of the film water itself, an effect which becomes more pronounced as the film depth increases.

An Estimation of the Depth of the Anomalously Conducting Film on the Mica Plate

Mica particles are quite suitable working materials for investigating the role played by surface effects. Water is almost unabsorbed by these particles but spreads over the particle surface. Grinding increases the total surface area. Thermal conductivity measurements on bodies consisting of mica particles of various sizes can therefore be used to bring out the role of surface effects. We have measured the thermal conductivities of systems consisting of mica lamellae which differed markedly in their measurement along the cleavage plane.

These measurements were carried by the previous regular regime method working in a copper jacketed α-calorimeter [21]. The results obtained are shown in Table 2 and Fig. 2. It is seen that the coefficient of

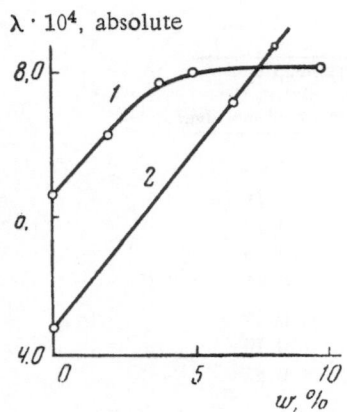

$\lambda \cdot 10^4$, absolute

Fig. 2. Thermal conductivity of the mica lamella and its variation with the humidity, W. 1) Lamella dimensions, 3-5 mm; 2) 2) lamella dimension 0.2-0.3 mm.

thermal conductivity of the dry lamellae was more than 1.5 times higher than the coefficient of thermal conductivity of the powders. This can be explained as the result of more regular ordering of the large particles and increased significance of the component of thermal conductivity parallel to the cleavage plane. In other words, the mean coefficient of thermal conductivity is higher for large mica particles than for small.

The thermal conductivity increases with the humidity, regardless of whether the body consists of large, or of small, particles. This increase takes place rapidly in the small-particle body and does not reach saturation even at 8% humidity, whereas saturation for the lamellae is observed at a humidity of 5%. This can be interpreted as indication that 5% humidity gives that limiting depth of coarse particle surface film at which the anomalous thermal conductivity disappears. The small particles have higher surface area, and films formed on them under these conditions are thin and highly conducting over the entire sectional area.

These experiments give a basis on which the water film depth, Δh, can be estimated for large mica particles (lamellar) from observations on the variation of thermal conductivity saturation with humidity. Here the bulk density of the body is expressed by the equation

$$\rho = \rho_{mi} S \cdot h \cdot n_0, \qquad (13)$$

ρ_{mi} being the bulk density of the mica, S and h, the respective area and thickness of the lamella, and n_0 of lamellae per unit volume.

If the water is assumed to be uniformly distributed over the particle surface and the small portion of surface included in the particle ends is neglected, the alteration of the total bulk density, $\Delta\rho$, with humidity resulting from the water cover of the particles can be expressed by the equation

$$\Delta\rho = 2\Delta h \rho_{wa} S n_0, \qquad (14)$$

Δh being the film depth and ρ_{wa} the bulk density of the water.

From (13) and (14) one obtains

$$\Delta h = \frac{\Delta\rho}{\rho} \cdot \frac{\rho_{mi}}{\rho_{wa}} \frac{h}{2}. \qquad (15)$$

From the data of Table 2, $\Delta\rho = 0.05$ g/cm^3, $\rho = 0.6$ g/cm^3, $\rho_{mi} = 2.7$ g/cm^3, $\rho_{wa} = 1$ g/cm^3, and h = 20 μ. The depth of the anomalously conducting film is obtained from (15) as $\Delta h \approx 2 \mu$. This result has only provisional significance since part of the water is distributed over the lamella ends and in the bridges between lamellae, and the value obtained is therefore too high.

This method of estimating the depth of the anomalously conducting water film leads to results which are conisistent with those obtained from the measurements on the mica pile.

I conclude with an expression of my sincere thanks to B. V. Deryagin, Corresponding Member, Acad. Sci., USSR, for his interest in this work, for his advice, and for a discussion of the results obtained here.

Literature

1. B. V. Deryagin. Zh. Fiz. Khim. 3(1):29 (1932).
2. B. V. Deryagin. Kolloidn. Zh. 1(5) (1935).

3. B. V. Deryagin and N. A. Krylov. Reports, Conference on the Viscosity of Liquids and Colloidal Solutions, Moscow (1944), p. 52.

4. B. V. Deryagin and V. V. Karasev. Dokl. Akad. Nauk SSSR 101(2):289 (1956).

5. B. V. Deryagin and I. I. Abrikosova. Zh. Fiz. Khim. 32(2):442 (1956).

6. B. V. Deryagin and E. Obukhova. Kolloidn. Zh. 1:385 (1935).

7. B. V. Deryagin. Izv. Sektora Fiz.-Khim. Analiza, Inst. Obshch. Neorgan. Khim., Akad. Nauk SSSR 9:55 (1937).

8. B. V. Deryagin. Zh. Fiz. Khim. 14:137 (1940).

9. B. V. Deryagin and M. Samygin. Reports, Conference on the Viscosity of Liquids and Colloidal Solutions, Moscow (1941), p. 69.

10. A. V. Bulgadaev. Dokl. Akad. Nauk SSSR 47(5):805 (1954).

11. D. L. Talmud. Surface Phenomena, Moscow, Gostekhizdat (1934).

12. A. V. Lukov. Transport Phenomena in Capillary Porous Bodies, Moscow, Gostekhizdat (1954).

13. A. Griffith. Phil. Trans. Roy. Soc. London, Ser. A. 2(21):163 (1921).

14. Bangham, Mosallam, and Saweris. Nature, No. 140:237 (1937).

15. S. I. Dolgov. A Study of the Mobility of Surface Moisture and Its Availability for Plants, Moscow—Leningrad, Izd, Akad. Nauk SSSR (1948).

16. P. I. Andrianov. The Heat of Wetting and the Specific Surface Area of Soils, Moscow, Izd. VASKhNIL (1937).

17. V. A. Bakaev, et al. Dokl. Akad. Nauk SSSR 125(4):831 (1959).

18. A. V. Dumanskii. Proceedings, May Session Acad. Sci. USSR, Akad. Nauk SSSR, Vol. 125 (1935).

19. H. H. de Boer. The Dynamic Character of Adsorption [Russian translation], Moscow, IL (1962).

20. M. S. Metsik. Tr. Irkutsk. Univ., Ser. Fiz.-Mat., Vol. 22 (1957).

21. G. M. Kondrat'ev. The Regular Thermal Regime, Moscow, Gostekhizdat (1954).

22. M. S. Metsik. Izv. Vysshikh Uchebn. Zavedenii, Fiz. No. 5:131 (1960).

23. E. K. Lashev, Mica, Physical Properties, Moscow—Leningrad, Promstroiizdat (1948).

24. D. Kei and T. Lebi. Handbook for Physical Experimenters [Russian translation], Moscow, IL (1940).

25. M. S. Metsik. In collection: Research in Surface Forces, Moscow, Izd. Akad. Nauk SSSR (1961), p. 66. [English translation: Consultants Bureau, New York (1963).]

26. N. N. Semenov and N. M. Chirkov. Dokl. Akad. Nauk SSSR, Nov. Ser. 51(1):37 (1946).

27. N. M. Chirkov. Zh. Fiz. Khim. 21:1303 (1947).

28. M. S. Metsik. Izv. Tomsk. Politekhn. Inst. 91:413 (1956).

29. M. S. Metsik. Izv. Vysshikh Uchebn. Zavedenii, Fiz. No. 4:29 (1958).

30. G. S. Zhdanov. Solid State Physics, Izd. MGU (1962).

31. B. V. Deryagin and L. M. Shcherbakov. Kolloidn. Zh. 23(1):40 (1961).

32. N. N. Fedyakin. Dokl. Akad. Nauk SSSR 138(6):1389 (1961).

THE EFFECT OF FILM WATER ON THE
DIELECTRIC PROPERTIES OF MICA
PART I

N. V. Afanas'ev, M. S. Metsik, and V. N. Popova

Irkutsk State University
Department of General Physics

The Effect of Adsorbed Films in Open Stratifications on the Electrical Properties of Mica-Phlogopite Crystals

Our studies on the effect of swelling on the electrical properties of mica have led us to the conclusion that the mica crystal contains open (to the atmosphere), and closed lens-shaped, stratifications whose surfaces are covered with adsorbed water films. Nothing is as yet known of the effect of these films on the electrical properties of the mica, study having been confined to measurements at right angles to the cleavage plane where the film effect is at a minimum. Study of effects in both directions, parallel and perpendicular to the cleavage plane, is of interest to a general description of the electrical properties of mica, and suggests, at the same time a method for investigating the properties of film water.

The present work made use of samples of phlogopite crystals of various degrees of hardness on which the electrical properties were studied in the two indicated directions.

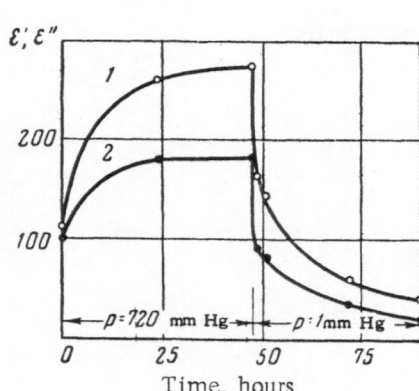

Fig. 1. Relation between the dielectric constant, ε' (1), and the loss factor, ε'' (2), of phlogopite, and the time of holding in moist air and in vacuum. Field parallel to the cleavage plane; frequency, 50 cps; Kuznetsov hardness, 100 sec.

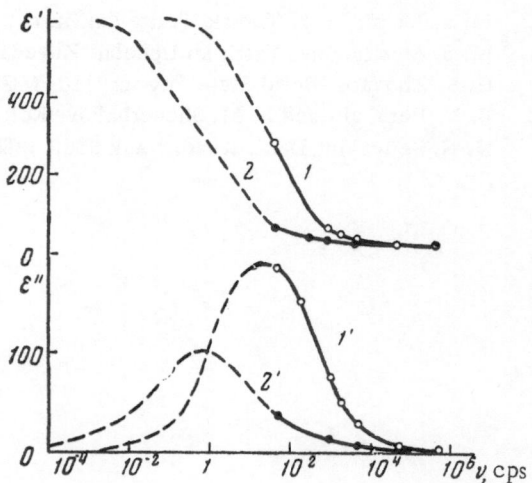

Fig. 2. The frequency (ν) dependence of ε' and ε'' for phlogopite. Temperature, 20°C; field parallel to the cleavage plane. 1, 1') ε' and ε'', respectively, after holding in vacuum for 48 hours; 2, 2') the same, after holding in vacuum for 19 hours.

TABLE 1

V.D.Kuznetsov hardness, sec	After drying		After moistening		Time of moistening, h	Change in mass, %
	ε'	ε''	ε'	ε''		
105	70	20	600	660	17	0.03
70	150	40	2,300	4,200	2	0.016
25	900	300	12,000	22,800	14	0.16

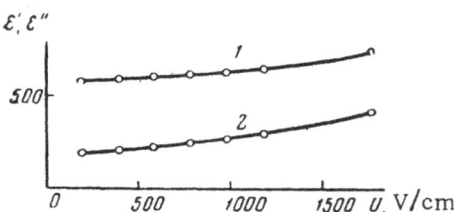

Fig. 3. The relation between ε' (1), and ε'' (2), for a soft phlogopite and the electric field strength, U, in the direction of the cleavage plane. Temperature, 25°C; frequency, 50 cps.

These experiments were carried out at temperatures ranging from -120 to $+350$°C, each involving measurement of the dielectric constant, ε', the loss coefficient, $\varepsilon'' = \varepsilon'\tan\delta$, and the specific resistance, at frequencies in the interval from 50 cps to 500 kc. Mechanical pressure was applied to prevent low temperature swelling of the phlogopite [1, 2].

The measurements in the perpendicular direction were carried out in such manner that the pressure on the sample was lowered to 0.2 kg/cm^2 in the second heating. This made it possible to study the effect of swelling on the electrical properties of phlogopite.

Figure 1 shows the relation between the 50 cps dielectric constant, ε', and loss factor, ε'', of solid phlogopite and the time of holding in moist air, and in vacuum at room temperature, the measurements being made along the cleavage plane at p = 720 mm Hg and p_{H_2O} = 6 mm Hg. Each sample was held in vacuum for two hours while the electrodes were being deposited, and then held in moist air for an additional three hours in preparation for the measurements. It is seen that the values of ε' and ε'' were quite high and increased with the time of holding in the air, the increase being rapid at first and then more gradual. The values of ε' and ε'' were found to fall after the sample had been put under vacuum, the fall-off being quite pronounced at first and comparatively slow at the end of the holding period.

The frequency (ν) relations for a phlogopite crystal which had been moistened and then held in vacuum are shown in Fig. 2. The dotted portions of these curves were developed by the Cole—Davidson extrapolation method [3]. These curves show a frequency maximum at 30 cps for the moistened state. Holding the sample in vacuum for 19 hours displaced the ε'' frequency maximum by almost 1 cps. The dielectric constant fell, at the same time, from 600 to 7. Thus the frequency maximum alters as the sample is moistened and dried. This observation is explained by the fact that muscovite and phlogopite mica crystals contain open stratifications which are parallel to the cleavage plane and in contact with the surrounding atmosphere. The surfaces of these stratifications are covered with conducting films of adsorbed water which have been taken up from the air. An alternating electrical field induces charge redistribution in these films, thus giving rise to interlayer dielectric polarization and energy loss.

The observation that the loss factor frequency maximum is displaced toward lower values on drying is explained by the fact that drying reduces the film depth in the stratifications and thus increases the surface resistance. This leads, in turn, to an increase in the time required for the establishment of polarization (relaxation time). The relaxation time is determined from the condition

$$\tau = \frac{1}{2\pi\nu_M}, \qquad (1)$$

ν_M being the frequency at which the loss factor shows a relaxation maximum in the frequency, or temperature relation.

Since drying brought about a 1:30 reduction in the value of ν_M, it must also have brought about a corresponding increase in both τ and the surface resistance (see Fig. 2). This makes it possible to employ the dielectric method for evaluating film depth and conductivity, as well as alterations in these parameters.

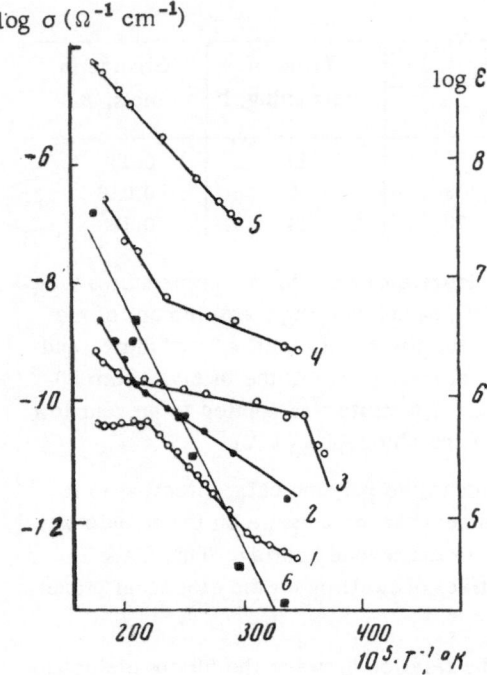

log σ (Ω⁻¹ cm⁻¹) — $\log \sigma \ (\Omega^{-1} \text{cm}^{-1})$

$\log \varepsilon$

$10^{5} \cdot T^{-1} \,^{\circ}K$

Fig. 4. Temperature dependence of the electrical properties of mica. 1) Surface conduction parallel to the cleavage plane; 2, 3, 4) bulk conduction parallel to the cleavage planes, for hard, medium, and soft phlogopites, respectively; 5) surface conduction at right angles to the cleavage plane; 6) dielectric constant in constant field.

Table 1 presents values of ε' and ε'' for phlogopites of various degrees of hardness which had been vacuum dried after moistening at 50% humidity, the measurements being made parallel to the cleavage plane at 50 cps. Reducing the hardness and moistening the sample resulted in an almost 10-fold increase in the values of ε' and ε''.

Figure 3 shows that an increase in the field strength, U, led to an increase in the low frequency values of both ε' and ε'', the effect being quite pronounced in the latter case. This is indication of a breakdown in Ohm's law in film conduction. This has already been shown to be the case with fresh cleavage [4].

An increase in the temperature displaces the ε'' frequency maximum toward higher values by reducing the surface resistance in the stratifications. Direct measurements of the relation between surface conduction and temperature in nonfreshly cleaved mica crystals confirm this conclusion, as can be seen from the data of Fig. 4. The relation between surface conduction and temperature is covered by the equation

$$\sigma = \sigma_0 \cdot e^{-\frac{U}{kT}}, \qquad (2)$$

where σ_0 can, according to [6], be set equal to $n_0 e^2 \delta^2 \nu / 6kT$, δ is the distance between ion bonding points, ν is the vibration frequency, n_0 is the number of ions per unit surface area, and U is the ion activation energy.

The activation energy proves to be approximately $7 \cdot 10^{-13}$ erg, and thus agrees rather well with the height of the potential barrier for water molecule transfer along the surface of the mica crystal [5]. It is worth noting that a high-temperature saturation effect is met in the temperature dependence of the surface conduction. If all of the surface ions are assumed to participate in the saturation conduction, n_0 can be determined from the σ_0 calculated from Eq. (2). Such calculations lead to a value of the order of 10^{15} cm^{-2}, which is in good agreement with the figure representing the maximum number of ions formed on the crystal surface after cleavage.

Table 2 also shows the variation of the dielectric constant, ε', of a sample of soft phlogopite with the humidity, the values in question here having been calculated from the adsorption capacity in a constant field directed along the cleavage surface. It is clear that the presence of open film conducting inclusions leads to

TABLE 2

Crystal state at time of measurement	ε'
After drying in vacuum at 10^{-4} mm Hg for several hours while contacts were being applied	7,400
After moistening for 2.5 hours in air at 50% humidity	85,000
After moistening for 50 hours under variable humidity (50%, desiccator, air)	1,470,000
After drying for 5 hours under evacuation	118,500
After drying for 16 hours under evacuation	26,100
After moistening for 9 hours in air at 50% humidity	6,430,000

TABLE 3

V. D. Kuznetsov hardness, sec	Activation energy, $U \cdot 10^{12}$, ergs		
	From bulk conduction		From surface conduction
	300-200°C	200-20°C	50-250°C
25	1.06	0.256	0.7
105	1.06	0.442	–
100	–	–	0.7

a pronounced increase in the dielectric constant, this increase amounting to several million units in moist samples at high temperatures. The value of ε'_s proves to vary inversely with the moisture content of the sample at temperatures up to 300°C, the two factors being related through an exponential equation. This is indication that the bonding of the ions to the surface is quite strong, the ions failing to dislodge even at 300°C.

Interesting conclusions concerning the properties of adsorbed films can be drawn from the results of studies of continuous bulk conduction along the cleavage plane. It is seen from Fig. 4 that there is a sharp and discontinuous increase in the conductivity in the neighborhood of 0°C. This discontinuity may be related to phase transitions in the film. Processes analogous to melting can obviously take place in the thicker films. Fusion would lead to an increase in film conduction with an accompanying decrease in the current carrier activation energy (Table 3).

Two factors affect the temperature dependence of the conductivity, σ, up to temperatures of the order of 200°C:

1) The conductivity tends to rise with the temperature because of an increase in the number of ions thermally torn from the surface, and

2) The conductivity tends to fall off with an increase in temperature because of water molecule evaporation and film thinning.

The slope of the straight lines applying to the various samples at these temperatures can vary over certain limits because of differences in the relative significance of these factors.

Polymolecular films do not form in open stratifications at temperatures in excess of 200°C, and the curve slope is then essentially determined by the potential barrier for surface ion transfer alone.

The data of Fig. 4 and Table 3 show this energy to be well-defined for mica.

Conclusions

1. Mica crystals contain open stratifications which are in contact with the surrounding atmosphere. Surface conduction in these stratifications leads to an interlayer polarization at sonic and supersonic frequencies.

2. The conduction of the open stratifications increases rapidly with an increase in temperature and adsorption of water. This conduction fixes the value of the dielectric constant, ε', for a field directed along the cleavage plane which is of the order of 12,000 at 50 cps, the companion value of $\varepsilon'\tan\delta$ being 23,000.

3. The surface conduction of the mica crystal rapidly increases as the temperature is raised to 200-250°C; it also increases with the field strength.

4. The activation energy for surface ions in the open stratifications is of the order of $7.0 \cdot 10^{-13}$ ergs, and is therefore close to the value of the potential barrier for transfer over a surface composed of water molecules and hydrated univalent ions.

Literature

1. N. V. Afanas'ev and M. S. Metsik. Izv. Vysshikh Uchebn. Zavedenii, Fiz.,No. 6, 132 (1961).
2. R. A. Zhidikhanov and M. S. Metsik. Izv. Vysshikh Uchebn. Zavedenii, Fiz., No. 3, 164 (1959).
3. R. H. Cole and D. W. Davidson, Chem. Phys. 9:1389 (1952).
4. M. S. Metsik. Izv. Tomsk. Politekhn. Inst. 91:415 (1956).
5. M. S. Metsik. In collection: Research in Surface Forces, Moscow, Izd. Akad. Nauk SSSR (1961), p. 66. [English translation: Consultants Bureau, New York (1963).]
6. G. I. Skanavi. The Physics of Dielectrics, Moscow—Leningrad, Gostekhizdat (1949).

THE EFFECT OF FILM WATER ON THE
DIELECTRIC PROPERTIES OF MICA
Part II
N. V. Afanas'ev and M. S. Metsik

Irkutsk State University
Department of General Physics

The Effect of Water Films in Closed Stratifications on the Electrical Properties of Mica-Phlogopite

The effect of closed stratifications whose surfaces are covered with film water on the dielectric properties of phlogopite can be brought out by calcining the sample at 100-300°C to minimize the effect of open stratifications and then making measurements parallel to the cleavage surface. These closed stratifications give rise to a maximum at about 10 kc on the loss factor, ε", curve at 20°C (Fig. 1 and Table 1), this maximum, and the corresponding frequency fall-off in the dielectric constant, increasing 60-fold as the hardness of the phlogopite is reduced. The ε" frequency maximum experiences a rapid displacement to higher frequencies with increase in temperature when the temperature is low, but no such displacement is observed in heating from 25 to 128°C. Increased surface conduction causes the effect of open stratifications on the values of ε' and ε" at sonic frequencies to become pronounced at these temperatures.

The temperature dependence of ε' and ε" for measurements parallel to the cleavage plane is shown in Fig. 2. The dielectric constant of the phlogopite is approximately 7, and the coefficient loss factor zero, at least to within the limits of experimental error, at temperatures in the interval from −180 to −100°C where the conducting films have no effect at the working frequencies. Increasing the temperature reduces the water film resistance in the closed stratifications, with the result that ε' also increases while each of the ε" curves at 50 and 500 cps, and 5 kc, passes through a maximum, the maxima being displaced toward higher temperatures as the frequency rises. The reduction of surface resistance makes the displacement toward higher frequencies of the dispersion and adsorption regions associated with open stratifications the principal determining factor for these ε' and ε" values above room temperature (see also, Fig. 1).

TABLE 1

V. D. Kuznetsov hardness, sec	Field parallel to cleavage plane				Field perpendicular to cleavage plane		
	$\varepsilon'_{\parallel}$ (static)	$\Delta\varepsilon'_{\parallel}$	$\varepsilon''_{M\parallel}$	$\tau_{\parallel} \cdot 10^5$ (after heating under-pressure) sec	$\varepsilon'_{\perp}$	$\varepsilon''_{M\perp}$	$\tau_{\perp} \cdot 10^5$ (after heating under-pressure) sec
105	14	7	2	3.2	6.0	0.007	1.20
70	45	38	16	1.6	5.9	0.021	0.94
25	447	440	120	1.6	6.0	0.008	1.60

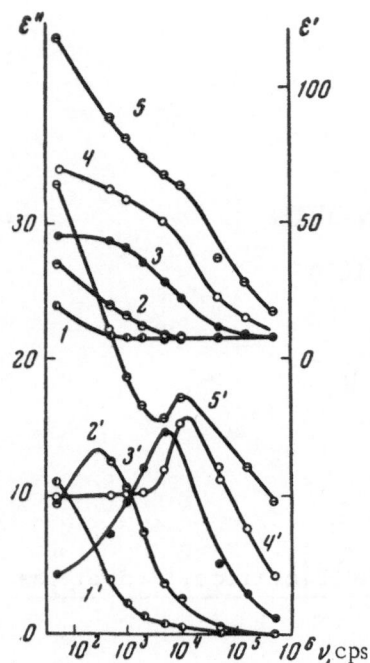

Fig. 1. The frequency dependence of the dielectric constant, ε', and the loss factor, ε'', of calcined phlogopite, at various temperatures: 1, 1') ε' and ε'', respectively, at $-75°C$; 2, 2') the same, at $-52°C$; 3, 3') the same, at $+25°C$; 5, 5') the same, at $+128°C$. Field parallel to cleavage plane; V. D. Kuznetsov hardness of phlogopite, 70 sec.

The temperature dependence of the relaxation time for interlayer polarization shown in curve 6 of Fig. 2 was developed from the ε'' frequency and temperature maxima displacements. A study of this curve shows the relaxation time and the water film resistance in closed stratifications to diminish rapidly with increasing temperature at low temperatures and then remain essentially constant. The curve break for the various samples comes at temperatures in the interval from -20 to $+20°C$.

The frequency and temperature dependencies of ε' and ε'' values measured perpendicular to the cleavage plane at 550 kg/cm^2 are similar to those observed for measurements parallel to this plane, but the values themselves are of considerably lower magnitude. Increase in the thickness of the sample brings about an increase of several factors in the loss factor frequency maximum. This result clearly traces back to an increase in the bulk concentration of open stratifications with increasing thickness of the mica sheet, this leading to field scattering and the appearance of a field component in the cleavage direction, with increased loss.

Special experiments were set up to study the effect of pressure on those dielectric properties perpendicular to the cleavage direction which are affected by stratification. Figure 3 shows the frequency dependence of the capacity and value of $\tan\delta = \varepsilon''/\varepsilon'$ for compressed samples which had been given a previous heat treatment. Increasing the pressure from 1.2 to 250 kg/cm^2 is seen to reduce the value of $\tan\delta$ to about one-fourth of its original value, this being the result of an irreversible compression of the closed stratifications. It should be noted that pressure had little effect on the electrical properties prior to heating, and this despite the fact that the sample obviously contained numerous stratifications about 0.1 mm in diameter. The enhanced effect of pressure is due to irreversible expansion of the closed stratifications which occurred during the first heating although the pressure was high (550 kg/cm^2).

Low pressure heating to temperatures in excess of 100°C, leads to evaporation of water from the closed stratifications and swelling. The crystal expands as a result of these changes, and the interlayer polarization and loss in the closed stratifications increases by a whole order. The time for establishment of polarization does not usually diminish, but increases with the temperature, this being the result of film water vaporization and increase in the surface resistance with increasing stratification diameter. The final result is a rise in the loss factor maximum as the crystal swells, with displacement toward lower frequencies (Fig. 4).

The above remarks give an understanding of the temperature dependence of ε' and ε'' in swelling (Fig. 5). Initiation of swelling and increase in the thickness of the sample (curve 6) is accompanied by a diminution in the high frequency ε' values, this resulting from the fact that the considerable film resistance prevents interlayer polarization in the closed stratification while gaseous inclusions reduce the capacity of the sample. These inclusions are, at first, shunted out by the conducting films, and have no effect on the low frequency capacity, ε' increasing as the thickness of the sample rises. Evaporation eventually reduces the effect of these shunting films, even at low frequencies, and ε' then begins to fall off. The low frequency loss is low at the beginning of swelling, charge displacement in the films occurring over only a small fraction of the half-period. Continued swelling leads to an increase in the film resistance and the time required for the establishment of polarization, and an active film current is then observed over the entire half-period, the loss and the loss factor, ε'', both maximizing at first and then falling off. A reduction in the frequency clearly displaces the ε'' maxi-

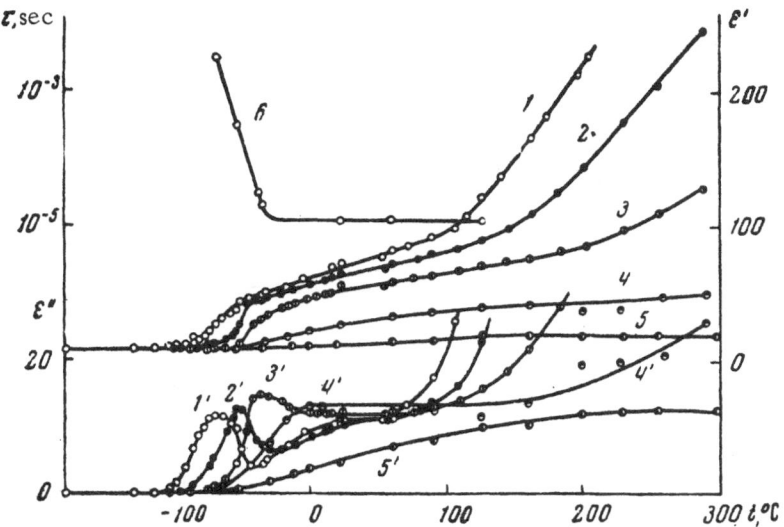

Fig. 2. The temperature dependence of the dielectric constant, ε', and the loss factor, ε", at various frequencies: 1, 1') ε' and ε", respectively, at 50 cps; 2, 2') the same, at 500 cps; 3, 3') the same, at 5 kc; 4, 4') the same, at 50 kc; 5, 5') the same, at 500 kc; 6) relaxation time.

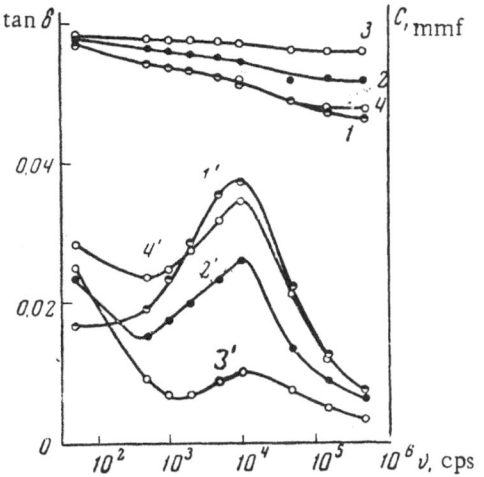

Fig. 3. The frequency dependence of the capacity, C (in microfarads), and the tangent of the loss factor, tan δ, of a phlogopite which had been previously heated to 300°C at 550 kg/cm², measurements being made at room temperature under various pressures. Field perpendicular to cleavage plane, V. D. Kuznetsov hardness, 70 sec; thickness of sample, 0.21 mm. 1, 1') C and tan δ, respectively, at 1.2 kg/cm²; 2, 2') the same, at 20 kg/cm²; 3, 3') the same, at 250 kg/cm²; 4, 4') the same, at 1.2 kg/cm², once more.

mum toward higher temperatures, and thus should lead to higher values of the film resistance. This is in accord with the actual observations (see curves 1 and 2, Fig. 5). The condition for a relaxation maximum $\omega\tau = 1$ (where $\omega = 2\pi\nu$) cannot be fulfilled at higher frequencies where the time required for establishment of polarization is extensive in comparison with the field period, the result being that the ε" temperature maximum is not a relaxation maximum at 50 and 500 kc and is not displaced by an alteration in the frequency. The appearance of this maximum can be explained from Fig. 4 in the following manner. The high frequency values of ε" increase at the beginning of swelling as a result of an increase in the dimensions of the closed stratifications in the direction of the field, which is to say, as a result of an increase in volume of the conducting component of the dielectric. The value of ε" then falls off because of an increase in the film resistance, and the properties are thereby improved.

The diameters of the closed stratifications in the phlogopite crystal, the depth of the aqueous electrolytic films covering the surfaces of these stratifications, and the specific film resistance, can be calculated from experimental data on the dielectric properties and other characteristics. These stratifications resemble compressed ellipsoids of revolution, the axis of revolution, c, being perpendicular to the cleavage plane, or spheroids in which the semiaxes are related by a = b > c. Mean

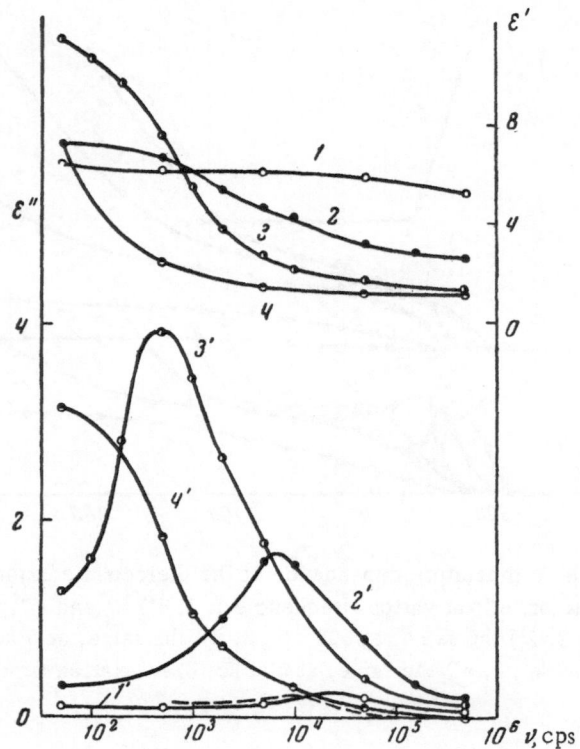

Fig. 4. The frequency dependence of the dielectric constant, ε', and the loss factor, ε'', of phlogopite in a second heating at 1.2 kg/cm², at various temperatures. Field perpendicular to the cleavage plane; V. D. Kuznetsov hardness, 70 sec; thickness of sample prior to swelling, 21 mm. The dotted curve applies to ε'' at 24°C. 1, 1') ε' and ε'', respectively, at 94°C; 2, 2') the same, at 121°C; 3, 3') the same at 149°C; 4, 4') the same, at 194°C.

hardness values for phlogopite crystals show that $2a = 0.015$ cm. Estimation of mean cross sectional diameters for closed stratifications in other crystals are less exact, since the stratifications themselves are more poorly defined, but the same value can also be adopted here as a first approximation. These stratifications contain both film water and gaseous inclusions, and can be looked on as double spheroids, i.e., as gaseous spheroids, each covered with a water layer.

The involved form of the equations makes for difficulties in calculating the complex dielectric constant of phlogopite. The problem is simplified if the major portion of the volume of the stratification can be considered as filled with gas. The stratification can then be treated as a gaseous spheroid covered with a thin film of aqueous electrolyte. The resulting equations can also be extended to the case in which the covering water film is rather thick. For this purpose, it is sufficient to replace the dielectric constant of the gas by the mean dielectric constant of the stratification [8]

$$\varepsilon_{me} = \varepsilon'_{me} - j\frac{2(1-N)}{\varepsilon_0 c \omega \rho_s}, \tag{1}$$

$$\varepsilon'_{me} = \varepsilon'_1\left[1 + \frac{\dfrac{d}{2c}(1-\varepsilon'_1)}{\varepsilon'_1 + \left(1 - \dfrac{d}{2c}\right)N(1-\varepsilon'_1)}\right], \tag{2}$$

184

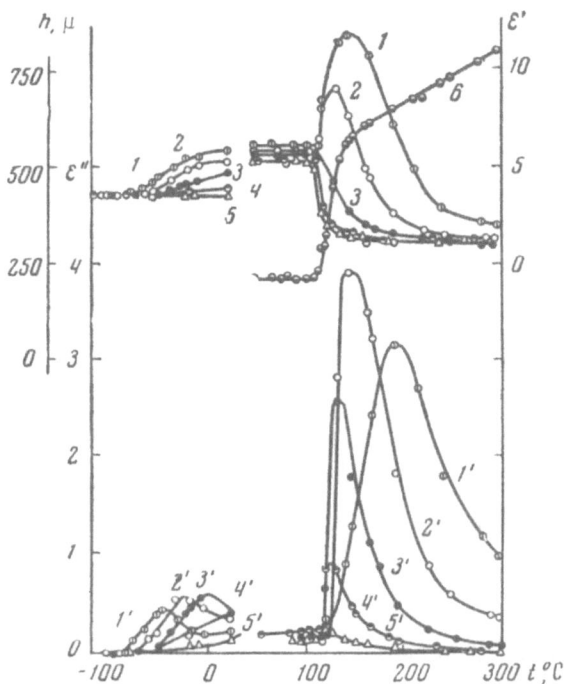

Fig. 5. The temperature dependence of the dielectric constant, ε', and the loss factor, ε'', of phlogopite in a second heating at a pressure of 1.2 kg/cm², at various frequencies. Field perpendicular to the cleavage plane; V. D. Kuznetsov hardness, 70 sec; thickness of sample prior to swelling, 0.21 mm. 1, 1') ε' and ε'', respectively, at 50 cps; 2, 2') the same, at 500 cps; 3, 3') the same, at 5 kc; 4, 4') the same, at 50 kc; 5, 5') the same, at 500 kc; 6) thickness of sample.

$$\frac{d}{2c} = \frac{(\varepsilon'_1 - \varepsilon'_{cp})\,[\varepsilon'_1\,(1-N) + N]}{(\varepsilon'_1 - 1)\,[\varepsilon'_1\,(1-N) + N\varepsilon'_{me}]}. \tag{3}$$

Here N is the depolarization coefficient of the inclusions, ε'_1 is the dielectric constant of the film water, assumed equal to 80 in what follows, ρ_s is the specific film resistance, d is the thickness of the gaseous inclusions between the films, and c is the smaller semiaxis of the ellipsoid.

The Lorenz–Lorentz equation for a double layer spheroid dielectric can be written as [2]

$$\varepsilon = \varepsilon_2\left[1 + \frac{V\,(\varepsilon_{me} - \varepsilon_2)}{\varepsilon_2 + (1-V)\,N\,(\varepsilon_{me} - \varepsilon_2)}\right], \tag{4}$$

V being the bulk concentration of stratifications. Substituting the dielectric constant of the mica $\varepsilon_2 = \varepsilon'_2$, and the expression for ε_{me} from (1), into Eq. (4), one obtains [8]

$$\varepsilon' = \varepsilon_\infty + \frac{\Delta\varepsilon'}{1 + \omega^2\tau^2}, \tag{5}$$

$$\varepsilon'' = \frac{\Delta\varepsilon'\omega\tau}{1 + \omega^2\tau^2}. \tag{6}$$

185

Here $\Delta\varepsilon'$, is given by the expression

$$\Delta\varepsilon' = \frac{V\varepsilon_2'^2}{[\varepsilon_2' + (1-V)\,N\,(\varepsilon_{me}' - \varepsilon_2')]\,(1-V)\,N}, \tag{7}$$

the relaxation time, τ, for interlayer polarization by

$$\tau = \frac{[(1-V)\,N\,(\varepsilon_{me}' - \varepsilon_2') + \varepsilon_2']\,\varepsilon_0\rho_s c}{2\,(1-V)\,N\,(1-N)}, \tag{8}$$

and the dielectric constant, ε_∞, for the crystal in the high frequency field by

$$\varepsilon_\infty = \varepsilon_2'\left[1 + \frac{V\,(\varepsilon_{me}' - \varepsilon_2')}{\varepsilon_2' + (1-V\ N\,(\varepsilon_{me}' - \varepsilon_2')}\right], \tag{9}$$

V being the relative volume of the inclusions in the mica, and $\varepsilon_0 = 8.85 \cdot 10^{-14}\ f/cm$.

Further development of the calculations requires an estimate of the spheroid depolarization coefficient which appears in the above equations.

Since the a and b semiaxes of the spheroids are equal, the depolarization coefficient in a field parallel to the cleavage plane is [2] given by

$$N_\parallel = \frac{a^2 c}{2}\int_0^\infty \frac{dS}{(S+a^2)^2\,(S+c^2)^{1/2}} = \frac{c}{2}\left(\frac{\pi}{2a} - \frac{c}{a^2} - \frac{1}{a}\,\text{arc tg}\,\frac{c}{a}\right).$$

When the second and third members of this expression (which make up 20% of the total value at $c/a = 0.1$, and 2% at $c/a = 0.01$) are neglected, one obtains

$$N_\parallel \approx 0.8\,\frac{c}{a}. \tag{10}$$

The depolarization coefficient, $N_\perp$, perpendicular to the cleavage plane can be expressed in terms of $N_\parallel$, through the equation [2]

$$N_\perp = 1 - 2N_\parallel = 1 - 1.6\,\frac{c}{a}. \tag{11}$$

The cross-sectional dimensions of the stratifications being much less than the linear ($c \ll a$), it follows from Eqs. (10) and (11) that $N_\parallel$ is small and $N_\perp$ essentially equal to unity.

We will now determine the mean value of the stratification dielectric constant perpendicular to the cleavage plane, our aim being to use this quantity in calculating $N_\parallel$ and V. Writing Eq. (8) for the parallel and perpendicular directions, dividing the resulting expressions, and neglecting low-magnitude terms, one obtains

$$\frac{\tau_\parallel}{\tau_\perp} = \frac{2\varepsilon_{2\parallel}'}{\varepsilon_{me\,\perp}'}. \tag{12}$$

At room temperature, the mean ε'' frequency maximum in phlogopite of average hardness is at about 10 kc for the parallel direction and 17 kc for the perpendicular direction, and $\varepsilon_{2\parallel}' = 7$, so that Eq. (12) gives $\varepsilon_{me\,\perp}' = 8.2$. The dielectric constant of water, free of relaxation polarization, is approximately 2-3 (see [2]). The fact that the value of $\varepsilon_{me\,\perp}'$ is considerably larger than this, is indication of relaxation polarization in the film water molecules. Knowing the value of $\varepsilon_{me\perp}$, the depolarization coefficient, $N_{\parallel}$, for the inclusions can be readily obtained from the dielectric properties of the crystal. Setting $\omega\tau = 1$ in Eq. (6), the expression for the loss coefficient, ε'', frequency maximum is obtained in the form

$$\varepsilon_{\text{M}}'' = \frac{\Delta\varepsilon'}{2}.$$

Taking account of (7), and of the fact that the values of V and $N_{\parallel}$ are low, one has

$$N_{\parallel} \approx \frac{\varepsilon_{\text{M}\perp}'' \varepsilon_{2\parallel}' \varepsilon_{\perp}'}{\varepsilon_{\text{M}\parallel}'' \varepsilon_{2\perp}'^{2}}. \tag{13}$$

The cross-sectional width (2c) of the stratifications can be obtained from $N_{\parallel}$ through Eq. (10). We will now determine V, the volume concentration of the stratifications. Setting up Eq. (7) for the parallel direction, and neglecting the V in comparison with unity in the numerator, one has

$$U = \frac{\Delta\varepsilon_{\parallel}' N_{\parallel}}{\varepsilon_{2\parallel}'}; \tag{14}$$

the number of laminations per 1 cm³ is then

$$n_0 \approx \frac{V}{\pi a^2 c}. \tag{15}$$

The specific film surface resistance in the closed stratifications can then be obtained from (8). Writing (8) for the parallel direction, and taking account of the fact that the values of V and $N_{\parallel}$ are low, one obtains

$$\rho_s \approx \frac{2N_{\parallel}\tau_{\parallel}}{\varepsilon_0 \varepsilon_{2\parallel}' c}, \tag{16}$$

the specific bulk resistance of the electrolyte film is

$$\rho = \rho_s h, \tag{17}$$

the film depth, h, being equal to $c - d/2$.

We have substituted the experimental data of Table 2 in the above equations to calculate the values of the principal characteristics of those lens-like conducting inclusions in the phlogopite crystal which contain film water. The results of these calculations for crystals of medium hardness are shown in Table 2. The values presented indicate that the stratification length in these phlogopite crystals was some 400 times greater than the width; the stratification volume represented about 1% of the whole and their water about 0.3% of the entire water mass. The concentration, n_0, was close to $3 \cdot 10^6$ cm⁻³. The water film depth was several hundred times greater than the depth of the monomolecular water layer. The cross-sectional dimensions of the closed stratifi-

TABLE 2

V.D. Kuznetsov hardness, sec.	$\varepsilon_{me\perp}'$	$N_{\parallel} \cdot 10^3$	$N_{\perp}$	$2c \cdot 10^5$, cm	$\dfrac{d}{2c}$	$d \cdot 10^5$, cm	$h \cdot 10^5$, cm	V, %	$n_0 \cdot 10^{-6}$, cm⁻³	$\rho_s \cdot 10^{-10}$, ohm	$\rho \cdot 10^{-4}$, ohm · cm
105	5.4	3.6	0.993	6.8	0.25	17	3.1	0.36	0.6	1.7	7
70	8.2	2.1	0.996	4	0.14	5.6	1.9	1.1	3.2	0.5	5.4
25	14.0	0.18	0.999	0.35	0.06	0.21	0.17	1.2	39	0.4	0.47

cations tended to decrease as the hardness of the phlogopite diminished, and the same is true of the depth and resistance of the water films; on the other hand, the volume concentration of stratifications, and the film water content tended to increase under the same conditions. These data furnish essentially new information on the properties of water inclusions and films in phlogopite crystals [3, 6].

Modern concepts [4, 5] indicate that the hydration of phlogopite involves replacement of potassium ions by oxonium ions; on this basis, the film should contain K^+ and OH^- ions, and these ions, in turn, fix the film conductivity. If all of the potassium ions in the stratification are assumed to be displaced and the ion mobility in the film is considered to be the same as in bulk water, the calculated specific resistance would be about 40 ohm $\cdot$ cm, i.e., three orders less than the $5.4 \cdot 10^4$ ohm $\cdot$ cm obtained experimentally. The same result is obtained in surface conduction measurements on electrolyte films deposited on a mica surface [6]. This effect clearly reflects an ion—crystal surface interaction.

The following points can be made in connection with the break in the $\tau = f(T)$ curve (see curve 6, Fig. 2).

When account is taken of possible experimental errors, it is seen that this break occurs in the neighborhood of 0°C, which is to say, near the freezing point of ordinary water. The water films are of finite depth, and it is therefore possible that a phase transition with structural alteration takes place in the neighborhood of this point. If such is the case, differences in the temperature coefficient of τ, and the film resistance, at positive and negative temperatures may trace back to differences in film structure. It is obvious that vaporization of the film could be significant above 0°C, since this would reduce the film depth, and thereby diminish the conductivity [6]. This diminution could partially, or completely, compensate for the increase in film conductivity with temperature which is commonly associated with electrolytes. From this it follows that the conductivity of the film water would be relatively insensitive to alterations in temperature, or possibly entirely independent of the temperature.

The activation energy for the conduction determining ions in the water film can be obtained from the $\ln \tau = f(1/T)$ or $\ln \rho = f(1/T)$ curves. The activation energies for the various phlogopites fall in the interval from $0.7 \cdot 10^{-12}$ to $1.0 \cdot 10^{-12}$ erg at lower temperatures and are therefore close to the activation energies for the transfer of water molecules and univalent ions over the mica crystal surface [7].

Conclusions

1. The phlogopite crystal contains two types of stratifications, the open which are in contact with the atmosphere, and the closed. Surface conduction in these stratifications leads to interlayer polarization, the open stratifications determining the range of dispersion and adsorption at supersonic frequencies and the closed the less interesting range of dispersion and adsorption at sonic and radio frequencies. The concentration of closed stratifications varies with the hardness and falls in the 10^5-10^7 cm^{-3} range.

2. The depth of the electrolytic aqueous film in the closed stratifications is several hundred times greater than the depth of the monomolecular layer; its specific conduction at room temperature is 10^{-4}-10^{-5} ohm^{-1} $\cdot$ cm^{-1}. Over the temperature interval from -100 to 0 ± 20°C and with various phlogopite samples, the specific conduction of the aqueous films in the closed stratifications was found to increase rapidly with heating and become essentially constant at higher temperatures.

3. The activation energy for surface ions in the closed stratifications is $(7\text{-}10) \cdot 10^{-13}$ erg.

4. The contribution to the dielectric constant, ε', of the phlogopite crystal arising from the presence of the closed stratification can be as large as 450 in the direction of the cleavage plane, at sonic frequencies while ε'' (loss factor) is equal to 120.

Literature

1. A. V. Lykov. Transport Phenomena in Capillary Porous Bodies, Moscow, Gostekhizdat (1954).
2. A. V. Netushil et al.. High Frequency Heating of Dielectric and Semiconductors, Moscow—Leningrad, Gosenergoizdat (1959).
3. M. S. Metsik. Izv. Tomsk. Politekhn. Inst. 91:415 (1956).

4.   V. S. Sobolev, Introduction to Silicate Mineralogy, L'vov, Izd. Gos. Univ. (1949).
5.   G. Brown and K. Norrish, "Hydrous Micas," Mineral. Mag. 29(218):929 (1952).
6.   N. M. Chirkov, Zh. Fiz. Khim. 21:1303 (1947).
7.   M. S. Metsik, "New Methods for Studying Surface Phenomena," Tr. Inst. Fiz. Khim., Akad. Nauk SSSR (1951), p. 66.
8.   N. V. Afanas'ev, Dissertation, Irkutsk. Gos. Univ. (1962).

SECTION IV

SURFACE PHENOMENA IN DISPERSED AND POROUS SYSTEMS

CAPILLARY HYSTERESIS IN THE RISE OF WETTING LIQUIDS IN SINGLE CAPILLARIES AND POROUS BODIES

M. M. Kusakov and D. N. Nekrasov

Institute for Petroleum Chemistry Syntheses
Acad. Sci., USSR

The quantitative aspects of capillary hysteresis have been only inadequately studied, although the phenomenon itself has been known for a long time [1-5]. Even a strict definition of the effect has been lacking up until recently. Hysteresis in capillary rise can be defined as the existence of quasi-equilibrium heights of rise corresponding to metastable equilibrium in the liquid. It should be emphasized that capillary hysteresis is frequently observed in conjunction with wetting hysteresis, an effect which traces back either to purely mechanical factors, or to the specific characteristics of the liquid and the solid surface. These two effects can be distinguished only with difficulty, and our discussion will, unless mention is made to the contrary, be limited to capillary hysteresis in wetting liquids which are entirely free of wetting hysteresis.

It will be proven in the present communication that capillary hysteresis leads to several different values for the height of rise in individual capillaries showing a variation of cross-sectional area with height. The number of such values of the height of capillary rise is determined by the capillary geometry and the physical properties of tne liquid. Capillary hysteresis does not appear when the cross-sectional area of the capillary is constant. This paper will also treat capillary rise of wetting liquids in porous bodies where hysteresis can be shown to lead to an infinite number of values of height of rise falling between a certain maximum and a certain minimum.

We will first consider hysteresis in the capillary rise of a wetting liquid in a single capillary of simple form.

This problem can be treated by three different and independent methods, the static, the kinetic, and the energetic.

The static method of determining the height of capillary rise [6] involves the solution of a system of two equations, the one being the usual expression for the equality of the forces of gravity and surface tension acting on the liquid, and the other a functional relation characterizing the capillary geometry.

This system of equations has the form

$$\left.\begin{aligned} \frac{2\sigma}{r} &= h\rho g, \\ r &= f(h), \end{aligned}\right\} \tag{1}$$

for a wetting liquid in a capillary of circular cross section, where σ is the surface tension at the air interface (more exactly, at the saturated vapor interface), ρ the density, and h the height of rise of the liquid, the latter measured from the meniscus to the plane, r is the capillary radius (r will also be taken, for simplicity, to be the radius of curvature of the meniscus assumed spherical in form), and g is the acceleration of gravity.

The height of liquid rise in capillaries of various forms can be found by analytical or graphic solution of this system of equations.

We note certain simple, but important, special cases

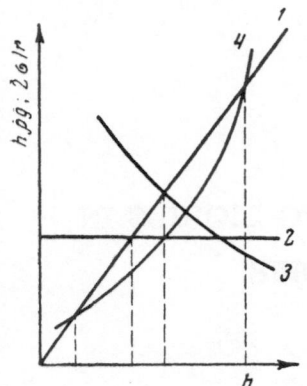

Fig. 1. Schematic representation of the variation of the hydrostatic (1), and capillary (2-4), pressure with the height, in cylindrical (2), diverging conical (3), and converging conical (4) capillaries.

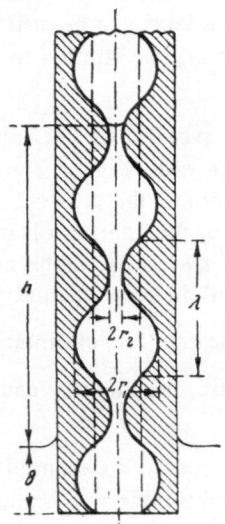

Fig. 2. Schematic representation of the sinusoidal capillary profile.

It can be readily shown that there is a single equilibrium height of capillary rise in cylindrical capillaries and in conical diverging capillaries [7]. Converging capillaries show two different values of the equilibrium height of capillary rise (h_1 and h_2), the smaller of the two (h_1) being the more stable. These conclusions are illustrated schematically in the graph of Fig. 1; here line 1 corresponds to the alteration of the hydrostatic pressure, $h\rho g$, with the height, h, while curves 2-4 correspond to the alteration of the capillary pressure, $2\sigma/r$, 2 being for cylindrical, 3 for conical diverging, and 4 for conical converging, capillaries. It will be shown later that the problem of stability can be solved on the basis of quite simple considerations.

It is interesting to note that any height will be an equilibrium height in a capillary satisfying the relation $r = 2\sigma/h\rho g$, i.e., in a capillary whose inner surface is generated by rotating the hyperbola $rh = 2\sigma/\rho g$ around its vertical axis.

It is clear that the frequently observed lack of reproducibility in capillary rise measurements traces back to this fact. Cylindrical capillaries often show alterations in cross section, and the height of rise will not be reproducible if the form of the capillary and the values of σ and ρ for the liquid are such that the relation $rh = 2\sigma/\rho g$ is approximately satisfied.

We will briefly consider the case of the sinusoidal capillary where the system of equations has the form

$$\left.\begin{array}{r} \dfrac{2\sigma}{r} = h\rho g, \\[2mm] r = \alpha + \beta \sin \gamma\,(h + \delta), \end{array}\right\} \qquad (2)$$

where

$$\alpha = \frac{1}{2}\,(r_1 + r_2), \quad \beta = \frac{1}{2}\,(r_1 - r_2), \text{ and } \gamma = \frac{2\pi}{\lambda}.$$

The significance of the remaining terms is clear from Fig. 2. This system must be solved by graphical means.

The stability problem for the height of rise can be solved on the basis of the following considerations [8]. Figure 3 makes it clear that the stable values of the height of capillary rise correspond to the abscissas of the points of intersection of the straight line $h\rho g = f(h)$ and the descending branches of the $2\sigma/r = f(h)$ curve. This conclusion follows from the fact that the meniscus will tend to pass over to the height corresponding to such an intersection point, the capillary pressure tending to pull it up from lower levels (arrow a), and the hydrostatic pressure tending to drive it down from higher (arrow b). Similar considerations lead to the conclusion that the heights of capillary rise corresponding to the abscissas of the points of intersection of the line $h\rho g = f(h)$ with an ascending branch of the curve $2\sigma/r = f(h)$ will correspond to unstable equilibria (arrows c and d).

This same line of reasoning can be applied to capillaries of arbitrary form, and is therefore applicable to the diverging conical capillary (see Fig. 1).

Decision concerning the number of equilibrium heights for a capillary of arbitrary form can be reached through a study of graphs of the functions $h\rho g = f(h)$ and $2\sigma/r = f(h)$.

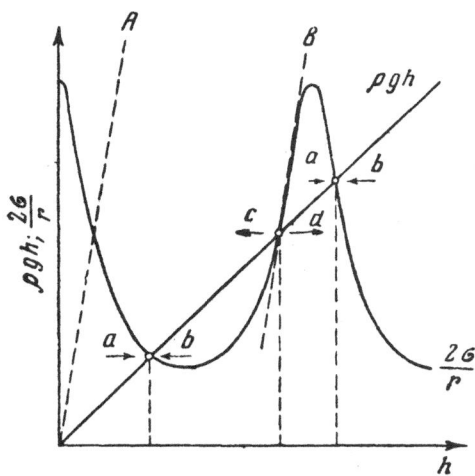

Fig. 3. Schematic representation of the functions $\rho gh = f(h)$ and $2\sigma/r = f(h)$ for the sinusoidal capillary [graphical solution of the system of equations (2)].

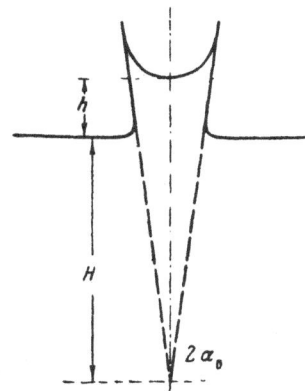

Fig. 4. Capillary rise in a diverging conical capillary.

It follows from Fig. 3 that there will be only one stable height of capillary rise when

$$\frac{d\,(\rho gh)}{dh} \geqslant \frac{d}{dh}\left[\frac{2\sigma}{\alpha + \beta \sin \gamma\,(h + \delta)}\right]$$

and the right hand member of the inequality is at a maximum. When this condition is fulfilled, the tangent to the $2\sigma/r = f(h)$ curve passes through the point of inflection on the ascending branch (see the dotted lines, A and B, of Fig. 3).

These calculations lead to the following inequality as the condition for a single equilibrium height independent of the position of the meniscus in the capillary (i.e., independent of the depth of capillary immersion):

$$\rho g \geqslant \frac{8\pi\sigma\,(r_1 - r_2)}{\lambda\,(r_1 + r_2)}.$$

It can be shown that there can be either one or several heights, depending on the depth of immersion, δ, and the capillary dimensions when

$$\rho g < \frac{8\pi\sigma\,(r_1 - r_2)}{\lambda\,(r_1 + r_2)}.$$

The height will be stable in the first of these cases. In the second case, the first of the several heights can be either stable or unstable. If this first height is stable, all of the odd heights (h_1, h_3, h_5, ...) will correspond to stable equilibria, and all of the even (h_2, h_4, h_6, ...) to unstable equilibra. The reverse situation holds when the first height is unstable, all the odd heights being then unstable and all of the even stable.

The principle of the kinetic method is that of setting up a simplified solution of the problem of the movement of the viscous liquid in the capillary tube (of exactly described form) under the action of the capillary and hydrostatic pressure difference, Δp, and then applying the equation of continuity to obtain a functional relation between the height, h, and the time, τ, of capillary rise.

The equilibrium height of capillary rise will then be that limiting value at which

$$\frac{dh}{d\tau} = 0. \tag{3}$$

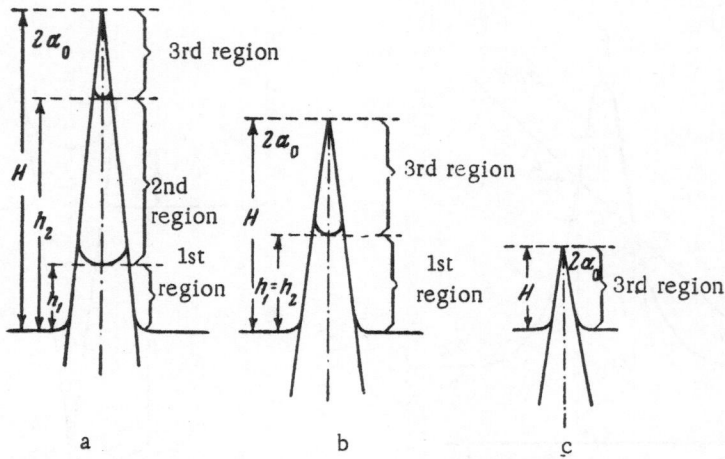

Fig. 5. Capillary rise in a converging conical capillary.
a) $H^2 > 4A$; b) $H^2 = 4A$; c) $H^2 < 4A$.

The following simplifying assumptions can be introduced in applying this method:

1) The rise of the viscous liquid in the narrow capillary is slow and the nonstationary process in question adequately described by steady state equations;

2) The capillary of the tube is narrow and the radius of the meniscus of the wetting liquid considerably less than the height of capillary rise, so that the height of rise can be measured from the lowest point on what is assumed to be a meniscus of spherical form;

3) The liquid wets the capillary walls and is incompressible;

4) The liquid movement is along an axis of symmetry.

V. V. Lebedev and M. M. Kusakov [7] have used this method to obtain a complete solution to the problem of the rise of a viscous liquid in converging and diverging conical capillaries of circular cross section. The calculations of [7] show a single equilibrium height for the diverging conical capillary (Fig. 4); this is given by the expression

$$h = -\frac{H}{2} + \frac{\sqrt{H^2 + 4A}}{2} , \qquad (4)$$

where

$$A = \frac{2\sigma \cos^2 \alpha_0}{\rho g \sin \alpha_0} ;$$

and the significance of the remaining symbols is clear from the figure. The converging capillary has two equilibrium heights given by

$$h_1 = \frac{H}{2} - \frac{\sqrt{H^2 - 4A}}{2} , \quad h_2 = \frac{H}{2} + \frac{\sqrt{H^2 - 4A}}{2} . \qquad (5)$$

Here $h_1 + h_2 = H$. It can be shown that there are three regions for liquid movement in the converging conical capillary under the action of the pressure difference, Δp, referred to above. These regions are indicated in Fig. 5.

The liquid will rise or fall to the level h_1 if the initial height is such that $h_0 < h_1$ or $h_1 < h_0 < h_2$; if $h_0 > h_2$, the liquid will rise until it fills the entire capillary ($h = H$). Both of the values of the equilibrium height can be obtained from the simple relation

$$\Delta p = \frac{2\sigma \cos^2 \alpha_0}{(H - h) \sin \alpha_0} - h\rho g = 0. \qquad (6)$$

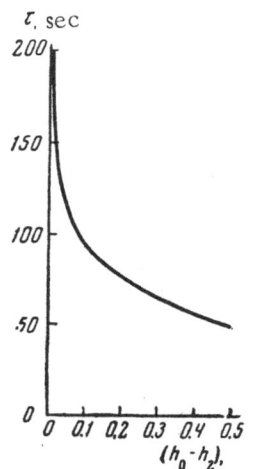

Fig. 6. The relation between the time of rise, τ, and the difference $h_0 - h_2$ in a converging conical capillary $H^2 > 4A$.

It might seem that the stable equilibrium would correspond to h_1, the lower of the two values since the rise of the liquid from h_1 to h_2 would require the expenditure of work by external forces. A more careful analysis of equilibria stability on the basis of energy considerations such as given below indicates, however, that maximum stability is associated with a rise to height $h = H$, which is to say, with complete filling of the closed tube and rise to the top of the open tube. It should be noted that the time required for rise to the end of the capillary is finite and strongly dependent on the difference $h_0 - h_2$, when $h_0 > h_2$ (Fig. 6).

Especial interest attaches to the case in which the geometrical dimensions of the capillary (H, α_0) and the properties of the liquid (σ, ρ) are such that

$$H^2 < 4A \quad \text{or} \quad H^2 < \frac{8\sigma \cos^2 \alpha_0}{\rho g \sin \alpha_0}.$$

The first and second regions are lacking here, and there is no equilibrium value for the height of capillary rise; the time required for rise to the height $h = H$ is finite and the liquid rises to the end of the capillary regardless of whether the latter is open or closed. The rise of soil moisture to considerable heights in plants is explained by the conical form of the capillary pores, the angle $2\alpha_0$ diminishing as the plant grows and becomes taller.

The more general energetic method treats the condition (minimum potential energy of the force of gravity and the surface tension) for liquid equilibrium in the capillary. The height of liquid rise, h, in a capillary of circular cross section can be obtained from the condition

$$\frac{\partial U}{\partial h} = 0. \tag{7}$$

Here U, the potential energy of the wetting liquid in the capillary, is given by

$$U = \pi \rho g \int_0^h r^2 h \, dh - 2\pi\sigma \int_0^h r \, dh, \tag{8}$$

the first term in the right-hand member representing the potential energy due to the force of gravity, and the second the potential energy due to surface tension. The capillary surface is once more determined as a surface of rotation around the capillary axis through the function $r = f(h)$.

For the cylindrical capillary ($r = r_0$)

$$U = \frac{1}{2} \pi r_0^2 \rho g h^2 - 2\pi r_0 \sigma h, \tag{9}$$

for the conical capillary [$r = (H - h) \sin \alpha_0$]

$$U = \pi \rho g \left(\frac{H^2 h^2}{2} - \frac{2Hh^3}{3} + \frac{h^4}{4} \right) \sin^2 \alpha_0 - 2\pi\sigma \left(Hh - \frac{h^2}{2} \right) \sin \alpha_0, \tag{10}$$

for the sinusoidal capillary [$r = \alpha + \beta \sin \gamma (h + \delta)$]

$$U = \pi \rho g \left\{ \frac{1}{4} h^2 (2\alpha^2 + \beta^2) - \frac{1}{4} \beta h \left[\frac{8\alpha}{\gamma} \cos \gamma (h + \delta) + \frac{\beta}{\gamma} \sin 2\gamma (h + \delta) \right] + \right.$$

$$+ \frac{4\beta}{\gamma^2}\left[\alpha \sin \frac{\gamma h}{2}\cos \gamma\left(\frac{h}{2}+\delta\right)+\frac{\beta}{16}\sin \gamma h \sin \gamma(h+2\delta)\right]\Big\}-$$

$$-2\pi\sigma\left[\alpha h + \frac{2\beta}{\gamma}\sin \frac{\gamma h}{2}\sin \gamma\left(\frac{h}{2}+\delta\right)\right]. \tag{11}$$

The value of $U = f(h)$ can be calculated for capillaries of still other forms if the function $r = f(h)$ is known.

The positive and negative roots, $h_1, h_2, \ldots, h_i, \ldots$ of Eq. (7) correspond to heights of stable equilibria when $(\partial^2 U/\partial h^2)_{h=h_i} > 0$, and to heights of unstable equilibria when $(\partial^2 U/\partial h^2)_{h=h_i} < 0$. Liquid rise in the capillary is over the interval of h values from 0 to h_1, or from h_i to h_{i+1}, if $\partial U/\partial h < 0$ in this interval; capillary rise can be realized over intervals in which $\partial U/\partial h > 0$ only through the expenditure of external work.

Thus solution of Eq. (10) for the converging conical capillary leads to three roots, and not two, as earlier. In fact, this solution leads not only to the two roots h_1 and h_2 which were obtained from (5), but gives a third root as well, namely $h_3 = H$. Calculation shows that there will be three roots (h_1, h_2, h_3) when $H^2 > 4A$, two roots ($h_1 = h_2$, h_3) when $H^2 = 4A$, and one root (h_3) when $H^2 < 4A$. It follows from (10) that

$$\left(\frac{\partial^2 U}{dh^2}\right)_{h=h_1} > 0, \quad \left(\frac{\partial^2 U}{dh^2}\right)_{h=h_2} < 0, \quad \left(\frac{\partial^2 U}{\partial h^2}\right)_{h=h_3} > 0.$$

Thus h_1 and h_3 correspond to stable equilibria, while h_2 is for an unstable equilibrium. Since $U_{h=h_1} > U_{h=h_3}$, the most stable equilibrium is that in which the liquid completely fills the converging conical capillary. The previous remarks are illustrated by the rise of water in a capillary $U = f(h)$ in which $H = 32$ cm and $2\alpha_0 = 4'$ (Fig. 7). Here the zero of potential energy has been located at $h = h_3 = H$.

These conclusions were checked by experiments on the rise of water and liquid hydrocarbons in convergent conical capillaries and in specially constructed periodically converging and diverging glass capillaries where the surface approximated sinusoidal form. The graph of Fig. 8 illustrates the results obtained by comparing direct

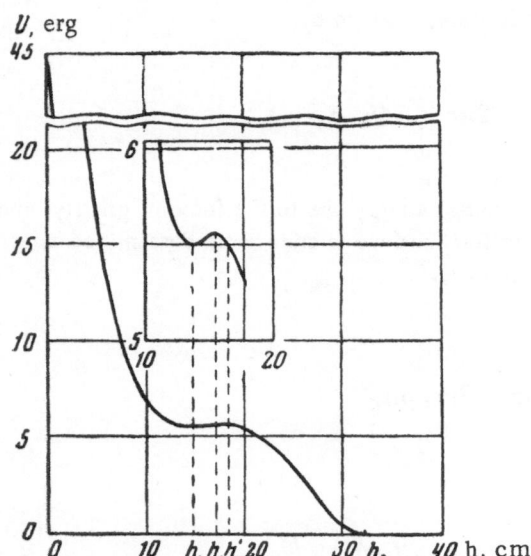

Fig. 7. The relation between the potential energy, U, and the height of rise, h, in a converging conical capillary. (H = 32 cm, $2\alpha_0 = 4'$).

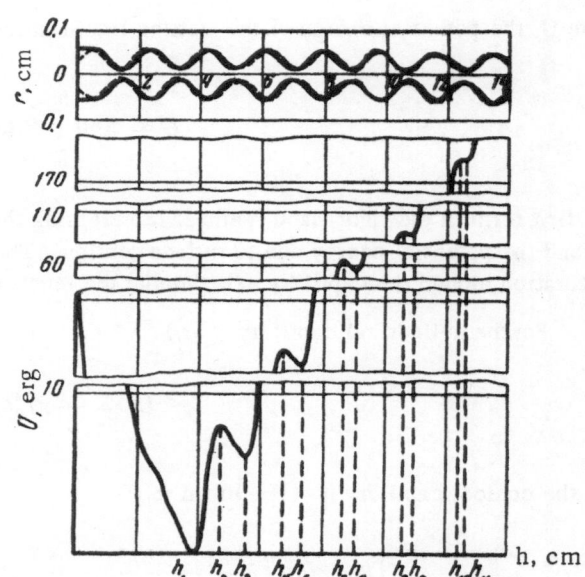

Fig. 8. The $r = f(h)$ and $U = f(h)$ relations for the rise of water in a capillary. Capillary characteristics (mean values): $r_1 = 0.0560$ cm; $r_2 = 0.0107$ cm; $\lambda = 1.8600$ cm; $\alpha = 0.0334$ cm; $\beta = 0.0227$ cm; $\gamma = 3.3763$ cm^{-1}.

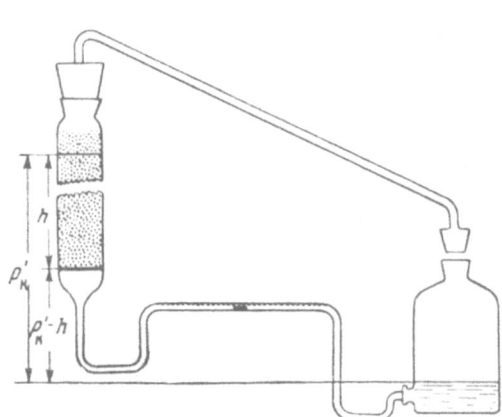

Fig. 9. Schematic representation of the apparatus for measuring liquid rise capillary hysteresis in a porous body.

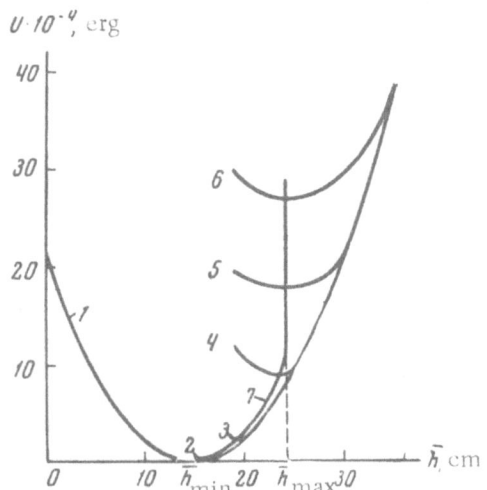

Fig. 10. The relation between potential energy of the forces of gravity and capillary attraction acting on the liquid in the porous body, and the mean height of capillary rise.

measurements on the capillary profile (full curve) with calculations based on Eq. (11) and data characterizing the capillary (dotted curve). The data referred to were obtained by direct microscopic measurement of the internal diameter of the capillary, the capillary being sectioned and ground at the end of the capillary rise measurements. Figure 8 also includes a curve showing the $U = f(h)$ function which was developed for the capillary through the aid of Eq. (11). All of the measurements in question here were carried out at zero immersion, $\delta = 0$. Six heights corresponding to stable equilibria were marked out for this capillary. The most stable of all these equilibria is the one associated with h_1. The table below presents values of the equilibrium height of capillary rise of water, as experimentally observed and as calculated from the $U = f(h)$ graphs:

Height, cm	h_1	h_2	h_3	h_4	h_5	h_6	h_7	h_8	h_9	h_{10}	h_{11}
Obtained from the graph	3.77	4.67	5.48	6.65	7.26	8.62	9.04	10.57	10.87	12.45	12.63
Observed experimentally	3.81	—	5.42	—	7.18	—	9.03	—	10.79	—	12.67

The agreement between values of stable height and capillary rise experimentally observed and calculated from the graph is good. Equally good agreement was obtained in other experiments involving variation of the immersion depth, and in all other cases in which experimentally observed values were compared with values calculated from the graph.

The problem of capillary rise cannot be solved analytically in the case of the porous body where an infinite number of complex capillaries of various forms and cross sections are in communication with one another. Reasoning in terms of an equivalent vertically oriented circular capillary in which the cross-sectional area varies with the height, and applying the energetic method as outlined above to data on the rise and fall of liquids in actual porous bodies, it can be shown that capillary hysteresis leads to an infinite number of values of the height of capillary rise, all included between two limiting values, a maximum and a minimum [9].

The system of equations (1), for the equivalent capillary has an infinite number of roots $(h_1)_{min} < h_2 < \ldots < (h_\infty)_{max}$. By analogy with (8), the potential energy of the column of wetting liquid which has risen in the equivalent capillary to the same height, h, as in the porous body is expressed by

$$U = \rho g \int_0^V (h - p_{\kappa}') \, dV, \tag{12}$$

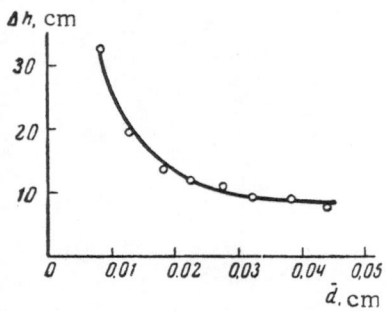

Fig. 11. The relation between the capillary hysteresis, Δh, in water and the mean quartz grain diameter, d.

p'_K being the capillary pressure in the same units as the height of liquid column (i.e., $p'_K = p_K/\rho g$, where p_K is the capillary pressure in pressure units), and V the volume of liquid in the porous body or equivalent capillary. Since $h - p'_K$ and V can be measured experimentally, the relation $V = f(h - p'_K)$, or $h - p'_K = f(V)$, can be developed and the potential energy, U, of the liquid column calculated for each value of the mean height $\bar{h} = V/s$, s being the transmission of the porous body, a quantity which can be considered as independent of the height. A graph of the $U = f(\bar{h})$ relation can be constructed in this way.

The transmission, s, can be set equal to the volume of liquid per 1 cm length of a column filled with quartz powder, measurement being made over that portion of the column in which this volume is independent of the height. The equilibrium height of rise can be obtained from this graph and the condition $\partial U/\partial h = 0$.

This method was experimentally checked by measurements on the rise of water in porous bodies composed of hydrophilic quartz powders with grain diameters in the 0.075-0.45 mm range. The apparatus is represented schematically in Fig. 9; movement of the air bubble in the graduated capillary tube made it possible to observe the rise and fall of liquid in the capillary and permitted accurate determination of the alteration in the volume of liquid taken up at various values of $p'_K - h$.

Figure 10 shows the $U = f(h)$ graph for a hydrophilic quartz powder with grain diameters in the 0.35-0.42 mm range. Curve 1 corresponds to the capillary rise. For convenience, the state corresponding to $U = 0$ has been arbitrarily set at minimum mean height of equilibrium rise of water, $(\bar{h}_1)_{min} = 15.42$ cm, in the porous body. The illustrative curves, 2-6, (an infinite number of such curves could be developed) correspond to water flow out of the porous body from heights such that $h > (\bar{h}_1)_{min}$. Liquid rise to heights greater than $(\bar{h}_1)_{min}$ is brought about by correctly altering the sign and magnitude of the difference $h - p'_K$.

It is seen from Fig. 10 that each of the curves 2-6 has a minimum corresponding to an equilibrium height of capillary retention of liquid introduced into the porous body. The geometrical locus of the points of minimum potential energy forms curve 7 which passes over to the line $h = (\bar{h}_\infty)_{max}$, where $(\bar{h}_\infty)_{max}$ is the maximum value of mean capillary rise in the porous body. Liquid rise to $h > (\bar{h}_\infty)_{max}$ with subsequent release, leads to retention at $h = (\bar{h}_\infty)_{max}$. The case of Fig. 10 is that in which $(\bar{h}_\infty)_{max} = 24.48$ cm. All other heights of capillary rise of water in the porous body correspond to the condition $(15.42 < h < 24.28)$ cm. The most stable minimum height $(\bar{h}_1)_{min}$ corresponds to the lowest value of the potential energy, U, of a liquid column drawn up into the porous body by capillary attraction.

From these remarks concerning the conditions for liquid equilibrium, and the experimental work, it is seen that capillary hysteresis of wetting liquids in porous bodies leads to an infinite number of values of the equilibrium height of capillary rise, all lying between a maximum and a minimum.

Construction of $U = f(\bar{h})$ curves for quartz powders composed of particles of various diameters makes it possible to develop a relation between the capillary hysteresis, $\Delta\bar{h} = \bar{h}_{max} - \bar{h}_{min}$, and the mean particle diameter. A curve covering this relation is shown in Fig. 11. When $\bar{h}_{min}$, or $\bar{h}_{max}$, is used as argument in place of the mean particle diameter, the hysteresis curve, $\Delta h = f(\bar{h}_{min})$ or $\Delta h = f(\bar{h}_{max})$, takes the form of a single straight line and applies, within the limits of experimental error, to all liquids (Fig. 12). From this it follows that the straight line slope will be determined by the properties of the solid phase alone, which is to say by the pore diameters and pore geometry.

A photographic method was used to obtain an idea of the liquid distribution within the porous body at various stages of rise and fall. This made it possible to determine the relative saturation of the porous body at various stages of capillary rise without destroying the body itself.

One of the most interesting cases of capillary hysteresis arises when the porous body contains two practically immiscible liquids. Differences in wetting results in the simultaneous appearance of wetting hysteresis

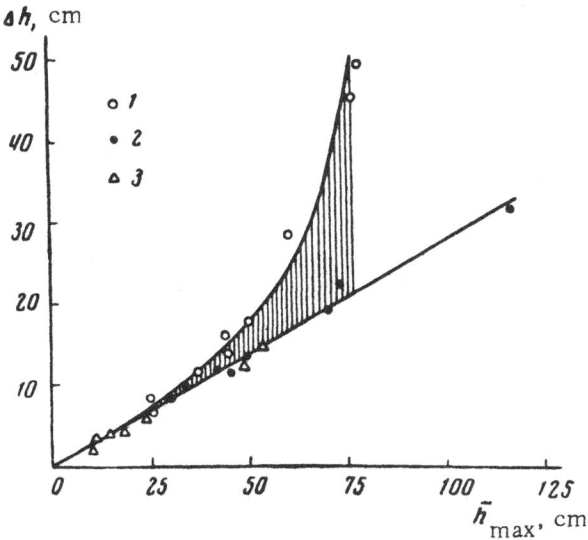

Fig. 12. The relation between $\bar{h}_{max}$ and capillary hysteresis, Δh, in various liquids. 1) Water—isooctane; 2) water—air; 3) kerosene—air.

and capillary hysteresis. The appratus of Fig. 9 was modified by connecting the upper portion of the column to a bottle through a glass tube and filling the entire space between the water level in the bottle and the water level in the porous body with isooctane, and then used in studying capillary hysteresis at the interface between two immiscible liquids in the porous body. The $U = f(\bar{h})$ curve for the liquid—liquid interface has the same form as the curve for the liquid—air interface. Liquid—liquid interface hysteresis also gives rise to an infinite number of heights of capillary rise falling between maximum and minimum values. Hysteresis is more pronounced here, however, wetting hysteresis being combined with capillary hysteresis. The $\Delta h = f(\bar{h}_{max})$ relation for this case is shown by the curve of Fig. 12. The hatched region corresponds to wetting hysteresis. An increase in h_{max} entails a reduction in the quartz grain diameter, which is to say, an increase in the total perimeter of the liquid—liquid—solid, three-phase interface, and thus results in a more pronounced wetting hysteresis.

Conclusions

1. Study of the general equilibrium conditions shows that capillary hysteresis in the rise of wetting liquids in capillaries of arbitrary form, or porous bodies, is an instance of metastable liquid equilibrium. Hysteresis gives rise to a finite number of heights of capillary rise in single capillaries, and an infinite number of such heights of rise in the porous body.

2. The problem of determining the number of heights of capillary rise and their values in the case of individual capillaries of simple form can be solved by three independent methods, the static, the kinetic, and the energetic.

The static method involves solution of a system of two equations, one covering the familiar relation of equality between the force of gravity and the surface tension, and the other a functional relation characterizing the capillary geometry.

The kinetic method involves solution of the problem of the movement of the viscous liquid in the capillary under the combined action of the force of gravity and the surface tension, with application of the condition of zero rate of liquid rise at equilibrium.

The very general energetic method involves a study of the conditions required for the establishment of equilibrium of the liquid in the capillary (minimization of the potential energy associated with the force of gravity and the surface tension).

3. Experimental test of the existence of multiple values of the height of capillary rise of water in glass capillaries of variable cross section (e.g., sinusoidal) shows agreement with the calculated values.

4. Study of the general conditions required for establishing equilibrium in the liquid together with the experimental work on these conditions, shows capillary hysteresis in the porous body to lead to an infinite number of values of the equilibrium height of capillary rise, these included between a certain maximum and a certain minimum.

5. Experimental study of capillary hysteresis at the interface between two immiscible liquids in a porous body shows that there will also be an infinite number of heights of capillary rise included between a maximum and a minimum. Capillary hysteresis is here complicated by the presence of wetting hysteresis.

201

Literature

1. A. Yu. Davidov. The Theory of Capillary Phenomena, Moscow (1851), p. 201.
2. H. Bouasse. Capillarité, phénomènes superficiels, Paris (1924), p. 189.
3. A. B. Haines. J. Agr. Sci. 20:97 (1930).
4. A. V. Lykov. Transfer Phenomena in Capillary Porous Media, Moscow, Gostekhizdat (1954), p. 32.
5. V. Cupr. Spisy Vydavane Prirodovedeckon Fak. Masary. Univ. 3: 137 (1956).
6. M. M. Kusakov and D. N. Nekrasov. Dokl. Akad. Nauk SSSR 119(1):107 (1958).
7. V. V. Lebedev and M. M. Kusakov. Izv. Vysshikh Uchebn. Zavedenii, Fiz. 1:15 (1958).
8. D. N. Nekrasov and M. M. Kusakov. Dokl. Akad. Nauk SSSR 133:1379 (1960).
9. M. M. Kusakov and D. N. Nekrasov. Zh. Fiz. Khim. 34:1602 (1960).

THE RELATION BETWEEN CAPILLARY PERMEATION AND DIFFUSIONAL EXTRACTION IN POROUS BODIES

M. A. Al'tshuler and B. V. Deryagin

Institute of Physical Chemistry, Acad. Sci., USSR
Laboratory for Surface Phenomena
UkrNIIGipronet'

There now exists a well-developed theory of the permeation of through capillaries; this has been repeatedly checked experimentally [1] and is carried over mechanically to the treatment of permeation processes in actual porous bodies.

The work of recent years has shown that both natural and artificial porous objects also contain dead-end pores, the volume and specific surface areas of these being comparable to the entire pore surface and volume [2, 3]. Cases are also met in both nature and technology in which the through-pored body behaves as if its pores were blocked. Thus solvent extraction of valuable materials from porous bodies can be hampered by the resistance offered to movement of the impregnation front by the air entrapped in the blocked pores.

When the porous body is completely immersed in a liquid, resistance of this kind will arise not only from air entrapped in the blocked pores but from air in the through pores, as well. It is clear that it would be helpful here to increase both the rate and density of permeation of the porous material.

Published work on the durability of building materials has indicated that the opposite effect of retarded permeation might be desirable in certain cases, since the longer a material can be made to hold its pore air the better its resistance to the effects of negative temperatures [4].

The effect of blocked pores in petroleum extraction (especially extraction by secondary methods), and on phase penetration of petroleum shales, has been discussed in other papers [2].

The present communication will treat the kinetics of permeation of blocked capillaries and porous materials considered to have the form of spherical granules, as well as diffusional extraction with capillary permeation. The rate of capillary permeation is determined by the dynamic pressure, p_{dy}, under which the liquid is driven into the capillary. For the through capillary

$$p_{dr} = p_K - \delta g l \sin \beta, \tag{1}$$

where $p_K = (2\sigma/r)\cos\theta$ is the capillary pressure, δ the specific gravity of the liquid, g the acceleration of gravity, l the depth of impregnation, β the angle of inclination of the capillary, σ the liquid surface tension, r the capillary radius, and θ the wetting angle.

The movement of liquid in the capillary is adequately described by the Poiseuille Equation [5]

$$Q = \pi r^2 \frac{dl}{d\tau} = \frac{\pi p_{dr} r^4}{8\eta l}, \tag{2}$$

where Q is the volume rate of liquid filtration, l is the length of the capillary, τ is the time of permeation, and $dl/d\tau$ is the mean linear filtration rate.

The Poiseuille Equation for the blocked capillary has the form [6]

$$\frac{dl}{d\tau} = \frac{r^2}{8\eta l}\left(p_{\text{к}} + p_0 - p_0\frac{l_0}{l_0 - l} - \delta g l \sin\beta\right),$$ (3)

where p_0 is the initial gas pressure, and l_0 is the total length of the capillary.

One method used to intensify capillary permeation is by preliminary evacuation of the sample. It should be noted that this procedure is effective only with blocked pores, or with through-pored materials which are completely immersed in the liquid during permeation. Here the rate of capillary permeation naturally increases because of the reduction in the pressure of the entrapped air, becoming equal to

$$\frac{dl}{d\tau_1} = \frac{r^2}{8\eta l}\left(p_{\text{к}} + p_0 - p_{\text{vac}}\frac{l_0}{l_0 - l} - \delta g l \sin\beta\right).$$ (4)

Since the relation

$$\frac{2\sigma\cos\theta}{r\delta g l \sin\beta} \gg 1,$$ (5)

is always applicable to pores of ordinary dimensions, Eqs. (3) and (4) can be respectively rewritten as

$$\frac{dl}{d\tau} \cong \frac{r^2}{8\eta l}\left(p_{\text{к}} + p_0 - p_0\frac{l_0}{l_0 - l}\right),$$ (6)

$$\frac{dl}{dt_1} \cong \frac{r^2}{8\eta l}\left(p_{\text{к}} + p_0 - p_{\text{vac}}\frac{l_0}{l_0 - l}\right).$$ (7)

The limiting depth of capillary permeation, l_∞, can be found from the condition $p_{\text{к}} + p_0 - p_0\, l_0/(l_0 - l_\infty) = 0$; it has the expression

$$l_\infty = \frac{p_{\text{к}}}{p_{\text{к}} + p_0}\, l_0 = \alpha l_0,\ \text{where}\ \alpha = \frac{p_{\text{к}}}{p_{\text{к}} + p_0}.$$ (8)

The blocked capillaries differ from the through capillaries in that the limiting depth of permeation depends on the total capillary length [1, 5]. This fact can be drawn on in studying the distribution of blocked pore radii [6, 8]. The time required for capillary permeation can be calculated from the expression

$$\tau = \frac{4\alpha l_0^2}{r\sigma\cos\theta}\left\{\frac{1}{2}\,\varphi^2 + (1 - \alpha)\left[-\alpha\ln\frac{\alpha - \varphi}{\alpha} - \varphi\right]\right\},$$ (9)

$\varphi = l\,/l_0$ being a dimensionless measure of the degree of permeation.

A similar expression for the time required for permeation of the previously evacuated capillary was derived in [6].

Air entrapment can occur in pores which are not blocked. Porous materials will behave in the same manner as a set of closed pores when completely immersed in a liquid, all or a part of the air being entrapped [6, 7]. The fraction of the air entrapped on immersion will clearly depend on the uniformity of pore structure, the form of the sample, and the properties of the liquid. Our experience has indicated that the quantity of air entrapped by immersion increases with the viscosity of the liquid. Air bubbles generally escape through the large fissures and pores where the capillary pressure is considerably less than the mean value. It will be found that all of the pore air will be entrapped if the sample has essentially uniform pore structure and is of such shape as to assure uniform permeation.

We will now consider capillary permeation of a porous body in the form of a single spherical granule, this granule completely immersed in the liquid. This case is interesting since silica gel, the most important adsorbent, is prepared commercially in granular form. The kinetics of capillary permeation of the spherical granule will be developed from the assumption of a quasi-stationary state [5, 11], with no bubbles of entrapped air left behind the impregnation front. This last is equivalent to assuming the granule to be free of pores which are blocked or difficultly accessible.

We write the Darcy Equation in the form

$$Q = kS \frac{\partial p}{\partial x} = 4\pi x^2 n \frac{\partial p}{\partial x}.$$

(10)

Since for spheres,

$$Q = 4\pi (R - l)^2 n \frac{dl}{dt},$$

(11)

$$\Delta p = p_\kappa + p_0 - p_0 \left(\frac{R}{R - l} \right)^3 = p_0 \left[\gamma - \left(\frac{R}{R - l} \right)^3 \right],$$

(12)

(10) can be integrated over the depth of impregnation layer, and (11) and (12) drawn on to obtain the following expression for the rate of permeation of the granule:

$$\frac{dl}{d\tau} = \frac{kRp_0 \left[\gamma - \left(\frac{R}{R - l} \right)^3 \right]}{l(R - l)}.$$

(13)

Here Q is the volume rate of filtration, k is the filtration coefficient, $\gamma = (p_K + p_\infty)/p_0$; $\partial p/\partial x$ is the pressure gradient, $S = 4\pi x^2 n$ is the filtration surface, n is the porosity, x the distance from the center of the granule to the impregnation front, R the radius of the granule, l the depth of permeation, and Δp the pressure difference between the liquid surrounding the granule p_0 and the gas entrapped in the granule, γ. Separating the variables in (13) and integrating, one obtains a relation between the time and the depth of capillary permeation, namely

$$M\tau = \frac{1}{3\gamma} (\psi^3 - 1) - \frac{1}{2\gamma} \quad (\psi^2 - 1) \quad \frac{1}{3\gamma^2} \ln \frac{\gamma \psi^3 - 1}{\gamma - 1} -$$

$$- \frac{1}{6\gamma^{5/3}} \ln \frac{(\psi - \gamma^{-1/3})^2 (1 + \gamma^{-1/3} + \gamma^{-2/3})}{(1 - \gamma^{-1/3})^2 (\psi^2 + \gamma^{-1/3} \cdot \psi + \gamma^{-2/3})} - \frac{\sqrt{3}}{3\gamma^{5/3}} \left(\text{arctg} \frac{2\gamma^{1/3}\psi + 1}{\sqrt{3}} - \text{arctg} \frac{2\gamma^{1/3} + 1}{\sqrt{3}} \right),$$

(14)

where $\psi = 1 - \varphi = 1 - l/R$ and $M = kp_0/R^2$.

It should be noted that the condition for cessation of capillary permeation, namely

$$\gamma - \left(\frac{1}{1 - \varphi_\infty} \right)^3 = 0,$$

(15)

can be drawn on, and the pore radius in the silica gel determined from the expression

$$r \approx \frac{2\sigma}{\left[\left(\frac{1}{1 - \varphi_\infty} \right)^3 - 1 \right] p_0}.$$

(16)

The depth of permeation of the silica gel can be conveniently observed under a low-power microscope. The kinetics of permeation can be studied from the change in the hydrostatic weight of the granule if the test material does not become transparent as impregnation proceeds.

The derivation of Eqs. (3)–(15) did not consider the physical chemical properties of the gas entrapped in the pore space. The effect of the viscous resistance of this gas on the kinetics of capillary permeation of through pores has been treated in [9], and shown to call for only an insignificant alteration in the description of the process. It was shown in [10] that the solubility of the entrapped gas has a pronounced effect on the rate of water uptake.

The fact that the solubility of the gas entrapped in the capillary increases under the action of the pressure gives rise to a diffusion current of dissolved gas, this current being directed from the meniscus where the concentration is high to the capillary mouth where the concentration is low. Permeation leads to dissolution of a part of the entrapped gas, so that the pressure, p, of the latter is always less than $p_0 \, l_0/(l_0 - l)$, and permeation does not cease at depth l_∞.

The movement of liquid in the capillary and the dissolution and diffusion of the entrapped gas are described by the equations

$$\frac{\partial C}{\partial \tau} = D \frac{\partial^2 C}{\partial x_1^2} + \frac{dl}{d\tau} \frac{\partial C}{\partial x_1},$$ (17)

$$\frac{\partial l}{\partial \tau} = \frac{r^2}{8 \eta l} \left(\frac{2\sigma}{r} \cos \theta + p_0 - p_\tau \right),$$ (18)

C being the concentration of the dissolved gas in the liquid which is permeating the capillary, and D the diffusion coefficient for the gas in this liquid.

Interest attaches to the effect of difficultly soluble gases on the permeation since this is the process which determines, to a degree, the durability of impregnated building materials, and to the effect of readily soluble gases since these can be used to study the intensification of permeation and related technological problems. Exact solutions of Eqs. (17) and (18) can be obtained only with difficulty, and discussion will be in terms of limiting cases only.

The difficultly soluble gas will dissolve to only a limited extent in the time required for the meniscus to penetrate to the depth l_∞. This situation does not persist for an extended period. Limiting the treatment to a rough analysis of the gas dissolution, which leads, none the less, to results in satisfactory agreement with experimental data, we write the equation for convective diffusion, (17), in a system of moving coordinates with origin fixed in the meniscus (this will be subsequently referred to as the "meniscus" system, thus distinguishing it from the "mouth" system which is fixed in the capillary walls):

$$\frac{\partial C}{\partial \tau} = D \frac{\partial^2 C}{\partial x^2}.$$ (19)

Equation (19) will be solved with the approximation boundary conditions formulated below.

The dissolved gas will be assumed to follow Henry's Law, $C = kp$, k being the Henry constant. It will also be assumed that the force of viscous resistance is so low at the lower permeation rates that the pressure of the entrapped gas can be considered constant and equal to $p_K + p_0$. The boundary conditions will then be

$$x = 0, \quad C = kp_\tau + k (p_K + p_0),$$ (20)

$$x = \infty, \quad C = kp_0.$$ (21)

Although $x = f(\tau)$ has been replaced by $x = \infty$ in (21), this has no essential influence on the results obtained [12].

The solution of (19) with the boundary conditions of (20) and (21) can be written as

$$C(x, \tau) = - kp_K \operatorname{erf} \left(\frac{x}{2 \sqrt{D\tau}} \right) + k (p_K + p_0).$$ (22)

Quantity	Radius, cm		
	10^{-3}	10^{-4}	10^{-5}
α	0.129	0.595	0.935
$1-\alpha$	0.871	0.405	0.065
τ_1	$2.9\cdot10^{-2}$	3.4	33
τ_2	0.276	2.76	27.6
τ_3	$4.4\cdot10^9$	$4.64\cdot10^7$	$4.83\cdot10^5$
τ_4	9.45	0.1	$1.04\cdot10^{-3}$

(with $\tau_1, \tau_2, \tau_3, \tau_4$ in sec)

The mass of the entrapped gas is determined from the expression

$$m = \frac{\pi r^2 (p_{\kappa} + p_0)(l_0 - l)}{RT},$$

while the reduction of this mass through dissolution and diffusion of the soluble portion of the gas at fixed pressure is given by

$$\frac{dm}{dt} = -\frac{\pi r^2 (p_{\kappa} + p_0)}{RT}\frac{dl}{d\tau}, \tag{23}$$

which can be expressed through Frick's First Law as

$$\frac{dm}{d\tau} = -DS\frac{\partial l}{\partial x}\bigg|_{x=0} = -\frac{\pi r^2 D_{\kappa} p_{\kappa}}{\sqrt{\pi D \tau}}. \tag{24}$$

By equating (23) and (24), one obtains the following expression for the rate of permeation due to dissolution and diffusion of the entrapped gas:

$$\frac{dl}{d\tau} = \frac{\alpha k D R T}{\sqrt{\pi D \tau}}. \tag{25}$$

Integration of (25) leads to

$$\Delta l^2 = 4\left(\frac{\alpha k\, D R T}{\sqrt{\pi D}}\right)^2 \tau, \tag{26}$$

R being the molar gas constant.

It should be noted that the kinetics of diffusional permeation follow a parabolic equation, just as do the initial stages of the permeation of through and blocked capillaries and spherical granules [5-7, 11, 12]. Special study will be given to the case in which the coefficient of proportionality depends on gas solubility and diffusion, as well as the properties of the capillary. Using the known values of the diffusion coefficient and the Henry constant for water at 293°K, one obtains

$$\text{for nitrogen entrapment}\qquad \frac{dl}{d\tau_{N_2}} = \frac{5\cdot10^{-5}\alpha}{\sqrt{\tau}},$$

$$\text{for ammonia entrapment}\qquad \frac{dl}{d\tau_{NH_3}} = \frac{1.08\alpha}{\sqrt{\tau}};$$

and accordingly,

$$\Delta l^2_{N_2} = 10^{-8}\alpha^2\tau, \tag{27}$$

$$\Delta l^2_{NH_3} = 4.65\alpha^2\tau. \tag{28}$$

The table shows calculated values of the time required for the permeation of blocked capillaries, 1 cm in length, and its variation with the radius (r, cm).

Here τ_1 is the time required for permeating a blocked capillary containing an insoluble gas to a depth of $0.99\,\alpha$ (it will be recalled that α is the limiting depth of impregnation), τ_2 the time for complete permeation of a through horizontally oriented capillary $[\tau_2 = (2\,\eta/r\,\sigma\cos\theta)^{l^2}]$; τ_3 the time for the dissolution of entrapped nitrogen, and τ_4 the time for the dissolution of entrapped ammonia.

The data presented make it clear that gas dissolution and diffusion are the slow steps when the solubility of the entrapped gas is low, and these processes therefore determine the time required for complete permeation of the capillary. The dissolution of a highly soluble entrapped gas occurs so rapidly that it can be entirely neglected, and the process considered as one taking place in a through capillary.

It is interesting to note that the time required for complete filling of the blocked capillary falls off with the pore radius when the solubility of the entrapped gas is low, whereas the time required for permeation of through capillaries increases under the same conditions; these facts have been experimentally confirmed [13].

We now pass to an analysis of diffusional extraction with capillary permeation in the case of capillaries and spherical porous granules.

There is a well developed theory of diffusional solvent extraction from porous bodies, but extraction by capillary permeation does not fall within its scope [14]. It is clearly necessary to consider permeation when the dissolved substance occupies only a part of the pore volume as a wall covering and the rest of the pore is occupied by air [7, 11].

Diffusional transfer of the dissolved substance from the meniscus to the capillary mouth is complicated by convective transfer in the direction of liquid movement in the capillary.

Passing from (17) to (19) in the "meniscus" system by the transformation $x_1 = x + \beta\sqrt{t}$, and then solving (19) with the boundary conditions

$$x = 0, \qquad D\frac{dC}{dx} = \frac{\beta C_S}{2\sqrt{\tau}}, \tag{29}$$

$$x = -\beta\sqrt{\tau}, \quad C = C_0, \tag{30}$$

one obtains

$$C(x_1, \tau) = C_0 + \frac{\beta C_S \sqrt{\pi}}{2\sqrt{D}}\left[\operatorname{erf}\left(\frac{\beta}{2\sqrt{D}}\right) + \operatorname{erf}\left(\frac{x_1 - \beta\sqrt{\tau}}{2\sqrt{D\tau}}\right)\right]. \tag{31}$$

Here C_S is the bulk concentration of the dissolved substance without diffusion, $\beta = \sqrt{r\sigma\overline{(\cos\theta)}/2\,\eta}$; and C_0 is the concentration of this same substance in the solution surrounding the capillary.

The extraction rate is given by the expression

$$j = -D\left(\frac{\partial C}{\partial x_1}\right)_{x_1=0} + C_0\frac{dl}{d\tau} = -\frac{\beta C_S}{2\sqrt{\tau}}\left(e^{-\frac{\beta^2}{4D}} - \frac{C_0}{C_S}\right). \tag{32}$$

It is clear from this equation that extraction will cease when $\beta_{cr}^2 = 4D\ln(C_S/C_0)$, a fact which makes it possible to find the radii of capillaries in which extraction can be expected at fixed surface tension, wetting angle, and concentration ratio:

$$r < r_{cr} = \frac{8\eta D}{\sigma\cos\theta}\ln\frac{C_S}{C_0}. \tag{33}$$

It is clear that extraction will begin earlier in blocked capillaries than in the through capillaries, the rate of permeation being lower and convective transfer less significant.

Extraction from the spherical granule is described by the equation

$$\frac{\partial C}{\partial \tau} = D \left(\frac{\partial^2 C}{\partial x_1^2} + \frac{2}{x} \frac{\partial C}{\partial x} \right) + \frac{dl}{d\tau} \frac{\partial C}{\partial x} ; \tag{34}$$

this, rewritten for the "meniscus" system, has the form

$$\frac{\partial C}{\partial z} = D \left(\frac{\partial^2 C}{\partial x_1^2} + \frac{2}{x_1} \frac{\partial C}{\partial x_1} \right) . \tag{35}$$

Thus our extraction problem has reduced to solution of the Stephan equation, which is possible (exactly) only when the movement of the liquid interface follows a parabolic or a linear equation.

Since a parabolic law is followed in the initial stages of permeation of the spherical granule, and $\partial^2 C/\partial x^2 \gg (2/x)(\partial C/\partial x)$ as $x \to R$, the results of extraction from the single capillary in the "meniscus" system can be drawn on and passage made back into the "granule" system by the transformation*

$$C(x, \tau) = C_0 + \frac{\beta C_S \sqrt{\pi}}{\alpha \sqrt{D}} \left[\mathrm{erf} \left(\frac{\beta}{\alpha \sqrt{D}} \right) \mathrm{erf} \left(\frac{x - R + \beta \sqrt{\tau}}{2 \sqrt{D\tau}} \right) \right] . \tag{36}$$

The extraction rate in the initial stages of the permeation of the spherical granule is given by

$$j = -S \left[D \frac{\partial C}{\partial x} + \frac{dl}{d\tau} C_x \right]_{x=R} = \frac{4\pi R^2 \beta C_S n}{2 \sqrt{\tau}} \left(e^{-\frac{\beta^2}{4D}} - \frac{C_0}{C_S} \right) . \tag{37}$$

Here, just as in the individual capillary,

$$\beta^2_{cr} = 4D \ln \frac{C_S}{C_0} .$$

It should be noted that this implies that extraction proceeds in a medium essentially free of the extracted substance, assuming the coefficients of viscosity and diffusion have their normal values; this agrees with the experiments of Musienko and Tovbin [15].

When the conditions are such that $\tau > h^2/D$, diffusion becomes quasi-stationary and is described by the equation

$$j = S \left[D \frac{\partial C}{\partial x} + \frac{dl}{d\tau_{(x,l)}} C_x \right] . \tag{38}$$

It follows from the condition of flux continuity that

$$\frac{dl}{dt_{x,l}} = \frac{kRp_0(R-l) \left[\gamma - \left(\frac{R}{R-l} \right)^3 \right]}{lx^2} = \frac{A}{x^2} . \tag{39}$$

---

*Here $\beta = \sqrt{2kRp_0\gamma}$.

Equation (38) can be solved with the boundary conditions

$$x = R, \quad C = C_0,$$

(40)

$$x = R - l, \quad C_S \frac{kRp_0 \left[\gamma - \left(\frac{R}{R-l} \right)^3 \right]}{l(R-l)} = -D \frac{\partial C}{\partial x}.$$

(41)

The result is the following expression for the extraction rate in the quasi-stationary stage:

$$j = 4\pi An \left[C_S e^{\frac{A}{DR} - \frac{A}{D(R-l)}} - C_0 \right].$$

(42)

By setting the bracketed term of (42) equal to zero, one can find the parameter values under which extraction becomes possible:

$$D \ln \frac{C_S}{C_0} = kp_0 \left[\gamma - \left(\frac{R}{R-l} \right)^3 \right].$$

(43)

Equation (43) makes it clear that the significance of convective transfer diminishes in the latter stages of the extraction, with the result that the extraction can be then observed under smaller concentration gradients. It is also possible to obtain an expression for the impregnation depth at which extraction is first observed:

$$l = R \left(1 - \frac{1}{\sqrt[3]{\gamma - \frac{D}{kp_0} \ln \frac{C_S}{C_0}}} \right),$$

(44)

but this is of significance only for rough estimates since extraction can begin even though the quasi-stationary stage has not yet been reached.

Conclusions

1. Air (gas) entrapment causes almost all of the capillaries in the spherical granule to behave as blocked capillaries during permeation.

2. The theory which has been developed here makes it possible to find the rate and depth of permeation. Depth measurements permit a determination of the pore radius.

3. Extension of the theory to permeation of the difficultly soluble gas leads to parabolic kinetics for diffusional impregnation. Here, the permeation rate diminishes as the pore radius increases, thus differing from the usual case. This explains the high durability of coarsely granular building materials.

4. A theory has been developed for diffusional extraction from individual blocked capillaries and spherical granules. An expression has been obtained for the extraction rate and a criterion set up for the possibility of extraction. Determination has also been made of the permeation depth at which extraction begins.

Literature

1. F. W. Washburn. Phys. Rev. 17:274 (1921).
2. C. P. Stewart, A. Lubinski, and K. A. Blenken. J. Petrol. Technol., No. 4:383 (1961).
3. B. V. Deryagin, N. N. Zakhavaeva, V. V. Filippovskii, and M. V. Talaev. Zh. Fiz. Khim. 32(8) (1958).
4. Proceedings of a Conference on the Frost Resistance of Ceramic Products, Riga (1957).
5. V. G. Levich. Physical Chemical Hydrodynamics, Moscow, Izd. Akad. Nauk SSSR (1959), p. 67.
6. M. A. Al'tshuler. Kolloidn. Zh. 83(6) (1961).
7. B. V. Deryagin and M. A. Al'tshuler. Dokl. Akad. Nauk SSSR 146:139 (1963).
8. B. V. Spektor, V. I. Ryazantsev, and M. A. Al'tshuler. Patent No. 125689.

9. J. R. Ligenza and R. B. Bernstein. J. Am. Chem. Soc. 73:4636 (1951).
10. I. N. Plaksin and D. M. Yukhtanov. Hydrometallurgy, Moscow, Metallurgizdat (1949).
11. B. V. Deryagin and M. A. Al'tshuler. Dokl. Akad. Nauk SSSR 152:651 (1963).
12. B. V. Deryagin and M. A. Al'tshuler. Dokl. Akad. Nauk SSSR 152:911 (1963).
13. F. G. Arutyunov. Izv. Akad. Nauk Arm.SSR 11(5) (1958).
14. G. A. Aksel'rud. The Theory of Diffusional Extraction from Porous Materials, Lvov (1959).
15. V. P. Musienko and M. V. Tovbin. Ukr. Khim. Zh. 28(3) (1962).

THE EFFECT OF SURFACE FORCES ON THE TRANSFER OF MOISTURE IN POROUS BODIES

M. P. Volarovich and N. V. Churaev

Kalinin Peat Institute

Interfacial surface phenomena have a considerable effect on the transfer of moisture in finely porous bodies such as peats and clays. These effects have been only inadequately studied, however, so that calculations on moisture transfer in highly dispersed bodies cannot be carried out and the development of optimal technological conditions for the treatment and application of such bodies is hindered. We have already proposed that radioactive labeled water be used in studying moisture transfer processes [1]. The present communication reports on the results of our continued studies in the Physics Department of the Kalinin Peat Institute.

We will first consider certain problems which arise in connection with moisture transfer in a two-phase porous body which is already saturated completely with water. It was early observed that the filtration of water failed to follow Darcy's Law, the permeability of the finely porous body increasing with the pressure gradient. This effect is now explained in terms of traces of shear rigidity in the pore water [2] and (possibly) an increase of the pore size in nonrigid colloidal structures [3]. Breakdown of filtrational flow can also be due to osmotic slip [4], or alteration in the depth of the liquid boundary layer with increased viscosity [5] as a result of changes in the composition of the filtering water.

Studies on the filtration of water with varying dissolved salt content have shown the alteration in the filtration rate (by a factor of 5-10) in peats containing a hydrophilic colloidal fraction to be essentially due to structural changes in the colloid, effects which might arise from surface phenomena being masked by these processes. When, however, a small quantity of labeled soluble substance which is negatively adsorbed at the peat interface and causes no appreciable alteration in the colloidal structure (sugar with C^{14}, or $S^{35}O_4$) is introduced into the filtering water, it is found that the concentration of this tracer solution changes markedly during passage through the sample (Fig. 1). In the figure, the ratio of the volume of the filtrate, V, to the volume of the sample, V_0, has been plotted along the axis of abscissas, and the ratio of the concentration of the tracer in the filtrate, C, to its concentration, C_0, in the solution undergoing filtration on the axis of ordinates. The difference between C and C_0 is reflection of the reduction in the concentration of the tracer substance in the neighborhood of the interfacial surfaces.

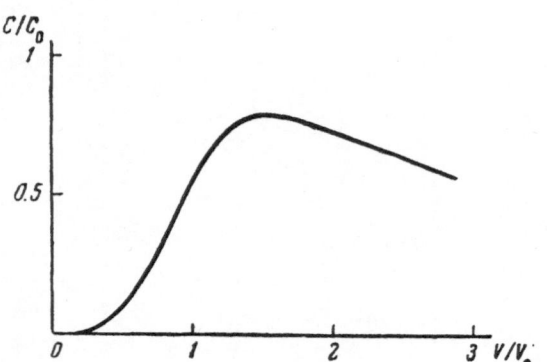

Fig. 1. Graph to illustrate the variation of the relative concentration, C/C_0, of the filtrate tracer during filtration through a sample of a finely porous peat.

This observation was first explained by B. V. Deryagin and his coworkers as indication that surface phenomena have a significant effect on the filtration of liquids through finely porous bodies [6]. Stripping a filtrate labeled both molecularly (sugar) and ionically shows that filtration in a finely porous peat can give rise not only to a streaming potential but also to a concentration gradient in the neutral

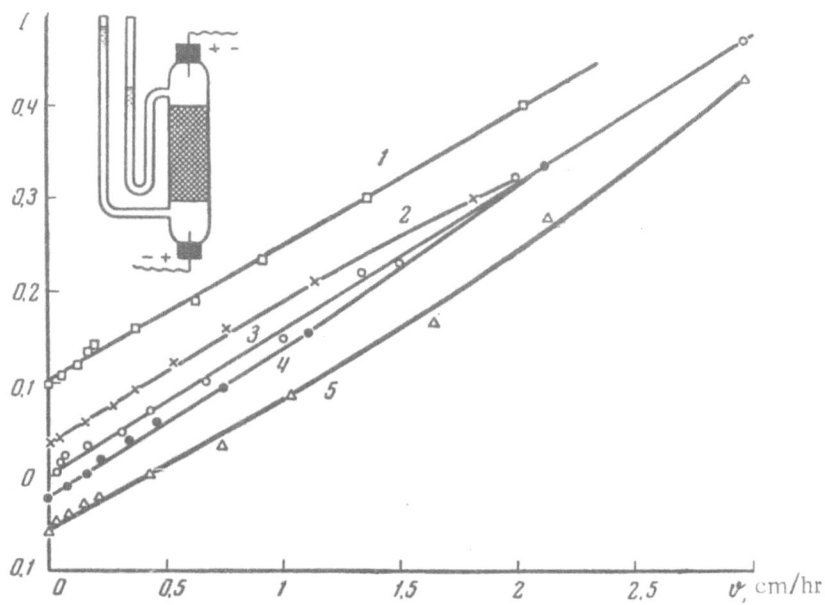

Fig. 2. The effect of electroosmosis on the filtration of water in a finely grained sand. 1) Pressure head + electroosmosis (H + E), i = 6 ma, U = 300 V; 2) (H + E), i = 2 ma, U = 100 V; 3) pressure filtration, U = 0; 4) pressure head − electroosmosis (H − E), i = 2 ma, U = 100 V; 5) (H − E), i = 6 ma, U = 300 V.

molecules. The latter factor induces capillary-osmotic slip of the moisture in a direction opposite to that of the filtrational flow, thus leading to that additional reduction in the tag concentration of the filtrate which is quite clearly marked in the graph.

The effect of an electric potential gradient on the filtration was separately studied on a fine grained sand (mean pore radius, R_c, 12 μ; porosity, m_0, 0.41) which was free of that colloidal fraction which readily undergoes alteration during flow, this being the factor which makes for difficulty in attempting to carry out similar experiments on clays and peats. These experiments were performed in a filtration tube, varying the pressure gradient, I, with fixed potential gradient to increase (pressure head + electroosmosis, condition H + E), or diminish (condition H − E), the flow rate, v, of the water. Figure 2 gives a schematic representation of the apparatus used here and shows the v(I) relations developed at two values of the potential difference, U, namely, 100 and 300 volts. It is seen that the overall rate of transfer is not the sum of the filtrational and electroosmotic flux rates. Thus the effect of electroosmosis is greater when (H + E) than when (H − E), a fact requiring special study. An increase in the pressure gradient diminishes the effect of electroosmosis, as noted earlier by others [7]. It is clear that high velocities of water movement impede the establishment of an electroosmotic flux.

Experiments show that departures from the Darcy Law in the porous body filtration of a liquid which follows this law when U = 0 become more pronounced as the potential gradient increases and the filtration rate diminishes (see Fig. 2).

The rheological properties of the dispersed medium are obviously controlling factors for the filtration of water or fixed composition which has come to equilibrium with the peat phase. Solutions of the problem of the filtration of structured liquids through model soils containing cylindrical pores of uniform diameter, can be obtained by drawing on the Buckingham equation [2, 8]. A more detailed picture of the filtration of liquids with limiting shear stress, θ, can be obtained from an analogous solution of the problem of filtration through a soil in which there is a definite distribution of pores over diameters. Here, the active porosity, m, will depend on the pressure gradient, I, and is calculated from the equation

$$ m = m_0 \left[1 - \int_{R_1}^{R_F} \varphi\,(R)\,dR \right], \tag{1} $$

213

m_0 being the limiting porosity, corresponding to the free water content, R the pore radius, and $\varphi(R) = dV/dR$ the function covering the distribution of pore volumes over diameters.

Integration is to be carried out between the minimum pore radius, R_1, and the flow radius, $R_F = 2\theta/I$, the latter being the pore radius at which liquid flow becomes possible at given θ and I. For a normal distribution of pores over radii, one has

$$m = m_0 [1 - \Phi(u)],$$

(2)

$\Phi(u)$ being a normalized integrated function, values of which are available in tabular form, and $u = (R_F - R_C)/\sigma$, where R_C is the mean pore radius and σ the mean square error.

These solutions can be used as a basis for the following description of structured liquid filtration. The initiation of filtration in the heteroporous body is determined by the condition $I \geq I_{01}$ (with $I < I_{01}$ $m = 0$), $I_{01} = 2\theta/R_2$ being fixed by the radius, R_2, of the largest pore since it is in these that flow is first observed. Increase in the value of I entails the participation of still finer pores in the filtration process, with an accompanying increase in the active porosity. The filtration rate and pressure gradient are not linearly related in this stage of the filtration [8]. Filtration through the entire pore space ($m = m_0$) becomes possible when $I \geq I_{02} = 2\theta/R_1$, the process being then described by an equation of the type

$$v = AI - B + \frac{C}{I^3},$$

(3)

in which A, B, and C are constants which depend on the pore radii, the type of pore distribution, the porosity of the sample, and the rheological properties of the liquid.

In this connection it should be noted that the equation $v = k_\Phi(I - I_0)$ which is frequently applied to the treatment of filtration problems is only of limited validity. The data presented here make it clear that such an equation can be used only when working with pressure gradients of considerable magnitude and materials which are uniform in regard to pore radii.

Figure 3 shows the results obtained in experiments in which several specimens of various peats were studied in the filtration apparatus under various pressure gradients. An equilibrated dispersed medium which had been previously pressed out of the peat was used as the filtering liquid. Analysis of the data indicates good agreement with calculations based on Eq. (3) with the limiting shear in the test water set equal to $\theta = 1 \cdot 10^{-2} - 3 \cdot 10^{-2}$ dyne/cm², this value being consistent with the results obtained in the studies of B. V. Deryagin, S. V. Nerpin,

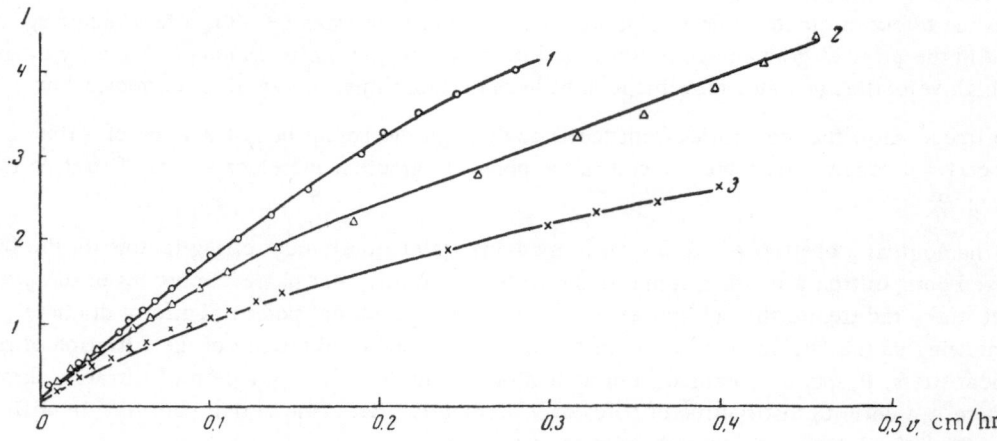

Fig. 3. The relation between filtration rate, v, and pressure gradient, I, in various peat samples: 1) Bog peat with 35% decomposition, mean pore radius, $R_C = 0.6$ μ; 2) pine-cotton grass peat with 65% decomposition, $R_C = 0.8$ μ; 3) the same, $R_C = 1.1$ μ.

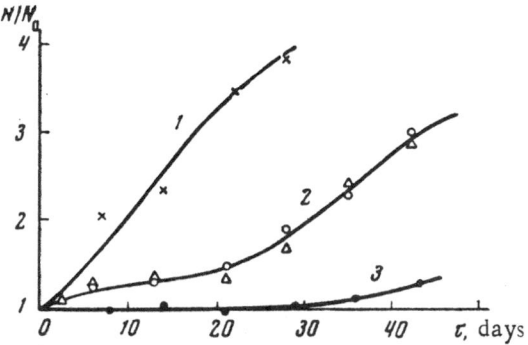

Fig. 4. Alteration of the specific surface activity, N/N_0, during evaporation of moisture from the sample at a relative humidity, W, of 95%: 1) Powdered silica gel, particle diameters, 0.1 mm; 2) quartz sand, particle diameters, 0.5-1.0 mm; 3) granular silica gel, particle diameters, 3-5 mm.

and their collaborators [2]. These peat experiments confirm the possibility that traces of liquid shear rigidity can affect the filtration of water through finely porous bodies. It should also be noted that the mean pore radii obtained from the graphs with the aid of Eq. (3) are consistent with the values obtained through the use of radioactive indicators [9]. These experiments also point to an increase in the active porosity with rising pressure gradient, thus supporting the validity of Eq. (1). It should be pointed out, however, that the filtration of structured liquids can be complicated by boundary slip, by retarded flow of damaged structures, and by the intersection of the pore canals, factors which require special study. Experiments with model media (sands, with filtration of sodium humate solutions with θ values in the 2-4 dyne/cm^2 interval) also yield results in satisfactory agreement with the theory. This work shows, in particular, that the observed effects do not arise from pore clogging, but from the rheological properties of the filtered liquid, the experimentally developed $v(I)$ relation being independent of the method of altering I and the elapsed time. It should, at the same time, be pointed out that the Darcy Equation gives a rather good description of the filtration of water through coarse peats when the mean pore radius is in excess of 3-4 μ, and flow is not accompanied by colloidal chemical interaction of the dispersed phase with the dispersing medium.

The effect of surface phenomena on moisture transfer in three-phase systems (porous bodies incompletely saturated with moisture) has been studied for the vaporization and transfer of water under the action of an electrical potential gradient.

It has been shown in [10] that the moisture transfer under vaporization results not only from vapor diffusion but also as from film movement of water. The theory of this process leads to the conclusion that significance attaches to water film transfer to the capillary mouth followed by vaporization from the film surface. The intensity of film transfer must increase as the radius of the capillary diminishes and the relative humidity, φ, of the surrounding medium increases.

These conclusions were tested in experiments on the vaporization of radioactive tagged water from various porous bodies (sand, silica gel, clays, peats). The samples were placed in test tubes and the latter set in a desiccator where the relative humidity, φ, was maintained at 95%. The sample was weighed during the drying, and the specific activity of the surface from which vaporization was taking place measured. A solution of Na_2SO_4 containing the radioactive isotope S^{35} was used as a tracer here. Self-adsorption of the soft β-radiation of the S^{35} made it possible to determine the amount of tracer in a layer less than 0.1 mm deep. Thus the alteration in the specific activity of the sample surface was solely due to liquid phase moisture transfer to the surface. Vapor phase transfer of moisture through the sample did not bring about any alteration in the specific activity of the surface.

Figure 4 shows the alteration of the specific activity, N/N_0, of the surface in the course of an experiment, N_0 being the activity at time $\tau = 0$. It is seen that the evaporation of water led to an increase in the surface activity in each sample. The role of film water movement was most significant in the finely porous silica gel where the particle diameter was less than 0.1 mm (curve 1), less significant in the sand where the grain diameters were in the 0.5-1.0 mm range (curve 2), and completely insignificant in the granular silica gel* (curve 3) where vapor phase transfer of moisture along the coarse pores predominated.

Figure 5 shows the variation of the specific activity of the water, C/C_0 (full curve), and moisture content, W (dotted curve), along the sample (H, the sample height, is measured from the upper surface), as indicated

---

*These were spherical granules with diameters in the 3-5 mm range.

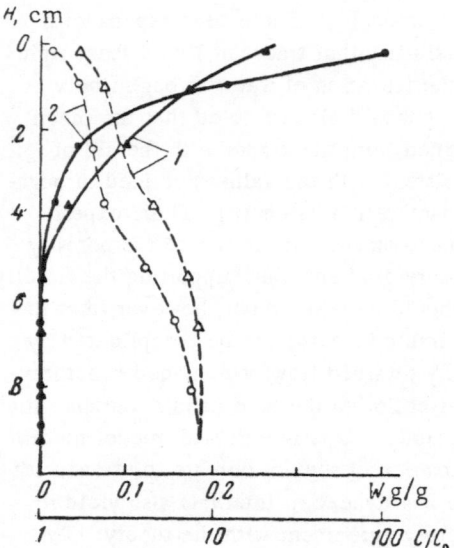

Fig. 5. Variation of moisture content, W (dotted curves), and specific activity, C/C_0 (full curves) of water, with depth after partial evaporation of water from a sand with grain diameters in the 0.25-0.5 mm range. Amount of water evaporated: 1) 1.5 g; 2) 2.6 g.

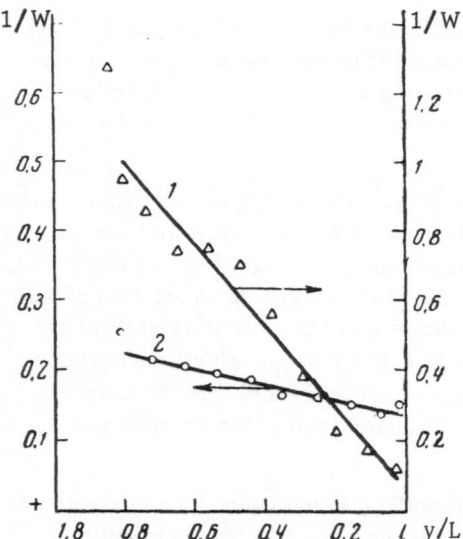

Fig. 6. Variation of the moisture content, W, along a sample in dynamic equilibrium under film electroosmosis. Initial moisture content: 1) $W_H = 0.027$ g/g, U = 1050 V; 2) $W_H = 0.046$ g/g, U = 1600 V.

by probe layer analysis at the end of the experiment. Here C is the tracer concentration, expressed as counts/min per 1 ml of water,* and C_0 is the tracer concentration in water which had been used to saturate the sample at the beginning of the experiment. In determining the value of C, the probe was weighed to fix the amount of moisture, and then washed free of tracer. The completeness of washing was checked through specific activity measurements on the water. A sample was prepared for analysis by evaporating the wash water in a standard cup. The value of C was obtained by dividing the activity, M, of the sample in counts/min by the water content of the layer which was under analysis. C/C_0 values have been plotted along the axis of abscissas on a logarithmic scale.

The experimental results shows that evaporation is actually accompanied by a liquid phase moisture transfer to the surface of the sample, where the largest amount of tracer accumlates. The fact that the C/C_0 ratio differed from unity at depths up to 4 cm is indication of internal vaporization of moisture from films and menisci, with subsequent diffusional vapor phase transfer along the pores. The dynamics of the vaporization process can be followed by analyzing experiments with two (see 1 and 2 of Fig. 5) samples of the same sand which were concluded at different times. The results indicate that film movement becomes more significant as drying proceeds. Thus these experiments with labeled water confirm the basic assumptions of the theory of transfer with accompanying film movement.

We now turn to the study of moisture transfer under an electric potential gradient in systems containing essentially film water alone. The "film electroosmosis" effect is illustrated by the data of Fig. 6. Here the porous body was a washed quartz sand in which the particle diameters fell in the 0.2-0.3 mm range; this was treated by the method of high column discharge to obtain a mean initial moisture content, W_i, of 0.027 g/g for the first experiment (curve 1) and 0.046 g/g for the second (curve 2). Saturation was carried out with a 10^{-4} N $CaCl_2$ solution labeled with radioactive Ca^{45}. This sand was loaded into a glass tube, 2.9 cm in diameter, and the tube then closed at both ends by rubber stoppers carrying platinum disk electrodes of the same diameter as the tube itself. The length of the sample, L, was 22 cm. The possibility of gas evolution at the electrodes was checked by introducing thin glass tubes into the sample and connecting these to a differential water manometer. Voltage was applied to the electrodes from a rectifier. The potential difference, U, was 1050 V in the first experiment, and 1600 V in the second. The current strength was automatically recorded and controlled with a microammeter. The current reached a steady state after 5-8 days, after which the tube was cut up and layer analysis

*The tracer concentration measurements were carried out under standard conditions, using dried samples without self-adsorption of radiation.

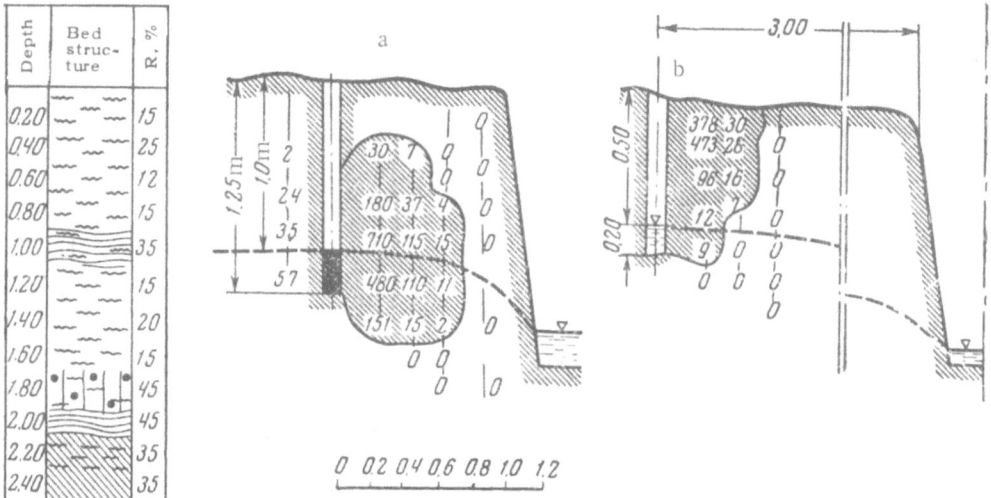

Fig. 7. Distribution of a radioactive tracer (NaI with I^{131}) near the drying channel in a peat bed. The numbers indicate the indicator concentration probe measured. 14 days after introduction of the tracer. a) Tracer added to well No. 58; b) tracer deposited in the sides of well No. 56.

carried out to determine the moisture content, W, and the specific activity of the sample. It should be noted that there was practically no evolution of gas in the course of the experiments, a fact which traces back to the very low values of the current strength (15-20 microamp).

Application of the potential difference, U, to the sand sets up an electroosmotic movement of moisture toward the cathode, thus establishing a splitting pressure gradient in the system which led, in turn, to a flow of film water in the opposite direction. Assuming a quasi-equilibrium stationary state, one can, in a first approximation, equate the rates of electroosmotic and film transfer (just as with electroosmosis in two-phase systems) drawing on the Deryagin—Nerpin Equation [11]

$$\frac{\varepsilon \zeta U}{4\pi \eta L} = \frac{h^2}{3\eta} \cdot \frac{d\Pi}{dy} \tag{4}$$

to obtain the second of these two expressions. In this equation, ε is the dielectric constant, and η the viscosity, of the water, h is the depth and π the splitting pressure of the film, ζ is the zeta-potential, and y the distance measured along the axis of the sample. Assuming with [11] that the splitting pressure isotherm can be represented by an equation of the form $\Pi = k/h^3$ in which the constant k depends on the properties of both liquid and solid phases, and setting $d\Pi/dy = -(3\pi/h)dh/dy$ in (4), one obtains a differential equation which can be solved to give

$$\frac{1}{h} = \frac{1}{h_0} + \frac{\varepsilon \zeta U}{4\pi k} (y/L). \tag{5}$$

Here h_0 is the film depth at the cathode where $y = 0$, and h is the film depth at the point of coordinate y ($y = L$ at the anode).

Equation (5) gives the variation of film depth along a sample in a state of dynamic equilibrium, the assumption being made that ε, ζ, and k are independent of the film depth and therefore constants for the experiment.

It is seen from Fig. 6 that the general validity of Eq. (5) is confirmed since the relation between $1/W$ and y/L is almost linear if the film depth, h, is assumed to be directly proportional to the moisture content, W. Pronounced deviations due to edge effects are noted in layers lying close to the anode ($y/L > 0.8$). It should be noted that the anode zone is strongly dried out ($W \leq 0.001$ g/g), even at the beginning of the experiment, and sharply delineated from the rest of the sample. An uncompensated transfer of double layer cations toward the

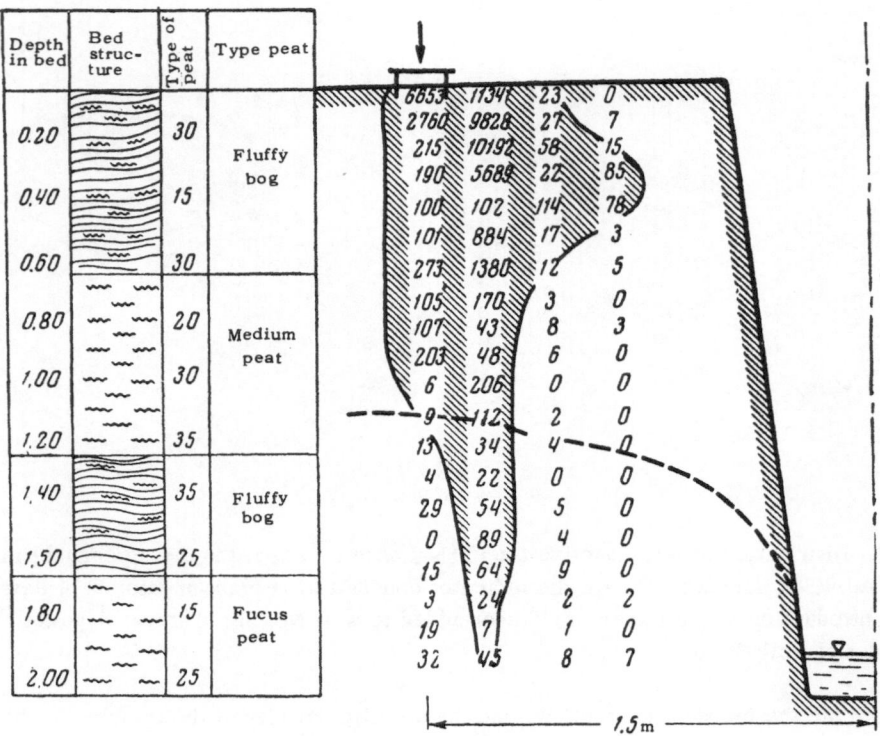

Fig. 8. Distribution of a radioactive tracer (S^{35}) deposited on gauze attached to an iron ring 2-3 cm from the surface of the peat bed as a result of tracer transfer by falling rain. The numbers indicate the tracer concentration as obtained by radiometric probe analysis of the bed 45 days after introduction of the ring.

cathode also occurs during the experiment, and the ζ-potential and the splitting pressure isotherm both alter as a result. Observations show that almost all of the Ca^{45} will have accumulated in the neighborhood of the cathode at the end of the experiment, the specific activity in the other parts of the sample being no greater than the background count. Thus a more exact check of the validity of Eq. (5) requires alteration of the experimental conditions so as to permit stationary moisture transfer through the sample.

Film electroosmosis can be considered as a new method for studying the double layer structures and other properties of thin liquid films. Experimental and theoretical studies of this effect should lead to a theory of electroosmosis in unsaturated porous bodies.

Moisture transfer in three-phase systems has considerable significance for various industrial processes. One example is the transfer of moisture during aeration of peat soils, an effect which has been observed in field experiments with radioactive tracers. It is seen from Fig. 7 that the rate of movement of water containing a radioactive indicator is approximately the same both above and below the ground water level indicated by the dotted line. The radioactive tracer was introduced directly into the well water in the first experiment (Fig. 7a), and deposited in the sides of the well in the second (Fig. 7b). The rate of horizontal transfer proved to be the same in both cases, namely 3.5-4.3 cm/day. Samples from the other side of the well (Fig. 7a) showed that the effect in question here was not due to tracer diffusion alone. These, and other observations outlined below, were checked by the method of coordinate measurements [9].

Figure 8 illustrates the movement of tagged rain water both downward, where they unite with the ground water, and along the gradient of the drying channel where evaporation takes place.

This intensive moisture transfer in the upper layers of the peat deposit under the action of temperature and splitting pressure gradients is not considered in hydrotechnical calculations, although experiment shows that makes a perceptible contribution to the over-all effect. Surface forces also have an appreciable effect on the moisture transfer which occurs during the drying of various highly dispersed materials [12].

Thus it becomes more and more clear that surface forces must be taken into account, especially in the technology of finely porous products. Since the rates of such technological process as drying, filtration, and vaporization are frequently determined by these forces, the study of the mechanism of surface effects acquires practical as well as theoretical significance.

Literature

1. M. P. Volarovich and N. V. Churaev. In collection: Research in Surface Forces, Moscow, Izd. Akad. Nauk SSSR (1961), p. 149. [English translation: Consultants Bureau, New York (1963).]

2. S. V. Nerpin and N. F. Bondarenko. Tr. Leningr. Inst. Inzh. Vodn. Transporta, No. 23:36 (1956); Dokl. Akad. Nauk SSSR 114(4) (1957).
   S. V Nerpin and B. V. Deryagin. In collection: Research in Surface Forces, Moscow, Izd. Akad. Nauk SSSR (1961), p. 156 [English translation: Consultants Bureau, New York (1963).]
   A. I. Kotov and S. V. Nerpin. Izv. Akad. Nauk SSSR, Otd. Khim. Nauk, No. 9:106 (1958).

3. B. F. Rel'tov. Proceedings, Conference on Soil Mechanics, Foundations, and Bases, Moscow, Gosstroiizdat (1956), p. 60; in collection: Geological and Engineering Properties of Rocks, Methods of Study, Moscow, Izd. Akad. Nauk SSSR (1962), p. 73.

4. B. V. Deryagin, M. K. Mel'nikova, and S. V. Nerpin. Reports, Sixth Congress on Soil Management, Soil Physics, Moscow, Izd. Akad. Nauk SSSR (1956), p. 101.
   B. V. Deryagin and M. K. Mel'nikova. Izd. Akad. Nauk SSSR (1956), p. 119.
   A. I. Kotov. Tr. Leningr. Inst. Inzh. Vodn. Transporta, No. 23:65 (1956).
   N. F. Bondarenko and S. V. Nerpin. In collection: New Methods of Measurement and Apparatus for Hydraulic Studies, Moscow, Izd. Akad. Nauk SSSR (1961), p. 268.
   S. A. Poza. Hydrotechnical Construction, No. 6:40 (1961).

5. B. V. Deryagin, N. N. Zakhavaeva, and A. M. Lopatina. Inzh.-Fiz. Zh., No. 3:66 (1960); In collection: Research in Surface Forces, Moscow, Izd. Akad. Nauk SSSR (1961), p. 175. [English translation: Consultants Bureau, New York (1963).]

6. B. V. Deryagin, et al.. Kolloidn. Zh. 9(5):355 (1947).

7. S. A. Yudovina. Tr. Leningr. Gidromet. Inst., No. 11:220 (1961).

8. A. M. Gutkin. Kolloidn. Zh. 23(3):350 (1961).

9. M. P. Volarovich and N. V. Churaev. Studies of the Properties of Peat by the Use of Radioactive Isotopes, Moscow, Izd. Akad. Nauk SSSR (1960).

10. N. V. Churaev. Dokl. Akad. Nauk SSSR 148(6) (1963).
    B. V. Deryagin, S. V. Nerpin, and N. V. Churaev. Kolloidn. Zh. 25(3):6 (1964).

11. S. V. Nerpin. Tr. Leningr. Inst. Inzh. Vodn. Transporta, No. 21:126 (1954).
    S. V. Nerpin and B. V. Deryagin. Dokl. Akad. Nauk SSSR 100(1):17 (1955).

12. N. V. Churaev. Inzh.-Fiz. Zh., No. 12:41 (1962); No. 2:31 (1963).

COMPLETE SPECULAR REFLECTION OF MOLECULES AT LOW ANGLES OF INCIDENCE AND ITS EFFECT ON THE MOLECULAR FLOW OF GASES THROUGH VERY NARROW CAPILLARIES

B. V. Deryagin and N. N. Fedyakin

Institute of Physical Chemistry, Acad. Sci., USSR
Laboratory for Surface Phenomena
Kostroma Technological Institute

Introduction

It is a well-known fact that molecular flow at high Knudsen number,

$$K = \frac{\lambda_0}{r_0},$$

(λ_0 is the mean free path of the molecule in the body of the gas and r_0 the radius of the capillary or tube) is characterized by rarity of collisions between molecules as compared with collisions between molecules and wall. By taking this fact into account and assuming complete accommodation with diffuse scattering of the molecules by the wall, it has been possible to develop the following expression [1, 2] for the density of molecular flux (in g/(cm^2 · sec)):

$$i = \frac{8}{3} \sqrt{\frac{RT}{2\pi M}} \frac{\Delta C r_0}{L}, \tag{1}$$

L being the length of the capillary, M the molecular weight of the gas, and ΔC the difference in the gas densities (concentrations) at the two ends of the capillary, R and T having their usual significance.

The necessary conditions for the applicability of Eq. (1) are

$$\lambda \gg r_0, \tag{2}$$

$$L \gg r_0. \tag{3}$$

It has been assumed up until recently that these were also sufficient conditions. We will, on the contrary, show that there is still another necessary condition, namely

$$\Omega \gg \frac{\lambda_0}{r_0}, \tag{4}$$

(Ω is a very large but well defined number), so that Eq. (1) is consistent with experiment only if the capillary radius is not too small.

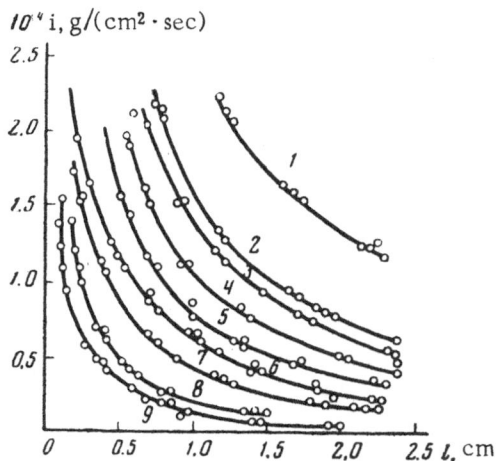

Fig. 1. The relation between the rate of vaporization of water in vacuum and the meniscus position, at 20°C. 1) $r_0 = 3.3\ \mu$; 2) $r_0 = 2.4\ \mu$; 3) $r_0 = 2\ \mu$; 4) $r_0 = 1.75\ \mu$; 5) $r_0 = 1.32\ \mu$; 6) $r_0 = 1\ \mu$; 7) $r_0 = 0.8\ \mu$; 8) $r_0 = 0.5\ \mu$; 9) $r_0 = 0.4\ \mu$.

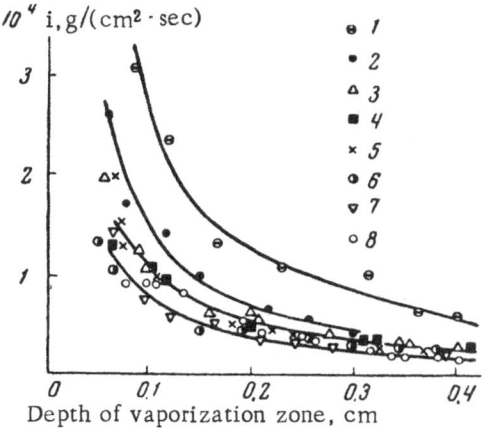

Fig. 2. The relation between the rate of vaporization of water in vacuum and the depth of the vaporization zone, for microcapillaries at 20°C. 1) $r_0 = 0.24\ \mu$; 2) $r_0 = 0.108\ \mu$; 3) $r_0 = 0.98\ \mu$; 4) $r_0 = 0.09\ \mu$; 5) $r_0 = 0.062\ \mu$; 6) $r_0 = 0.05\ \mu$; 7) $r_0 = 0.029\ \mu$; 8) $r_0 = 0.024\ \mu$.

Equation (4) fixes an upper limit for λ_0 at given r_0, or a lower limit for r_0 at given λ_0. Deviations from (1) will also be explained and equations derived which are applicable when

$$\frac{\lambda_0}{r_0} \gg \Omega, \quad \frac{L}{r_0} \gg \Omega, \tag{5}$$

which is to say, when the capillary is quite narrow.

Experimental Methods

We have undertaken the study of the vaporization of water from very narrow capillaries with a view to establishing the relations applying to molecular flow in capillaries of this kind. If a capillary is filled with liquid and introduced into an evacuated space where it can be held at fixed temperature, the rate of liquid evaporation from it at various meniscus levels will essentially depend on the laws of capillary vapor flow. It is possible to obtain some idea of the laws applying to this flow if the diameter of the capillary, the pressure of the saturated vapor, and the rate of vaporization at given depth (i.e., the quantity of material delivered per unit time) are known. By carrying out measurements on capillaries of different radii, information concerning the dependency of the flow laws on the L/r_0 ratio can also be obtained.

We have measured the rate of vaporization of water from capillaries whose radii varied over an interval of two orders of magnitude (3-0.02 μ). Methods of measuring capillary radii and observing liquid columns in microcapillaries have been described in [3, 4].

The capillary filled with the liquid was inserted into a glass tube, which was then evacuated to 0.1 mm Hg and sealed. This tube was set into a beaker of water and the latter placed in an air thermostat. The variation of temperature was less than 1°C.

Alteration in the liquid level in the capillary was followed with a cathetometer which could be accurately read to ±10 μ. The linear rate of vaporization at given depth was obtained from the time required for the meniscus to sink 20-30 units in the measuring eyepiece of the cathetometer. These data were used to develop the relation between the vaporization rate (vapor flux density, i) and the distance from the meniscus to the end of the capillary (depth of the vaporization zone).

Experimental Results

Results obtained in one series of microcapillary experiments are shown in Fig. 1. The curves were constructed from the experimental points. It is seen that the rate of vaporization, i, diminishes as the radius is reduced.

Figure 2 shows the results of experiments involving further reduction of the capillary radius to values below 0.24 μ. Such reduction at first leads to a decrease in the intensity of vaporization, but only in capillaries

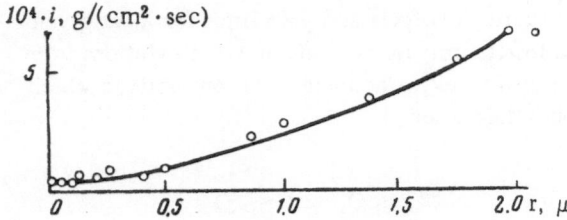

$10^4 \cdot i$, g/(cm$^2 \cdot$ sec)

Fig. 3. The relation between the water vapor flux
density and the capillary radius.

having radii in excess of 0.09 μ. The rate of vaporization is constant, and therefore independent of the
radius, in more narrow capillaries.

A more clear-cut idea of the relation between
flow type and capillary radius is furnished by Fig. 3.
Here it is a matter of the relation between the flux
density (i) of water vapor and the radius, in capillaries
of fixed length. The figure makes it clear that diminution of the radius entails a reduction in the slope of
the curve, with passage to a horizontal section. The
curve for radii in excess of 1.5 μ is described by the equation for viscous flow with slip. The relation is almost
linear over the radius interval from 0.5 to 1.5 μ, the slope of 3.2 g $\cdot$ cm$^{-3} \cdot$ sec^{-1} lying within the limits of experimental error and being almost identical with the 3.4 g $\cdot$ cm$^{-3} \cdot$ sec^{-1} calculated from Eq. (1) for water vapor
with $\Delta C = 1.72 \cdot 10^{-5}$; T = 293°K; L = 0.2 cm. The flux density is less and less sensitive to the radius as the radius
diminishes, and finally becomes completely independent of it. The limiting value of the rate of vaporization
is $0.42 \cdot 10^{-4}$ g $\cdot$ cm$^{-2} \cdot$ sec^{-1}.

Theory

It is necessary to review the theory of molecular flow in order to understand the results obtained here.
We will consider the derivation of the expression for the mass of gas, i, passing in unit time through unit cross
section of a capillary of length L. Taking account of the fact that the individual molecules move through a
series of mutually independent broken line paths in Knudsen flow, and applying the method developed by Einstein in his theory of the Brownian movement, one finds the following expression for the mass of gas, i, passing
in unit time through unit cross-sectional area of a capillary of radius r_0:

$$i = - D \frac{dC}{dz},$$

(6)

dC/dz being the gas density gradient along the capillary, and

$$D = \frac{1}{2} \frac{\overline{\Delta_z^2}}{\tau}$$

(7)

the diffusion coefficient.

Here $\overline{\Delta_z^2}$ is the mean square displacement of the molecule along the z-axis (which is identical with the
capillary axis) in time τ (the subscript z will be omitted in what follows).

If the molecule undergoes n collisions with the wall in time τ, $\overline{\Delta^2}$ can be calculated from the equation

$$\overline{\Delta^2} = \left(\sum_{i=1}^{i=n} \lambda_i \right)^2 = \sum_{i=1}^{n} \lambda_i^2 + 2 \sum_{i=1}^{n} \sum_{k=i+1}^{n} \overline{\lambda_i \lambda_k},$$

(8)

in which λ_i and λ_k are the displacements along the capillary axis in the interval between the i-th and (i + 1)-th,
or between the k-th and (k + 1)-th, collisions.

When the Knudsen hypothesis of diffuse molecular reflection is adopted, λ_i and λ_k become independent,
and the summation over $\overline{\lambda_i \cdot \lambda_k}$ products equal to zero, so that

$$\overline{\Delta^2} = n \overline{\lambda_i^2}.$$

(9)

A somewhat different result is obtained if the rather artificial assumption is made that a fraction, f, of
the molecules is diffusely reflected and the remainder, $(1 - f)$, reflected specularly, f being independent

222

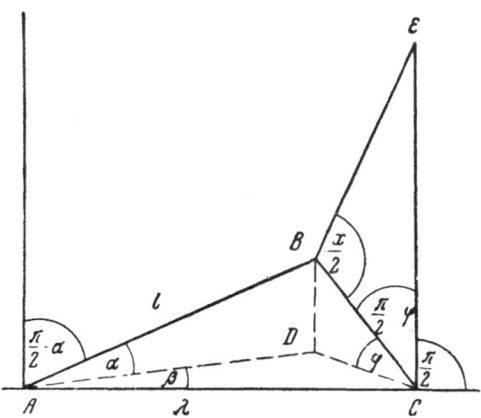

Fig. 4. Calculation of the mean free path and molecular flux in capillaries.

of the angle of incidence. If f is not too low, and n sufficiently large, (3) can be rewritten as

$$\overline{\Delta^2} \cong n\overline{\lambda_i^2} + 2\sum_{i=1}^{n}\sum_{m=1}^{\infty}\lambda_i\lambda_{i+l}.\tag{10}$$

Averaging over the products $\lambda_i\lambda_{i+m}$ must take account of the fact that the only nonzero terms are those associated with paths i and (i+m) which differ in specular reflections; for this one has

$$\lambda_i = \lambda_{i+m}.$$

The probability of specular reflection being $(1-f)$, the probability of n successive reflections of this type will be given by $(1-f)^m$. On this basis, (10) can be rewritten as

$$\overline{\Delta^2} = \overline{n\lambda_i^2}\left[1 + 2\sum_{m=1}^{\infty}(1-f)^m\right].$$

Since the second member in the bracket is the sum of a geometrical progression, the expression itself can be written as

$$\overline{\Delta^2} = \overline{n\lambda_i^2}\left[1 + \frac{2(1-f)}{f}\right].\tag{11}$$

Since

$$\tau = \frac{n\overline{l}}{\overline{c}},\tag{11'}$$

$\overline{l}$ being the mean free path of the molecule between collisions with the wall, and $\overline{c}$ the mean molecular velocity, one obtains from (7)

$$D = \frac{1}{2}\overline{c}\frac{\overline{\lambda_i^2}}{\overline{l}}\left(\frac{2-f}{i}\right).\tag{12}$$

The mean free path is calculated from the formula

$$\overline{l} = \frac{\int lq(\alpha)\,d\omega}{\int q(\alpha)\,d\omega}.\tag{13}$$

Here $q(\alpha)$ is the number of molecules emitted, or reflected, by unit area of wall surface within an infinitesimally small solid angle, $d\omega$, which is directed at an angle $(\pi/2 - \alpha)$ with the normal.

We will determine $q(\alpha)$ from Fig. 4 in which AD is the projection of the free path AB on the tangent plane at the point of reflection, and BCD is the normal cross section of the capillary:

$$q(\alpha)\,d\omega = k\sin\alpha\,d\omega = k\sin\alpha\cos\alpha\,d\alpha\,d\beta.\tag{14}$$

Here k is a factor of proportionality, and β is the azimuth of the reflection plane relative to the generatrix of the capillary surface which passes through the point of impact.

It is seen from Fig. 4 that

$$l = \frac{2r_0 \sin \alpha}{1 - \cos^2 \alpha \cos^2 \beta}.$$ (15)

By substitution of (14) and (15) into (13), one obtains

$$\bar{l} = \frac{2r_0 \cdot \displaystyle\int_{-\frac{\pi}{2}}^{+\frac{\pi}{2}} \int_{0}^{+\frac{\pi}{2}} \frac{\sin^2 \alpha \cos \alpha \, d\alpha \, d\beta}{1 - \cos^2 \alpha \cos^2 \beta}}{\displaystyle\int_{-\frac{\pi}{2}}^{+\frac{\pi}{2}} \int_{0}^{+\frac{\pi}{2}} \sin \alpha \cos \alpha \, d\alpha \, d\beta}.$$

Evaluation of these integrals gives *

$$\bar{l} = 2r_0.$$ (16)

Thus the mean free path of the molecule becomes equal to the capillary diameter when molecular collisions are neglected.

Proceeding in a similar fashion, one determines $\bar{\lambda}_i^2$, the mean square of the projection of the mean free path on the capillary axis:

$$\bar{\lambda}^2 = \frac{\displaystyle\int \lambda^2 q(\alpha) \, d\omega}{\displaystyle\int q(\alpha) \, d\omega}.$$ (16')

Evaluating the integrals, one obtains

$$\bar{\lambda}^2 = \frac{8}{3} r_0^2.$$ (17)

In view of (11), Eq. (12) can now be written as

$$D = \frac{2}{3} \bar{c} r_0 \left(\frac{2-f}{f} \right).$$ (18)

The case $f = 1$ corresponds to Knudsen scattering. Setting the result of (18) into (6) and assuming dc/dz to be constant over the entire length of the capillary, one obtains

$$i = \frac{8}{3} \sqrt{\frac{RT}{2\pi M}} \frac{\Delta C}{L} r_0,$$ (19)

the Knudsen equation, (1).

This expression can be made more precise if f is assumed to be a function of the angle of impingement onto the wall. Equation (11) can then be replaced by

$$\overline{\Delta^2} = n \left\{ \overline{\lambda^2 \left[1 + \frac{2(1-f)}{f} \right]} \right\}.$$ (20)

---

*The same result is obtained from the equation of B. V. Deryagin [5], $l = 4\,V/S$, where V/S is the volume to surface ratio for the cavity.

Leaving (11') unaltered, one then has

$$D = \frac{1}{2} \frac{c}{l} \cdot \overline{\left[\lambda^2 \left(\frac{2-f}{f} \right) \right]}.$$

(21)

If f is never equal to zero, regardless of the angle of incidence, the value of D given by (21) will differ from that of (12) only in the numerical factor k.

In each of these cases, one has to do with a diffusional Knudsen flux in which the density is proportional to the r_0/L ratio. This proportionality entails a flux density tending to zero with diminishing r_0, and is therefore in contradiction with our own experiments on liquid vaporization from microcapillaries.

An entirely different result is obtained if one assumes that

$$f = 0 \text{ when } \alpha < \alpha_0,$$

(22)

α being the angle of incidence and α_0 a critical value of this angle which may be quite low. Equation (21) gives an infinite value for D with this assumption, thus implying that the molecular movement has ceased to be diffusional.

Each molecule which collides with the capillary wall at an angle less than α_0 will clearly retain this angle of incidence, regardless of many impacts (all specular) it experiences, passing through the entire capillary if the length of the latter satisfies the condition

$$L \ll \lambda_0$$

(23)

(λ_0 is the mean free path in the bulk vapor).

The capillary flux due to molecules which impinge on the capillary walls at angles such that $\alpha < \alpha_0$ (glancing molecules) can be obtained by studying the contribution from molecules which have been reflected within the solid angle $d\omega$ from a section, ds, of the capillary surface. Taking account of the fact that the reflected molecules, as well as the incident, are isotropically directionally distributed according to a sine law, one obtains

$$dq = \frac{C\bar{c}}{4} ds \frac{\sin\alpha \cos\alpha d\alpha \, d\beta}{\int\limits_{-\frac{\pi}{2}}^{+\frac{\pi}{2}} \int\limits_{0}^{+\frac{\pi}{2}} \sin\alpha \cos\alpha d\alpha \, d\beta},$$

$C\bar{c}/4$ being the flux density due to all of the molecules* impinging on this surface area (this density is independent of the position of the area). Replacing ds by $2\pi r_0 L$, where $L \gg 2r_0/\alpha_0$, allows for the impingement of molecules on other portions of the capillary surface but results in each molecule being considered L/λ times. It is seen from Fig. 4 that λ is the projection on the capillary axis of the path traced out by the molecule between collisions.

This error is corrected by dividing dq by L/λ, which is to say, by replacing ds by $2\pi r_0 \lambda$.

One then has

$$dq = \frac{\pi}{2} C\bar{c} r_0 \cdot \frac{\lambda \sin\alpha \cos\alpha d\alpha \, d\beta}{\int\limits_{-\frac{\pi}{2}}^{+\frac{\pi}{2}} \int\limits_{0}^{\frac{\pi}{2}} \sin\alpha \cos\alpha d\alpha \, d\beta},$$

---

*The expression for C covers only those molecules with flow directed velocity components along the capillary axis.

or integrating and passing to the flux density,

$$i = \frac{C\bar{c}}{2} \cdot \frac{\displaystyle\int_{-\frac{\pi}{2}}^{+\frac{\pi}{2}} \int_0^{\alpha_0} \frac{\lambda}{r_0} \sin\alpha \cos\alpha \, d\alpha \, d\beta}{\displaystyle\int_{-\frac{\pi}{2}}^{+\frac{\pi}{2}} \int_0^{\frac{\pi}{2}} \sin\alpha \cos\alpha \, d\alpha \, d\beta}.$$

Expressing λ/r_0 in terms of elementary geometrical and trigonometrical functions, one obtains

$$i = C\bar{c} \cdot \frac{\displaystyle\int_{-\frac{\pi}{2}}^{+\frac{\pi}{2}} \int_0^{\alpha_0} \frac{\sin^2\alpha \cos^2\alpha}{1 - \cos^2\alpha \cos^2\beta} \cos\beta \, d\alpha \, d\beta}{\displaystyle\int_{-\frac{\pi}{2}}^{+\frac{\pi}{2}} \int_0^{\frac{\pi}{2}} \sin\alpha \cos\alpha \, d\alpha \, d\beta}.$$

Integration gives

$$i = \frac{C\bar{c}}{2} \cdot \left[1 - \left(1 - \frac{2\alpha_0}{\pi} \right) \cos 2\alpha_0 - \frac{\sin 2\alpha_0}{\pi} \right].$$

With $\alpha = \alpha_0$, one obtains the expression for the rate of vaporization in vacuum, just as was to be anticipated, since $C = C_S/2$ (C_S is the density of the saturated vapor) and the presence of a counter current of the same density is both a necessary and sufficient condition for equilibrium. If the angle α_0 is low, the rate of vaporization is given by

$$i \approx C\bar{c}\,\alpha_0^2 = \frac{C_S}{2}\,\bar{c}\alpha_0^2.$$

When a vapor of density C_S' exists at the open end of the capillary, a counter current is set up in the capillary itself, and the expression for the overall rate of vaporization, i, becomes

$$i = \frac{(C_S - C_S')}{4}\bar{c} \cdot \left[1 - \left(1 - \frac{2\alpha_0}{\pi} \right) \cos 2\alpha_0 - \frac{\sin 2\alpha_0}{\pi} \right], \tag{24}$$

$$i \approx \frac{(C_S - C_S')}{2}\bar{c}\alpha_0^2 \quad \text{when} \quad \alpha_0 \ll 1. \tag{24'}$$

Thus the flux density is independent of both the length and the radius of the capillary.

The movement of the nonglancing molecules is also diffusional, or, more exactly expressed, random, but somewhat different from that which led to Eq. (1). The first point of difference is that the number of these molecules is somewhat less than the total number; this is equivalent to introducing a factor $(1 - \varepsilon)$ into the expression for the flux density, ΔC, ε being the relative concentration of the glancing molecules in the capillary. In calculating a value for ε, account must be taken of the fact that molecules reflected from the wall within the solid angle $d\omega$ contribute to the relative concentration in proportion to their number times $\sin\alpha \cos\alpha \cdot d\alpha \, d\beta$, and therefore

$$\varepsilon = \frac{\int_0^{\alpha_0} \sin\alpha \cos\alpha \, d\alpha}{\int_0^{\frac{\pi}{2}} \sin\alpha \cos\alpha \, d\alpha} = \sin^2\alpha_0.$$

Thus the value of ε is low when α_0 is small.

The second point of difference goes back to the fact that the principle of microscopic reversibility requires an alteration of the Knudsen law of diffuse evaporation as a result of specular reflection of the glancing molecules. This alteration will be such that adsorbed molecules are not permitted to evaporate at angles such that $\alpha < \alpha_0$, thus giving an additional reduction in the flux density (evaporation rate) which is also small for small values of the limiting angle α_0. This alteration in Eq. (1) is insignificant since we are only concerned with the relative order of the diffusional component of the molecular flux.

Although the ordinary diffusional component of the Knudsen flow of Eq. (1) falls off as the length of the capillary increases and its radius diminishes at fixed pressure difference, the flow is almost completely described by Eqs. (24) and (24') when the conditions are such that

$$\frac{r_0}{L} \ll \alpha_0^2. \tag{25}$$

These equations are correct, however, only if the condition of (23) is fulfilled and there are no collisions with other molecules. The glancing molecules then pass through the capillary without resistance. On the other hand, the opposite condition, i.e.,

$$L \gg \lambda_0, \tag{26}$$

applied over most of the capillary in our own experiments. Over every section of the capillary distant ΔL from the end and satisfying the condition

$$L \gg \Delta L \gg \lambda_0 \tag{27}$$

there will be only a negligibly small number of glancing molecules which fail to experience a collision with another molecule after evaporation; all such molecules are the result of collisions between nonglancing molecules, in which one glancing molecule is accidentally generated. This reestablishes isotropic equilibrium distribution of velocities over the glancing molecules, at least to a first approximation. A second approximation shows that the distribution of directions of motion in the glancing molecules differs from the equilibrium distribution (characterized, in particular, by the fact that the numbers of oppositely directed molecules are identical) by a factor which is proportional to λ grad C. The numbers of glancing molecules leaving the two ends of the capillary is determined by the difference ΔC, when $L \ll \alpha_0$. Although the gas flux is radically reduced in the case under discussion, it is again completely diffuse and quite analogous to a Knudsen flow, being directly proportional to the concentration gradient and inversely proportional to L.

The flux is determined quantitatively by the expression for the projection of the displacement of the molecule along the capillary axis in time $\Delta\tau$:

$$\Delta = \sum_{i=1}^m \mu_i + \sum_{i=1}^m \nu_i. \tag{28}$$

Here μ_i is the total axial displacement of the glancing molecule prior to the first collision which interrupts the i-th series of specular reflections, and ν_i is the total axial displacement prior to initiating a new, $(i+1)$-th, cycle of specular reflections. Since the axial displacements, both μ_i and ν_i, are independent of one another, it follows that one will have

$$\overline{\Delta^2} = \sum_i \left(\overline{\mu_i^2} + \overline{v_i^2} \right) = \sum_i \overline{\mu_i^2} + \sum_i \overline{v_i^2}. \tag{29}$$

The contribution of the second summation in the right-hand member of (29) to the coefficient D is approximately equal to the result obtained from Eq. (18) with $f = 1$ when the angle α_0 is small. We will now estimate the contribution from the first summation of (29).

It is clear from Fig. 4 that

$$\mu_i = p\,\lambda = p\,l \cos \alpha \cos \beta = l_i \cos \alpha \cos \beta,$$

p being the number of ricochets and l_i the path of the glancing molecule from formation to collision with another molecule, a process which will, generally speaking, remove it from the "glancing" class. It is obvious that

$$\sum_{i=1}^{i=m} \mu_i^2$$

can be set equal to the integral

$$\sum_{i=1}^{i=m} \mu_i^2 = \int a \overline{l^2} \cos \alpha \cos \beta \, d\omega$$

for large values of m, $ad\omega$ being the number of paths directed in the solid angle $d\omega$ inclined at the angle α to the wall per time τ. In view of the isotropic distribution of molecular velocities, $a = k \sin \alpha$, where

$$\int_{-\pi}^{+\pi} \int_0^{\frac{\pi}{2}} \sin \alpha \cos \alpha \, d\alpha \, d\beta = n,$$

n being the total number of collisions of the molecule in question with other molecules in time τ. From this it follows that when $\alpha_0 \ll 1$

$$\overline{\Delta^2} = \frac{n}{\pi} \int_{-\pi}^{+\pi} \int_0^{\alpha_0} \overline{l_i^2} \cos \beta \cos^3 \alpha \sin \alpha \, d\alpha \, d\beta = n \overline{l_i^2} \cdot \frac{\alpha_0^2}{2}.$$

For a Clausius distribution of path length, l_i, $[\sim e^{-(l_i/\lambda_0)}]$, and $\overline{l_i^2} = 2\lambda_0^2$. It follows by analogy with (11') that one will also have $\tau = n\lambda_0/c$. The contribution to the coefficient D is then obtained from (7) as

$$D_0 = \overline{c}\lambda_0 = \frac{\lambda_0^2}{2}. \tag{30}$$

It is clear that one can neglect the contribution of the nonglancing molecules to D and the vaporization rate when the conditions are such that

$$L \gg \lambda_0 \gg \frac{r_0}{\alpha_0^2}, \tag{31}$$

a situation met in very narrow capillaries (λ is not very large for water vapor and λ_0 rather small).

Thus even if condition (26) is fulfilled, the flux density of the gas, i, will cease to be proportional to r_0 if the capillary is narrow enough for (31) to be satisfied, and the flux will tend to a fixed value which will depend only on λ_0 and $\overline{c}$, which is to say, only on the temperature and the gas density. The dependence on the capillary length, L, remains as before. Since $\bar{\lambda}_0$ is inversely proportional to C, (6), (30), and the continuity condition for stationary flux indicate that

$$i = \frac{\beta \bar{c}}{C} \frac{dC}{dL} = \text{const,} \tag{32}$$

β being a constant defined by the expression

$$\beta = \frac{\alpha_0^2 \gamma}{2} \tag{32'}$$

(here $\gamma = \lambda_0 \cdot C$ is a constant for the gas (vapor) which depends on the molecular diameter and is equal to $3.2 \cdot 10^{-9}$ for water vapor). Integration then leads to

$$C = C_0 e^{-\frac{L_i}{\beta \bar{c}}}, \tag{33}$$

$$\rho \frac{dL}{d\tau} \equiv i = \frac{\beta \bar{c}}{L} \ln \frac{C_0}{C_1}, \tag{34}$$

where ρ is the density of water, C_0 is the vapor concentration at the meniscus, and C_1 the concentration near the mouth of the capillary, but still inside of it. The relation between flux and gas concentration at the extremities of the capillary is seen to be quite different from that met in Knudsen flow.

Equations (32) and (34) are not directly applicable to vaporization in vacuum, since condition (27) is no longer fulfilled when $C_1 = 0$ at the free end; in addition, this condition is incompatible with gas flow through the capillary. Thus we have not used $C_1 = 0$ as a boundary condition for the first approximation determination of C_1, but have attempted to base this determination on the expression for the density of effusional discharge into vacuum

$$i = \frac{C_1 \bar{c}}{4}. \tag{35}$$

From this we find that

$$C_1 = \frac{4i}{\bar{c}}. \tag{36}$$

Substituting the expression of (34) for i, and dividing both sides of the equation by C_0, we obtain

$$\frac{C_1}{C_0} = \frac{4\beta}{LC_0} \ln \frac{C_0}{C_1}. \tag{37}$$

Let us write

$$\ln \frac{C_0}{C_1} = u. \tag{38}$$

The value of u is to be determined from the transcendental equation

$$ue^u = \frac{LC_0}{4\beta}, \tag{39}$$

which can be solved graphically. The result is a functional relation of the form

$$u = f\left(\frac{LC_0}{4\beta}\right). \tag{40}$$

229

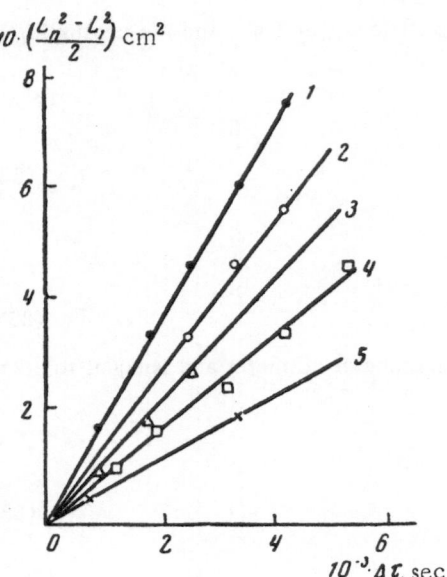

Fig. 5. The kinetics of water vaporization from capillaries with 200-300 A radii, evaporation being into a water vapor atmosphere at the relative humidity, W: 1) $W = 0.343$; 2) $W = 0.458$; 3) $W = 0.552$; 4) $W = 0.657$; 5) $W = 0.745$.

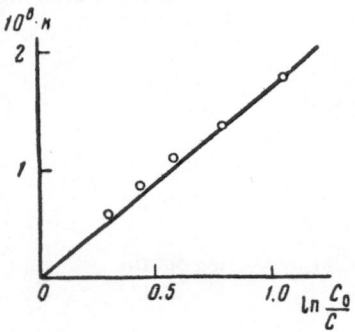

Fig. 6. The relation between the slopes of the curves for water vaporization from microcapillaries and the logarithm of the ratio of the vapor densities, at the meniscus and in the surrounding space.

Substitution of this expression for u into Eq. (34) gives i. Since C_1 enters Eq. (34) logarithmically, this method of determining i is much more accurate than Eq. (36):

$$i = \frac{\beta \bar{c}}{L} u. \qquad (41)$$

It is seen from (39) that this expression for the density of gas flux from capillary to vacuum is unusual in predicting only a very weak dependence of i on the molecular concentration, C_0, at the meniscus, or capillary mouth, and its complete independence of the capillary radius. Furthermore, i is seen to be almost exactly inversely proportional to L.

Test of the Theory

The theoretical results obtained here were tested by experiments on the evaporation of water from microcapillaries of 200-300 A radius into an atmosphere of unsaturated water vapor.

The experimental procedure differed from that described above in that the time required for vaporization from one position to another was measured, rather than the evaporation rates at various depths. The capillaries were held in the thermostat for 7-10 days, being taken out for 1-2 minutes each day to permit determination of the position of the meniscus.

By integrating Eq. (34), one obtains

$$\frac{1}{2}\left(L_n^2 - L_1^2\right) = \beta \bar{c} \Delta \tau \ln \frac{C_0}{C_1}, \qquad (42)$$

where L_1 is the meniscus position corresponding to the first measurement with the zero reading taken at the free end of the capillary, L_n is the position at the n-th measurement, and $\Delta \tau$ is the time interval between readings.

Figure 5 shows the relation between $(1/2)\,(L_n^2 - L_1^2)$ and $\Delta \tau$ for various values of the relative vapor pressure, in the surrounding atmosphere and at the liquid meniscus. Figure 6 shows that there is a clear-cut linear relation between the slopes, k, of the lines of this graph and $\ln C_0/C_1$. This confirms the validity of Eq. (34) and thus supports the assumption of specular reflection on which the description of microcapillary vapor movement has been made to rest. Since the slope of the line of Fig. 6 is equal to $\beta \bar{c}$, one finds the value $\beta = 2.7 \cdot 10^{-11}$ for water vapor. A further check on the theory is obtained by comparing values for vaporization into vacuum calculated from Eq. (41) with those directly measured. These measurements show the vapor flux density from capillaries 2 mm long (L) and of radii less than 800 A to be $0.42 \cdot 10^{-4}\,\mathrm{g \cdot cm^{-2} \cdot sec^{-1}}$ at 20°C. Solution of Eq. (39) gives $u = 8.25$. Setting this value back into (41) gives $i = 0.63 \cdot 10^{-4}\,\mathrm{g \cdot cm^{-2} \cdot sec^{-1}}$. The difference between these two results is not large, especially in view of the assumptions involved in the calculations and the errors of measurement. The limiting angle for specular reflection (in radians) can be calculated from Eq. (32') and the value of β obtained from Fig. 6:

$$L_0 = \sqrt{\frac{2\beta}{\gamma}} = 0.13.$$

Thus the limiting angle for specular reflection of water vapor molecules from a glass wall is approximately 7°.

Conclusions

1. Study has been made of the vaporization of water from glass capillaries with radii, r_0, ranging from 0.02 to 3 μ. The data obtained with capillaries ranging from 0.5 to 1.5 μ in radius are in agreement with the results of calculations based on the Knudsen equation.

2. The work with narrower capillaries showed departures from the Knudsen equation, the flux being proportional to r_0^2 rather than to r_0^3. This indicated the necessity of introducing a new hypothesis into the theory of molecular flow.

3. At angles of incidence less than a certain critical value, α_0, all water molecules are specularly reflected from the glass surface.

4. As a result, the adsorbed water molecules cannot evaporate from the glass surface at angles less than a certain critical value.

5. The existence of this limiting angle for specular reflection leads to an alteration in the laws of molecular flow in very narrow capillaries.

6. When the conditions are such that $\lambda_0 \gg L \gg r_0/\alpha_0^2$ (L is the length of the capillary, and λ_0 is the mean free path of the gas or vapor molecule), the gas flux is proportional to $(C_0 - C_1)r_0^2\alpha_0^2$ and independent of L, C_0 and C_1 being the gas densities at the two ends of the capillary.

7. When the conditions are such that $L \gg \lambda_0 \gg r_0/\alpha_0^2$, the gas flux is proportional to $\lambda_0 C_0 (r_0^2\alpha_0^2/L) \ln(C_0/C_1)$, ($\lambda_0$ is the mean free path at a gas density C_0), a result which has been experimentally confirmed.

8. Knowing α_0, one can theoretically calculate the rate of vaporization from capillary to vacuum, obtaining results which are in satisfactory agreement with experiment.

9. Comparison with experiment leads to the value $\alpha_0 = 7°$ for water vapor.

Literature

1.   M. Knudsen. Ann. Physik. 28:759 (1909).
2.   M. Smoluchovski. Ann. Physik. 33:1559 (1911).
3.   N. N. Fedyakin. Dokl. Akad. Nauk SSSR 138:1389 (1961).
4.   N. N. Fedyakin. Zh. Fiz. Khim. 36:1450 (1962).
5.   B. V. Deryagin. Dokl. Akad. Nauk SSSR 53:627 (1946).

THE EFFECT OF CERTAIN SURFACE ACTIVE SUBSTANCES ON THE COAGULATION OF SILVER IODIDE HYDROSOLS

Yu. M. Glazman and I. P. Sapon

Kiev Technological Institute for Light Industry

The distinguishing characteristic of lyophobic sols is their extraordinary sensitivity to foreign electrolytes. These systems are stabilized by the ionic double electric layers which form around the colloidal particles.

Lyophobic sol stability to coagulation under the action of small quantities of added electrolyte is, in our opinion [1], largely due to the molecular component of the splitting pressure [2]; this arises, according to B. V. Deryagin [3-6], through the formation of a polymolecular liquid layer of peculiar properties on the dispersed phase surface. It is clear that the state of this boundary phase will be closely dependent on the nature of the surface and the type of molecular forces which emmanate from it; more particularly, it must be profoundly altered by the adsorption of surface active materials, this adsorption increasing the depth of the lyosphere and and the splitting pressure effect of the solvated layers, as well [4].

These considerations (see, also, [1]) lie at the basis of our studies on the effect of surface active materials on the stability of hydrophobic sols to coagulation under the action of electrolytes. We have limited our initial investigations to the simple relations which apply when nonionogenic surface active materials are added to typical hydrophobic colloidal systems.

The present work was carried out with positively and negatively charged silver iodide hydrosols; these contained the dispersed phase at a concentration of 10 mmole/liter and were used without prior purification. These sols were prepared in the usual manner by reacting $AgNO_3$ and KI with either a 5% excess of the potassium salt or a 10% excess of the silver nitrate. The surface active substances were polyglycol ethers which differed in the length of the ethoxy chain.* The proposed formulas for these compounds were

$C_{18}H_{37}O(CH_2CH_2O)_{12}H$ (mol. wt.798),
$C_{18}H_{37}O(CH_2CH_2O)_{27}H$ (mol. wt.458),
$C_{18}H_{37}O(CH_2CH_2O)_{48}H$ (mol.wt.2382),
$C_{18}H_{37}O(CH_2CH_2O)_{98}H$ (mol.wt.4582).

The coagulating electrolytes used here were lithium chloride and sulfate, ammonium nitrate and sulfate, and calcium chloride and nitrate. Critical electrolyte concentrations were determined with a

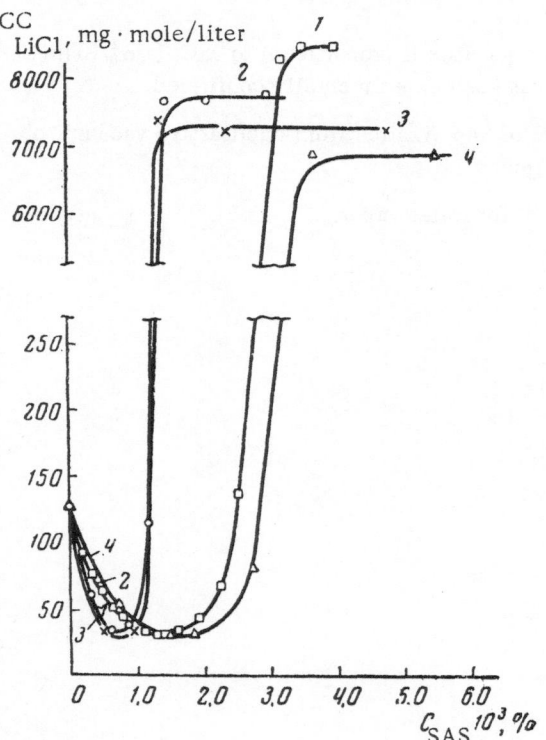

The effect of surface active substances (SAS) on the coagulation of a negatively charged silver iodide sol by lithium chloride. 1) $C_{18}H_{37}O(CH_2CH_2O)_{12}H$; 2) $C_{18}H_{37}O(CH_2CH_2O)_{27}H$; 3) $C_{18}H_{37}O(CH_2CH_2O)_{48}H$; 4) $C_{18}H_{37}O(CH_2CH_2O)_{98}H$.

---

* These materials were kindly furnished by O. K. Smirnov and T. M. Moshkina, to whom we express our deep thanks. .

TABLE 1. The Effect of Surface Active Substances on the Stability of Negatively Charged Sols

Electrolyte	Coagulating concentration, mmole/liter			Electrolyte	Coagulating concentration, mmole/liter		
	Original sol	Sol with SAS			Original sol	Sol with SAS	
		CC_{min}	CC_{max}			CC_{min}	CC_{max}
I. $C_{18}H_{37}O(CH_2CH_2O)_{12}H$				**III. $C_{18}H_{37}O(CH_2CH_2O)_{48}H$**			
LiCl	125	31	8500	LiCl	125	32	7300
$CaCl_2$	1.4	1.4	>4200	$CaCl_2$	1.4	1.4	>4200
				Li_2SO_4	75	18	980
II. $C_{18}H_{37}O(CH_2CH_2O)_{27}H$				**IV. $C_{18}H_{37}O(CH_2CH_2O)_{98}H$**			
LiCl	125	33	7700	LiCl	125	32	6900
$CaCl_2$	1.4	1.4	>4200	$CaCl_2$	1.4	1.4	>4200

FÉK-M photoelectric colorimeter, using a curve showing the relation between optical absorption and amount of electrolyte added to the sol [7]. Whenever possible, the effect of the surface active substance was studied all the way from the very smallest amounts up to a plateau, which is to say, a region in which further increase in the concentration did not appreciably alter the observed effect. The effect of these surface active substances on the stability of the test colloidal solutions is illustrated by the figure which shows the coagulating of lithium chloride on a negatively charged silver iodide sol and its variation with the concentration of added surface active substance. All of the surface active substances studied here behaved in the same way, a pronounced reduction in the stability at very low concentrations ($\approx 0.001\%$) of the surface active substance in which the coagulating concentration (CC) of lithium chloride was reduced by a factor of 1:3.5 being followed by a sharp almost 200-fold increase, in the sol stability. It should be noted that though the stabilizing effect is general and always appears, sensitization may be completely lacking, especially when coagulation is by an electrolyte containing divalent counterions (see Table 1).

Another fact stands out in studying the data of the table. Sol stability with respect to an electrolyte containing a divalent coagulating ion (calcium chloride) is increased more rapidly by SAS addition (more than 3000 times) than is stability with respect to lithium chloride. A silver iodide sol protected by a surface active substance cannot, in general, be coagulated by $CaCl_2$, the limit of solubility being reached before one enters the coagulation region. Thus though the critical concentrations of LiCl and $CaCl_2$, are markedly different in the original sol (roughly conforming to the Z^6 Deryagin—Landau law), the two become of the same order of magnitude after a surface active substance is added to the AgI solution.

These results can be interpreted on the basis of the concepts which we have advanced to account for the stability of lyophobic and lyophilic sols [1]. The adsorption of surface active substances on the colloidal hydrophobic particles of silver iodide clearly leads to hydrophilization of these surfaces. Coagulation of the resulting hydrophilic sol requires a pronounced reduction in the molecular component of the splitting pressure, the factor responsible for the stability of the colloidal system to aggregation. This result can be achieved only at high concentrations of electrolyte where breakdown of the solvated boundary layers occurs. It is obvious that the valence of the counterion of the electrolyte is of no significance here. The data presented show that such is actually the case.

These considerations should apply, regardless of the charge carried by the colloidal particles in the original sol. Thus it was of interest to study the effect of surface active substances on positively charged AgI hydrosols. Even here (see Table 2), the system stability was found to be increased markedly by the addition of a surface active substance. Now, however, the critical concentrations of electrolytes containing monovalent and divalent coagulating ions (NH_4NO_3 and $(NH_4)_2SO_4$) were markedly different (approximately six-fold) * even when the sol was protected by an SAS.

---

*It will be shown below that NH_4^+ is the counterion when the system contains an added surface active substance; for this reason, the coagulating concentrations of ammonium nitrate and sulfate differ only by a factor of three.

TABLE 2. The Effect of Surface Active Substances on the Stability of Positively Charged AgI Sols

Electrolyte	Initial sol	Coagulating concentration, mmole/liter			
		Sol with SAS			
		$C_{18}H_{37}O(CH_2CH_2O)_{12}H$ $(CC)_{max}$	$C_{18}H_{37}O(CH_2CH_2O)_{27}H$ $(CC)_{max}$	$C_{18}H_{37}O(CH_2CH_2O)_{48}H$ $(CC)_{max}$	$C_{18}H_{37}O(CH_2CH_2O)_{98}H$ $(CC)_{max}$
NH_4NO_3	20	5400	5900	5900	6300
$(NH_4)_2SO_4$	1	900	1000	1000	1000
$Ca(NO_3)_2$	10	>3500	>3500	>3500	>3500

The assumption could be made that this result does not follow from the charge on the counterion, but is due to an increase in the solvating effect of ammonium sulfate, tracing back to the position of the SO_4^{2-} ion in the lyotropic series, and it therefore seemed desirable to study the coagulating action of lithium sulfate on a negatively charged silver iodide sol.

Although the coagulating concentrations of lithium sulfate and chloride differed by only 15-20% for the original sol (see Table 1), the critical concentrations of these electrolytes for the SAS-stabilized colloidal solution differed by almost 3.5-fold. Thus these data confirm the validity of the above assumption, at least in its qualitative aspects.

Other, quite simple, experiments also indicate that the stabilizing action of the surface active substances in the cases under study here involves hydrophobization of the dispersed phase surface through adsorption of SAS molecules on the colloidal particles. Thus, coagulation of the SAS-protected sol is reversible and therefore differs from the coagulation of the original sol; in fact the silver iodide is completely peptized when the electrolyte responsible for flocculation is washed away. Experiments on selective wetting show convincingly that the coagulate obtained from the original colloidal solution is hydrophobic, while a sol to which a surface active substance has previously been added is hydrophilic; on shaking the aqueous suspension with benzene, the precipitate passed into the nonpolar liquid in the first case, and remained entirely in the aqueous phase in the second.

Certain information appearing in the literature [8], would seem to imply that particles stabilized by a nonionogenic substance would be free of electrical charge. Our own data do not confirm this assumption, however. On the contrary, the particles of such sol undergo displacement in electrophoresis, and remarkably enough, prove to be negatively charged after SAS stabilization, regardless of their charge sign in the original sol. These results are quite consistent with the ideas of [9] concerning the quasi-ionogenic nature of the hydrated molecules of those members of the polyethoxy series which we used as surface active substances. In explaining the observed effect it must be remembered that the molecules of most surface active substances orient in the same manner at interfaces, the negative end of the dipole being directed toward the aqueous phase [10-12]. The result of this orientation is to produce a rather considerable positive potential gradient, which can amount to several tenths of a volt [10, 11]. Thus in the presence of a surface active substance, the total potential difference at the interface between the colloidal particles and the dispersing medium (φ) is composed of two additive terms, the ionic potential of the double electric layer, φ_i, and a purely adsorptional potential difference, φ_a, which arises in an oriented layer of polar organic molecules situated in the liquid phase. Since it is the total potential difference ($\varphi = \varphi_i + \varphi_a$) at the phase interface which fixes the activity of the potential-determining ions in the solution [13], the appearance of an adsorptional potential difference, φ_a, must invariably lead to an alteration in φ_i. If φ_i and φ_a are of opposite sign, the appearance of φ_a must give a more or less pronounced increase in the absolute value of φ_i; if the two potential differences are of the same sign, orientation of the polar SAS molecules at the surface of the dispersed phase can not only reduce the value of φ_i but change in the sign of the latter, as well. This last situation is the one more likely to be met, since the value of φ_a is rather large, as has been pointed out above. Such charge inversion has been observed in our own experiments on the addition of surface active substances to positively charged silver iodide sols. Here the necessity arose for supplementary studies of the coagulating action of electrolytes with divalent cations on an SAS-supercharged AgI sol. The critical concentrations of ammonium and calcium nitrates proved to be about the same (see Table 2), just as was to be expected (the coagulation concentration of $Ca(NO_3)_2$ could not be determined be-

TABLE 3. Critical Concentrations for Micell Formation

SAS	Critical concentration for micell formation, %	
	c_{K_1}	c_{K_2}
$C_{18}H_{37}O(CH_2CH_2O)_{12}H$	0.004	0.25
$C_{18}H_{37}O(CH_2CH_2O)_{27}H$	0.004	0.23
$C_{18}H_{37}O(CH_2CH_2O)_{48}H$	0.006	0.38
$C_{18}H_{37}O(CH_2CH_2O)_{98}H$	0.011	0.73

cause of the limited solubility ot the salt). In this respect the positive and negative charged silver iodide sols behave similarly, addition of an SAS to the colloidal solution leading to the formation of a hydrophilic sol which is quite insensitive to electrolyte action. On the other hand, the peculiarities of the positive charged AgI sol appear when the added quantity of surface active substance is very small. The degree of sensitization reaches its maximum here, surface active substance bringing about coagulation of the colloidal solution even without the addition of the electrolyte. It is clear that this effect is related to sol supercharging under the action of the SAS, this process inevitably carrying the colloidal particle potential through the null point.

The fact that the presence of electrically charged particles has been demonstrated in an AgI solution stabilized with surface active substances is indication that there must be two factors contributing to sol stability here. In order to clearly demonstrate this point, we have repeated what is essentially the classic experiment of Kruyt and de Jong [14]. Electrolyte was added to a colloidal solution of silver iodide which had been SAS stabilized in the plateau region, addition being continued to the threshold concentration for coagulation of the unprotected colloid. This, of course, brought about no appreciable change in the system. Coagulation was then brought about by addition of ethyl alcohol, which broke down the hydrated boundary layers in the system once a certain concentration had been reached. The result obtained here was independent of the order of mixing, being the same if the alcohol was first added to the earlier concentration and an insignificant amount of electrolyte then introduced.

The surface active substance studied in this work are semicolloidal compounds, and it was therefore necessary to determine whether the SAS stabilization was related to the appearance of colloidal micells in the solution. For this purpose, the critical concentration (C_{cr}) for micell formation of each of these compounds was determined colorimetrically [15].

The results obtained are shown in Table 3, from which it is seen that micell formation is associated with a range of concentrations extending from C_{cr_1} to C_{cr_2}, just as indicated by the literature [16]. The region from initiation to completion of micell formation in solution is rather extensive. It should be emphasized that the point of sharp increase in sol stability (plateau region) is reached at lower concentrations of the surface active substance than would correspond to C_{cr_1}. This is indication that the SAS stabilization of lyophobic colloids which we have observed is not related to direct micell formation in solution.

The sol sensitization observed at very low concentrations of the surface active substance can be said to result from interaction between the oriented polar SAS molecules on the colloidal particles. Theoretical calculations have shown that this interaction must give rise to an attraction [17]. A purely descriptive explanation of this effect can be given if an oriented dipole on the one surface is assumed to project into the space between two oriented dipoles on the other.

On the basis of the Deryagin theory of lyophobic sol stability [18], it can be shown that the "critical gap," i.e., the distance of closest particle approach for unimpeded mutual attraction, is not the same for coagulation with various electrolytes, and is small (of the order of several molecular layers) only in the case of monovalent counterions [19]. Since dipole interaction rapidly diminishes with the distance of separation, it is understandable that sol sensitization by surface active substance can be observed in coagulation by univalent, but not divalent, counterions (see Table 1).

Increasing the SAS concentration of the solution entails an increase in the degree of coverage of the colloidal particles and leads to predominance of effects from hydrophilization of the dispersed phase surface. This lead, in turn, to a sharp increase in the stability of the colloidal solution.

Conclusions

1. Study has been made of the effect of ionogenic surface active substances (polyglycol ethers) on the stability and coagulation of silver iodide.

2. Very small additions of surface active substances ($\approx 0.001\%$) either increase the sensitivity of a negative charged silver iodide sol to electrolytes (in the case of univalent coagulating ions), or leave this stability unaltered (coagulation by electrolytes with divalent counterions); they themselves lead to the coagulation of a positive charged AgI sol. Sol sensitivization can be explained in terms of orientation of the polar SAS molecules at the surface of the particles of the colloidally dispersed phase.

3. The stability of a silver iodide sol with respect to electrolytes increases sharply (by a hundred or even a thousand fold) as the concentration of the surface active substances rises.

4. Pronounced sol stabilization sets in at concentrations of surface active substance much lower than the critical concentration for micell formation.

5. The results obtained here have been interpreted in terms of hydrophilization of the colloidal particle surface as a result of the adsorption of molecules of the surface active substance. This explanation has been confirmed by other experimental data.

The authors wish to express their deep thanks to B. V. Deryagin, Corresponding Member, Acad. Sci., USSR, for valuable advice, discussion of the results obtained, and a critical review of the manuscript.

Literature

1. Yu. M. Glazman. Kolloidn. Zh. 24:275 (1962).
2. B. V. Deryagin and S. V. Nerpin. Dokl. Akad. Nauk SSSR 99:1029 (1954).
   B. V. Deryagin, N. I. Moskvitin, and M. F. Futran. Proceedings, Third All-Union Conference on Colloidal Chemistry, Moscow, Izd. Akad. Nauk SSSR (1956), p. 285.
   B. V. Deryagin. Proceedings, Conference on the Engineering and Geological Properties of Rocks and Their Methods of Study, Moscow, Izd. Akad. Nauk SSSR 1:45 (1956); Dokl. Akad. Nauk SSSR 109:967 (1956).
   T. H. Voropaeva, B. V. Deryagin, and B. N. Kabanov. Dokl. Akad. Nauk SSSR 128:981 (1959); in collection: Research in Surface Forces, Moscow, Izd. Akad. Nauk SSSR (1960), p. 143. [English translation: Consultants Bureau, New York (1963).]
3. B. V. Deryagin. Zh. Fiz. Khim. 3:29 (1932); 5:379 (1934).
   B. V. Deryagin and E. Obukhov. Kolloidn. Zh. 1:385 (1935).
   B. V. Deryagin and M. M. Kusakov. Izd. Akad. Nauk SSSR, Ser. Khim. Vol. 741 (1936).
   B. V. Deryagin, M. Kusakov, and L. Lebedeva. Dokl. Akad. Nauk SSSR 23:670 (1939).
   M. M. Kusakov and A. S. Titievskaya. Dokl. Akad. Nauk SSSR 28:333 (1940).
   B. V. Deryagin, M. M. Kusakov, and A. S. Titievskaya. Kolloidn. Zh. 6:304 (1940).
   B. V. Deryagin. Priroda, No. 2:23 (1943); Proceedings, All-Union Conference on Colloidal Chemistry, Kiev, Izd. Akad. Nauk SSSR (1952), p. 26.
   B. V. Deryagin and M. M. Kusakov. Zh. Fiz. Khim. 26:1536 (1952).
   B. V. Deryagin and A. S. Titievskaya. Dokl. Akad. Nauk SSSR 89:1041 (1953); Kolloidn. Zh. 15:416 (1953).
   N. I. Moskvitin, M. F. Futran, and B. V. Deryagin. Dokl. Akad. Nauk SSSR 105:758 (1955).
   B. V. Deryagin. Kolloidn. Zh. 17:207 (1955).
4. B. V. Deryagin. Miner. Syr'e 9(2):33 (1934).
   B. V. Deryagin and M. M. Kusakov. Izd. Akad. Nauk SSSR, Ser. Khim. No. 5:1119 (1937).
5. G. I. Fuks. Zavodsk Lab., No. 12:1455 (1955); Proceedings, Third All-Union Conference on Colloidal Chemistry, Moscow, Izd. Akad. Nauk SSSR (1956), p. 301.
6. N. N. Fedyakin. Dokl. Akad. Nauk SSSR 138:1389 (1961); Zh. Fiz. Khim. 36:1450 (1962); Kolloidn. Zh. 24:497 (1962).
7. Yu. M. Glazman. Kolloidn. Zh. 15:334 (1953).
   Yu. M. Glazman and E. F. Zhel'vis. Proceedings, Third All-Union Conference on Colloidal Chemistry, Moscow, Izd. Akad. Nauk SSSR (1956), p. 341.

8. K. Meguro and T. Kondo, J. Chem. Soc. Japan, Pure Chem. Sect. 76:642 (1955).

9. T. V. Karabinos, G. E. Kapella, H. T. Ferlin, and D. L. Sawhill. Euclides (Madrid) 15 (174–175):253 (1955).

10. M. Gouy, Compt. Rend. 146:612 (1908); Ann. Chim. Phys. 7(9):129 (1917).

11. A. Frumkin, Z. Phys. Chem. 111:190 (1934).

12. D. N. Strazhesko. Doctoral dissertation, Kiev (1951).

13. A. Frumkin, Z. Phys. Chem. 103:55 (1923).

14. H. R. Kruyt and H. G. Bungenberg de Jong. In collection: The Coagulation of Colloids (translated under the editorship of A. I. Rabinovicha and P. S. Vasil'eva), ONTI (1936); p. 193.
    H. R. Kruyt and H. G. Bungenberg de Jong. Kolloid.—Beih. 28:1 (1928).

15. A. B. Taubman, V. V. Konstantinova, and A. S. Kryukova. The Chemistry and Technology of Fuels and Oils, No. 3:61 (1960).

16. A. B. Taubman and S. A. Nikitina. Dokl. Akad. Nauk SSSR 135:1179 (1960).

17. G. A. Martynov and V. P. Smilga. Kolloidn. Zh. (in press).

18. B. V. Deryagin and L. D. Landau. Zh. Éksperim. i Teor. Fiz. 11:802 (1941); 15:663 (1945).

19. Yu. M. Glazman. Kolloidn. Zh. 15:334 (1953).

THE BEHAVIOR OF SOLS AND SUSPENSIONS
IN THE MAGNETIC FIELD

E. E. Bibik, I. F. Efremov, and I. S. Lavrov

Lensovet Leningrad Technological Institute

Although magnetic properties find wider and wider application in the most diversified fields of human endeavor, the effect of the magnetic field on dispersed systems has received little study. The effect of the rotating magnetic field on certain liquid crystals and colloidal solutions has been investigated through magneto-optical measurements, analysis being on the basis of the theory developed by P. Langevin [1] and V. N. Tsvetkov [2]. Here the form and dimensions of the colloidal particles were determined from the lag angle between the optical axis of the solution and the field strength vector. The optical and magnetic properties of suspensions have also been treated by Elmor [3].

Lourila [4] has considered the effect of nonferromagnetic inclusions (or pores) on the magnetic permeability of ferromagnetic substances, and discussed the behavior of ferromagnetic powders in the rotating magnetic field. A method for determining the dimensions of highly dispersed particles from the alteration in the intensity of transmitted light caused by the magnetic field has been proposed by Rose [5]. Tranklin and others [6] have determined the dimensions of finely dispersed ferromagnetic powders by using electron microscopy, x-ray scattering, and nitrogen adsorption. A. P. Soboleva [7] has used an electron-optical method to study the microfields in ferromagnetic materials.

Phoresis (movement) of suspended particles under the combined action of the electric current and a magnetic field has been studied in various papers [8], as well as the magnetophotophoretic effect in ferromagnetic particles [9]. Bagguley [10] has studied resonance effects in colloidal systems containing ferromagnetic substances.

Harvey [11] has studied the effect of the magnetic field on the rheological properties of suspensions of ferromagnetic substances. By using a rotating viscosimeter with a narrow cylinder gap, it was shown that suspensions of iron and iron oxide have enhanced viscosity in the magnetic field. Studies which M. P. Volarovich [12] carried out on various molecular and colloidal solutions in the capillary viscosimeter failed to disclose any alteration of the viscosity with the time of holding in a strong magnetic field.

The literature also contains reference to an effect of the magnetic field on the crystallization of various substances (gypsum, sugar, etc.) from solution, a possibility that could find application to industrial processes [13]. G. L. Mikhnevich and others [14] have studied the effect of the magnetic field on the kinetics of betol crystal nucleation in supercooled solutions and the structure of the wall layer in the betol solution.

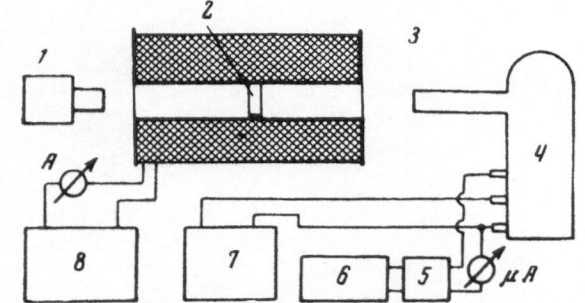

Fig. 1. Schematic representation of the appratus for measuring the intensity of light transmitted through the sol. 1) Light source; 2) cell with sol; 3) solenoid; 4) photomultiplier; 5) direct current amplifier; 6) loop oscillograph; 7) power source for the photomultiplier; 8) power source for the solenoid.

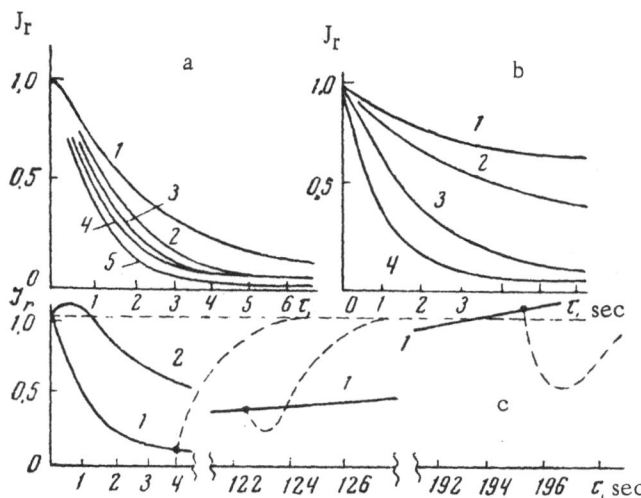

Fig. 2. The relation between the relative intensity of the light transmitted through the sol ($J_r = J_{mag}/J_{wf}$) and the exposure time. a) Curves 1, 2, 3, and 4, solenoid current, I, 1, 2, 4, and 8 amp, respectively; curve 5, sol with electrolyte, solenoid current strength, 8 amp; b) curves 1, 2, 3, and 4, same, for sols of respective concentrations 0.01, 0.04, 0.1, and 0.4%; c) curve 1 for various times of interaction of field with sol containing sodium oleate; dotted curve, the same, but after cutting out the field; curve 2) the same, but for sol without sodium oleate.

N. G. Koltusheva and others have found [15] that the magnetic field affects the character of the particles obtained in coprecipitation of the hydroxides of iron, aluminum, and other materials. A monograph is available [16] on the magnetic properties of very small ferromagnetic particles. Interest attaches to work on the technological and physical chemical properties of ferrites [17].

The works enumerated here have dealt with individual, largely physical, problems which lie outside the scope of the physical chemistry of the dispersed system.

In this first communication we present the results of our studies on sols and suspensions of ferromagnetic materials.

Magnetite (Fe_3O_4) Sols

Preparation of the Sols

Dvořák and Špaček [18, 19] obtained low concentration magnetite sols ranging up to 0.2 weight % by condensation from equimolecular solutions of divalent and trivalent iron with peptization of the precipitate with 0.1 N HCl and addition of soap as a stabilizer. Sols of concentrations as high as 8-9% could have practical value these we have prepared by ultrasonic peptization of precipitates at $\approx$20 kc. These precipitates were obtained from equimolar $FeCl_3 + FeSO_4$ mixtures containing an excess alkali, separated from the solution, and then washed and peptized with 0.1 N HCl. The sol particles were positively charged and invisible under the ordinary microscope; they coagulated when exposed to the magnetic field for 5-6 hours.

Supercharging resulted from the introduction of 1% sodium oleate, and the stability of the sol diminished somewhat, both in time and to the action of the magnetic field. Most of the experiments were carried out with sols which contained the oleate, the action of the magnetic field being more pronounced in them.

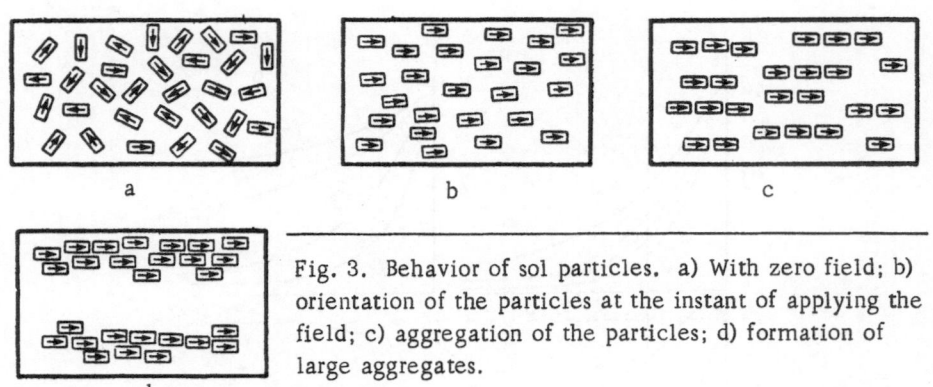

a   b   c

d

Fig. 3. Behavior of sol particles. a) With zero field; b) orientation of the particles at the instant of applying the field; c) aggregation of the particles; d) formation of large aggregates.

Optical Properties of the Sol

The study of optical properties in the uniform magnetic field was carried out in the appratus shown schematically in Fig. 1.

Alteration in the intensity, J_r, of the light transmitted through the sol was recorded on the oscillograph (Fig. 2). Application of the magnetic field reduced the intensity (J_{mag}) in the sols containing the sodium oleate, the reduction increasing with the field strength (Fig. 2a) and the sol concentration (Fig. 2b). The intensity remained at its minimum for several minutes, and then rose, eventually exceeding the original intensity for passage through the sol under zero field (J_{wf}) if the exposure was extended long enough, as can be seen from curve 1 (Fig. 2c).

The sol became turbid as the result of particle aggregation when the field was turned on, clouding following the Rayleigh Law.

The subsequent increase in the intensity of the light might be explained in terms of interaction of the particles with one another and the field, with formation of coarse particles oriented along the lines of force and reduction in the optical density (Fig. 3).

This supposition was confirmed by direct microscopic examination, the field of view being free of particles (particle diameter less than 0.2 μ) before the field was turned on, and the number of dark "spots" increasing rapidly when the field was activated with the incident light directed along the lines of force (Fig. 4a). These dark spots had a mean diameter of 3-4 μ and were in continual Brownian movement; they represented the ends

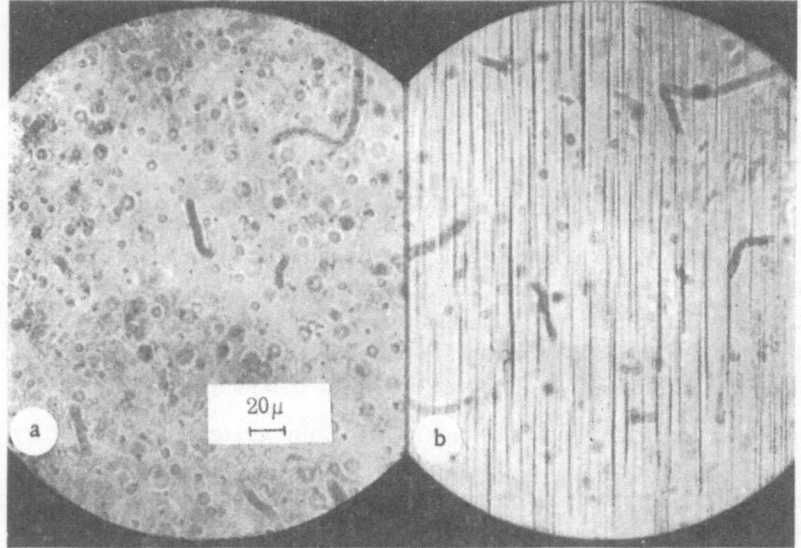

20 μ

Fig. 4. Fe_3O_4 sol viewed along (a), and at right angles to (b), the lines of force.

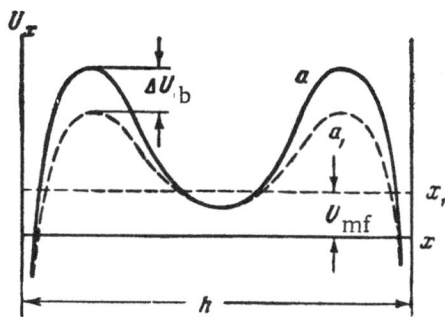

Fig. 5. Potential energy curves for sol particles, without (a), and with (a_1), the magnetic field.

of vertically directed thread-like aggregates which could be readily seen when the system was viewed at right angles to the lines of force (Fig. 4b).

The orientation of the magnetic particles resulting from application of the field (see Fig. 3b) should increase the intensity of the transmitted light up to the instant at which aggregation set in. There was actually a slight increase in the intensity of the transmitted light in the more stable sols which were free of sodium oleate (see curve 2, Fig. 2c). It is clear that particle aggregation and orientation occurred almost simultaneously in the less stable Fe_3O_4 sols which contained the sodium oleate.

Thus the optical properties of magnetite sols in the magnetic field can be drawn on to estimate differences in the splitting action of the protective envelopes and the ability of envelopes to undergo relaxation. This follows from the fact that addition of a quantity of electrolyte insufficient to bring about coagulation to a sol containing sodium oleate resulted in more rapid reduction in the light intensity (see curve 5, Fig. 2a, for the sol with added HCl, and curve 4, Fig. 2a, for this same sol at the same field strength but without addition of HCl).

When the field is turned off, the magnetite sol completely returned to its original state, at least for field strengths up to ~5000 oersteds, and times of exposure of 2-3 minutes such as were used in our experiments. The dotted curves of Fig. 2c show the alteration in the light intensity after cutting out the field at various stages of the process. It is seen that the sol traced out in reverse (and often in a shorter time) all the changes it experienced when the field was applied.

The reversal time was essentially the same for all cases and amounted to 5-6 seconds.

The reversibility of aggregation is indication that particle approach involves an energy barrier. The difference in the rates of aggregation of positive and negatively charged sols with all other conditions held constant, clearly points to differences in stability and the initial energy barriers, U_b^0. Orientation in the field increases the attraction between particles, thus leading to approach with reduction of the energy barrier to the value U_b.

One can write, to an approximation,

$$\Delta U_b = U_b^0 - U_b = U_m, \tag{1}$$

U_m being the increase in the energy of sol particle attraction resulting from oriented approach. Since such approach results from particle-field interaction, one can neglect the entropy factor and express the interaction energy by

$$U_{mf} \approx U_m. \tag{2}$$

Here the magnetic field functions as a "trap" for the sol particles, so that U_{mf} is equivalent to the energy of the potential well established between two widely separated particles when the force of attraction is greater than the force of repulsion. This last factor fixes the possibility of forming a quasi-crystalline lattice in gelatinization [20]. With limited volume, gelatinization becomes possible even though a potential well is not established at large distances of separation, which is to say, even though the force of repulsion exceeds the force of attraction [21]. This type of lattice formation is reversible, the gel spontaneously passing back into the sol when the volume is increased, through addition of a dispersing medium, for example, [22].

The interaction between the sol particles in the aggregate can be characterized by an overall potential curve which is analogous to the curve for the colloidal particles which constitute the quasi-crystalline gel lattice (Fig. 5).

Fig. 6. Behavior of a Fe_3O_4 sol at the interface with water in a nonhomogeneous magnetic field with gradually increasing field strength. a) Horizontal surface of sol; b) puckered surface of sol; c) formation of drop; d, e) separation of drop with tail.

Thus the aggregate of magnetite particles behaves as an irreversible gel (curve a_1, axis x_1) when a magnetic field is imposed, and as a reversible gel with unlimited volume (curve a, axis x) when the field is removed.

Extended application of the magnetic field should lead to an irreversible aggregation, or coagulation, if the protective coatings of the sol particles which fix the interparticle potential barrier are dissipated with the passage of time. The magnetite sol remained reversible, however, even after being held for 5-6 hours in a field ~1000 oersteds. It is clear that barriers which can be maintained for such extended periods must be essentially solvational in character.

In experiments which we have carried out with V. B. Murav'ev, poorly stable V_2O_5 and MnO_2 sols were subjected to extended (~17 hours) treatment in a constant magnetic field (H ≈ 1000 oersteds). It was observed that aggregation was markedly accelerated; the MnO_2 sol and a low concentration V_2O_5 sol (~0.005%) coagulated, while a 0.1% V_2O_5 sol underwent irreversible gelatinization. The maganese dioxide sols were freshly prepared and capable of undergoing spontaneous coagulation in the course of 2-3 days, while the vanadium pentoxide was two years old. Freshly prepared V_2O_5 sols would not undergo coagulation. Both freshly prepared iron oxide sols and very old (~5 years) iron oxide sols refused to coagulate, probably as a result of hydration.

Viscosity

The viscosity of the magnetite sols differed but little from the viscosity of the dispersing medium, the viscosity of the 0.1% sol being 2% higher than that of water. The viscosities of these sols were practically the same, both with and without the magnetic field. Measurements were carried out in a vertical capillary viscosimeter. The accuracy of measurement was obviously too low to disclose the details of the viscous-elastic behavior of the sol in the magnetic field.

Magnetophoresis

When placed in a nonuniform magnetic field, it was observed that the sol particles migrated in the direction of the field gradient, the migration varying with the distribution of the ferromagnetic substance in the magnetophoretic cell. For example, the particle movement extended throughout the entire cell when the latter was completely filled.

Equating the driving force

$$f_1 = V\chi H \frac{dH}{dx} \tag{3}$$

(V is the volume of the particles and χ the susceptability) and the force of viscous resistance $f_2 = 6\pi\eta rU$, one obtains an expression for the migration rate of spherical particles, namely

$$U = \frac{2r^2}{9\eta} \cdot \chi H \frac{dH}{dx} . \tag{4}$$

Difficulties are met, however, in attempting to apply this expression, since the particles tend to aggregate under the action of the field, as pointed out above.

If water, or some other liquid was superposed on the sol, it was observed that there is no movement of particles from sol to liquid when the field was applied. Increase in the field strength led to swelling of the sol surface (Fig. 6) and caused tailed drops to break away from the sol surface, if carried far enough. These effects can be explained by interaction of the sol particles with the field and with one another, the system functioning as a unit.

Suspensions of Ferromagnetic Materials

Study was made of the effect of particle magnetization on the stability of suspensions of ferromagnetic materials in the absence of an external magnetic field. Aqueous suspensions of ferromagnetic ZIL and T-6 powders which are used in the manufacture of magnetic recording tapes were prepared by stirring with water in a graduated cylinder to an overall concentration of 20-25%. After stirring had been completed, note was made of the position of the precipitate-suspension, or suspension-pure medium, interface at various times.

The suspension was then magnetized by being placed in a field of ~2000 oersteds. The suspension was carefully stirred after magnetization and observation again made of the movement of the precipitate-suspension interface.

The data (Fig. 7a) make it clear that magnetization altered the precipitation kinetics, and, in certain instances, the character of the precipitate, as well. A precipitation curve passing through the origin is known to be indication of the precipitation of a relatively stable suspension.

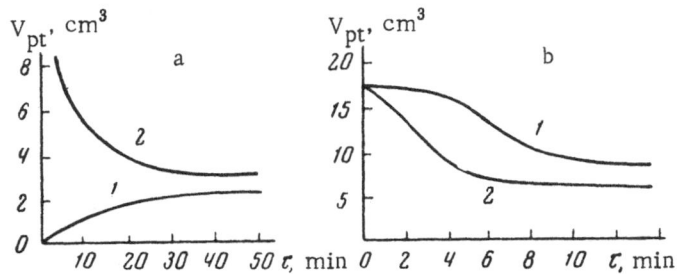

Fig. 7. The precipitation kinetics in suspensions of ZIL (a) and T-6 (b) powders. 1) Before magnetization; b) after magnetization.

Precipitation of an unstable suspension is characterized by a curve passing from top to bottom.

Magnetization caused the ZIL suspension (particle diameter, ~2 μ) to become unstable to aggregation. Magnetization of an unstable T-6 suspension (elongated particles, ~2 μ in length) caused the suspension to become still more unstable (curve 2, Fig. 7b). The almost horizontal segment at the beginning of the precipitation curve for the unmagnetized suspension (curve 1, Fig. 7b) is characteristic of the state of gel formation associated with long-range attractive forces weaker than those met in the preceding case. Rapid fall-off of the protective barriers and approach of the particles causes the externally homogeneous system to separate into flock regions, which compact to accelerate the precipitation process.

Comparatively rapid processes of gelatinization and syneresis are known to take place in various classical sols and suspensions (Berlin purple, etc., [22]). The gelatinization of T-6 powder suspensions is interesting since a quasi-crystalline structure is established there as the result of interaction of the magnetic forces of attraction between particles separated by potential barriers. It is clear that the effect of these attractive forces is manifest in the second stage of the process which leads to syneresis.

In concluding we note that the use of the magnetic field for studying dispersed system stability has the advantage over other methods (use of the electrical field, addition of electrolytes and dehydrators, etc.) in that it leaves the electrical structure of the colloidal particles completely, or almost completely, unaltered.

The result is that more clear-cut conclusions can be obtained in those cases in which it is possible to apply the magnetic field, a fact which is to a certain degree brought out by the material presented above.

Application of the magnetic field is, moreover, not limited to ferromagnetic suspensions. Thus we have prepared organic suspensions of ferromagnetic substances which have been emulsified in water. Separation of individual drops of the dispersed phase from the emulsion surface occurred when the latter was placed in a non-uniform magnetic field at relatively low H dH/dχ values. These emulsions coalesce rather rapidlly in the magnetic field and are useful systems for stability studies since the forces acting between the droplets can be varied over wide limits.

The range of applicability of ferromagnetic substances in surface property studies can be extended by using adsorption, hydrophobization, or coverage with a compact layer of some material (e.g., velon, etc.) to alter the particle surface.

Conclusions

1. The effect of an external magnetic field on the properties of sols and suspensions of ferromagnetic materials has been studied experimentally.

2. Use of ultrasonics for precipitation dispersion has made it possible to obtain high-concentration Fe_3O_4 sols (8-9%).

3. Direct microscopic observation and study of the optical properties of the Fe_3O_4 suspensions show that the external magnetic field orients the colloidal particles, causing them to approach one another more closely, and form ordered aggregates. The colloidal particles which enter into the quasi-crystalline lattice of the aggregate are separated one from the other by potential barriers. The sol returns to its initial stage when the field is cut off.

4. It has been shown that the rate of aggregation varies with the sol stability and can be used to characterize the latter.

5. It was found that the magnetization of ferromagnetic suspensions tended to reduce the suspension stability. Cases were observed in which gelatinization and syneresis proceeded rapidly during precipitation from suspension.

6. Study of the effect of the magnetic field on sols, suspensions, and emulsions makes it possible to investigate the stabilization of these systems, and effects arising from colloidal particle interaction.

Literature
_____

1. P. Langevin. Selected Works, Moscow — Leningrad, Izd. Akad. Nauk SSSR (1960).
2. V. N. Tsvetkov. Kolloidn. Zh. 41:197 (1949).
3. W. C. Elmor. Phys. Rev. 60:593 (1941).
4. E. Lourila, Ann. Acad. Sci. Fennicae, Ser. A Vol. 34 (1959); Vol. 70 (1961).
5. H. E. Rose. Chem. Prod. 18:375 (1955).
6. A. D. Tranklin. J. Appl. Phys. 24:1040 (1953).
7. L. P. Soboleva. Author's summary, dissertation, Moscow Energetics Institute (1954).
8. A. Kolin. J. Appl. Phys. 25:1065 (1954); V. A. Myamlin, et al., Dokl. Akad. Nauk SSSR 137:1405 (1961).
9. O. Preining. Acta Phys. Austriaca 7:147 (1953).
10. D. M. S. Bagguley. Proc. Phys. Soc. A 66:765 (1953).
11. E. N. Harvey. J. Colloid. Sci. 8:543 (1953).
12. M. P. Volarovich and D. M. Tolstoi. Zh. Fiz. Khim. 8:619 (1936).
13. A. F. Komarov. Spirt. Prom. No. 2:46 (1962).
14. G. L. Mikhnevich, et al.. Kolloidn. Zh. 24:49 (1962).
15. N. G. Koltusheva, et al.. The Biological and Medical Action of the Magnetic Field and Strictly Periodic Vibrations, Perm' (1948).
16. S. V. Vonsovskii. The Modern Theory of Magnetism, Moscow, Gostekhizdat (1953).
17. L. I. Rabkin. S. A. Sosnin, and B. I. Epshtein. Ferrite Technology, Moscow, Gosenergoizdat (1962).
18. J. Dvořák. Česk. Časopie Fys. 5:472 (1955).
19. L. Špaček. Česk. Časopie Fys. 5:395 (1955).
20. I. F. Efremov. Kolloidn. Zh. 18:276 (1956).
21. I. F. Efremov and S. V. Nerpin. Dokl. Akad. Nauk SSSR 113:846 (1958).
22. I. F. Efremov. Tr. Leningr. Tekhnol. Inst. im. Lensoveta, No. 38 (1957).

SECTION V

SURFACE PHENOMENA IN ADHESION AND FRICTION

STUDY OF THE GENERAL RULES APPLYING
TO THE ELECTROSTATIC COMPONENT OF THE FORCE
OF ADHESION FOR SEMICONDUCTORS WITH SURFACE STATES

V. P. Smilga and B. V. Deryagin

Institute of Physical Chemistry, Acad. Sci., USSR
Laboratory for Surface Phenomena

A study of the electrostatic component of the adhesion force for semiconductors with surface states has been reported in one of our papers [1]. For the model used there, the surface states were shown to be often determinant in fixing the electrostatic component of the force of adhesion at the interface between the semiconductor and a metal. It proved possible to extrapolate descriptively from this model to other systems so as to obtain a rather clear-cut concept of the effect of the surface on the electrostatic component of the force of adhesion.

Interest attaches, however, to the general study, one leading to equations that might be used for evaluating the electrostatic component of the adhesion force for any semiconductor, regardless of its band structure and surface state spectrum.

The present paper is devoted to this problem. The possibility of obtaining a general solution here traces back to one very broad assumption concerning the charge density in the semiconductor, namely, that both the bulk charge density, ρ, and the surface state charge are functions of the electrostatic potential, φ, alone, so that the Poisson Equation (which fixes the distribution of electrostatic potential in the semiconductor and thereby determines the electrostatic component of the force of adhesion) can be brought to integrable form.

Solution to a similar problem has already been obtained in [2].

It is to be noted that almost all actual systems satisfy the above requirement.

We will consider a semiconductor separated from a metal by a gap, d, in which the dielectric constant is ε_g. The field within this gap determines the force of adhesion. This semiconductor has surface states and we will assume that the surface can be charged relative to the bulk as a result of the charge distribution, even if the semiconductor is "taken by itself," i.e., insulated from the metal. The presence of this charge naturally alters the work function of the semiconductor. Further analysis by the method of [1] will make use of the quantity V_c^0, the difference in contact potentials between the metal and a semiconductor similar to our own in every respect except that of the total charge due to states on the free surface is equal to zero. The semiconductor with no surface states would be a special instance here.

It follows from what has been said that

$$eV_c^0 = \varphi_m - \varphi_{sc}^0 \, ,$$

φ_m being the work function of the metal and φ_{sc}^0 the work function of the semiconductor with zero surface state charge.

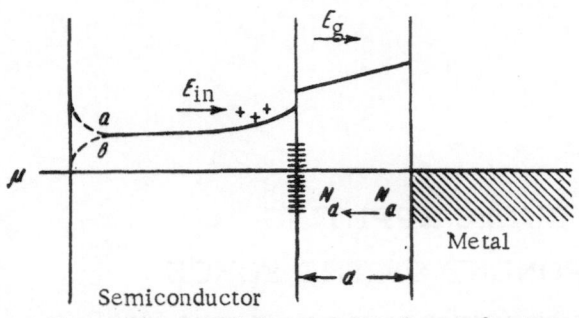

Electronic potential energy, in the conduction band and in the gap.

It is easily seen that the work function of the actual semiconductor, φ_{sc}, would be related to φ_{sc}^0 by

$$\varphi_{sc} = \varphi_{sc}^0 + e\varphi^f,$$

φ^f being the potential of the free surface of the insulated semiconductor relative to a point lying within the semiconductor and beyond the space charge layer (the superscript, f, designates the free surface, both here and in what follows).

For the sake of definiteness, it will be assumed that $V_c^0 > 0$. The earlier assumption leads to the following form of Poisson Equation for the potential inside the semiconductor:

$$\frac{d^2\varphi}{dx^2} = -\frac{4\pi\rho\,(\varphi)}{\varepsilon_{sc}}.$$

(1)

Here ε_{sc} is the dielectric constant of the semiconductor. To be more exact, we will now suppose the entire charge density within the semiconductor to be generally determined by the dimensionless electronic potential energy, $\widetilde{\varphi} = -e\varphi/kT$, rather than the electrostatic potential, φ, itself, and then rewrite the Poisson Equation as

$$\frac{d^2\widetilde{\varphi}}{dx^2} = +\frac{4\pi e}{\varepsilon_{sc}\,kT}\,\rho\,(\widetilde{\varphi}).$$

(1')

Separating the factor en_i (n_i is a quantity characterizing the carrier concentration in the semiconductor) out of $\rho(\widetilde{\varphi})$, we finally obtain

$$\frac{d^2\widetilde{\varphi}}{dx^2} = L^{-2}\,\rho_1\,(\widetilde{\varphi}),$$

(1")

where

$$\rho_1\,(\widetilde{\varphi}) = \frac{\rho\,(\widetilde{\varphi})}{en_i}\,,\quad L = \sqrt{\frac{4\pi e^2 n_i}{\varepsilon_{sc}\,kT}}$$

(L is the Debye length of the semiconductor).

The choice of characteristic concentration, n_i, depends on the particular type of semiconductor. For example, n_i is simply the carrier concentration in the case of semiconductors with a single type of carrier, it can be conveniently chosen as the carrier concentration for intrinsic semiconduction in the case of germanium and silicon, etc.

In addition, the concentrations of charged donor and acceptor surface levels are given by the well-known expressions:

$$p_d = \sum_\alpha \frac{N_d^\alpha}{1 + \exp\,[(\mu - e\varphi - E_d^\alpha)/kT]}\,,$$

(2)

$$n_a = \sum_\alpha \frac{N_a^\alpha}{1 + \exp\,[(E_a^\alpha + e\varphi - \mu)/kT]}\,.$$

(3)

Summation extends here over all the energy levels. N_d^α and N_a^α are, respectively, the total numbers of donor and acceptor levels with energies, E_a^α and E_p^α, and μ is the Fermi level of the semiconductor. The top of the normal band below the surface charge layer of the semiconductor is chosen as the zero point for the energy scale.

The surface charge is given by

$$\sigma\left(\widetilde{\varphi}\right) = p_d\left(\widetilde{\varphi}\right) - n_a\left(\widetilde{\varphi}\right). \tag{4}$$

The free surface charge is accordingly defined by the expression

$$\sigma^f = p_d^f - n_a^f. \tag{4'}$$

We will now formulate the boundary conditions for the problem.

The overall potential drop, in the gap between the metal and the semiconductor, and up to that part of the semiconductor which lies beyond the surface charge level, is equal to V_c^0. The electronic potential energy in semiconduction band and gap is shown schematically in the figure. The dotted curves a and c correspond, respectively, to negative and positive charging of the semiconductor surface.

Thus,

$$eV_c^0 = e\varphi + e\,dE_g. \tag{5}$$

It will be agreed that the positive direction of the field is from the semiconductor to the vacuum.

Condition (5) must be supplemented by the condition for continuity of the electrostatic induction vector at the semiconductor interface, namely

$$\varepsilon_g E_g - \varepsilon_{sc}\,E_{in} = 4\pi\sigma\left(\widetilde{\varphi}\right). \tag{6}$$

We will further assume the depth of the semiconductor to be much greater than the Debye length, so that Eq. (1") can be integrated by setting $\varphi \to 0$ and $d\varphi/dx \to 0$ when $\chi \to \infty$.

One then obtains

$$\frac{eE_{in}}{kT} = \frac{d\widetilde{\varphi}}{dx} = L^{-1}\sqrt{2\int_0^{\widetilde{\varphi}}\rho_1\left(\widetilde{\varphi}\right)d\widetilde{\varphi}}. \tag{7}$$

It should be noted that the integral of (7) can be calculated for any actual system by elementary means since $\rho_1(\widetilde{\varphi})$ is always a simple sum of constant terms and functions of the form $\exp\widetilde{\varphi}$. In addition, the choice of characteristic concentrations is such that all of the numerical coefficients appearing in the function $\rho_1(\widetilde{\varphi})$ are close to unity.

Using (5) and (6), one can obtain a system of equations for the surface potential, $\widetilde{\varphi}$, and the gap field, E_g.

Substituting (7) into (6), one has

$$\frac{eE_g}{kT} = \frac{\varepsilon_{sc}}{\varepsilon_g}L^{-1}\sqrt{2\int_0^{\widetilde{\varphi}}\rho_1\left(\widetilde{\varphi}\right)d\widetilde{\varphi}} + \frac{4\pi e}{kT\varepsilon_g}\sigma\left(\widetilde{\varphi}\right). \tag{8}$$

Substitution of the expression for E_g from (8) into (5), one obtains an equation for determining the surface potential, $\widetilde{\varphi}$:

$$\widetilde{V}_R^0 = \widetilde{\varphi} + d\left[\frac{\varepsilon_{sc}}{\varepsilon_g}L^{-1}\sqrt{2\int_0^{\widetilde{\varphi}}\rho_1\left(\widetilde{\varphi}\right)d\widetilde{\varphi}} + \frac{4\pi e}{kT\varepsilon_g}\sigma\left(\widetilde{\varphi}\right)\right]. \tag{9}$$

Once the band structure and surface state spectrum of the semiconductor are known, the electrostatic component of the force of adhesion can be calculated through Eqs. (8) and (9).

On the other hand, a qualitative study of the system (8) and (9) can be carried out with specifying the precise band structure of the semiconductor to obtain a number of significant conclusions.

First, we offer some remarks of general character. It is clear that the force of adhesion, $F = \varepsilon_g E_g^2 / 8\pi$, will increase with E_g at fixed $\tilde{V}_c^0$, i.e., will increase with the potential drop in the gap. Thus it becomes obvious that the problem here is that of the surface state screening of the interior of the semiconductor from the external field. This problem for a specific model was first treated by Bratten and Bardeen [3]. This is a very real problem for semiconductor physics, a fact which should, in itself, increase the interest attaching to the equations obtained.

It must also be pointed out that Eqs. (8) and (9) give a solution to an extremely significant problem of the physical chemistry of colloids and double layer theory. In fact, these equations make it possible to determine the double electric layer potential at the interface of an electrolyte of any degree of complexity with specific adsorption of the various types of ions. Here, the adsorbed ions function as semiconductor surface states, and $\sigma(\tilde{\varphi})$ is determined by the ionic adsorption isotherm.

We now pass to a quantitative analysis of the problem, limiting ourselves to the case of semiconductors.

A. Free Surface Potential Equal to Zero. Here it is clear that $\sigma(0) = 0$ and $V_c^0 = V_c$ under experimental measurement of the difference of contact potentials. Incidentally, it should be pointed out that V_c^0 is a more fundamental characteristic of the metal-semiconductor junction than is V_c, and the use of V_c^0 throughout the theory is therefore justified. Interest attaches to the conditions required for minimizing the surface potential, $\tilde{\varphi}$, at fixed $\tilde{V}_c^0$ and d. This problem has already been solved [1] for the case of the semiconductor with one type of carrier and two types of surface states which are widely separated from the Fermi level.

Here $\sigma(\tilde{\varphi}) \approx C \operatorname{sh} \tilde{\varphi}$, C being the number of states of given type which are ionized prior to contact.

Calculations show that the interior of the semiconductor will be almost fully screened and the entire difference of contact potentials concentrated in the gap when $C \approx 10^{11}$–10^{12} cm^{-2}; $d = 5 \cdot 10^{-8}$ cm; $L = 10^{-4}$ cm, and $\varepsilon = 10$. Detailed tables characterizing this effect have been presented in [14]. Estimates can, however, be set up for the general case. We first note that the expression

$$\sqrt{2 \int_0^{\tilde{\varphi}} \rho(\tilde{\varphi})\, d\tilde{\varphi}} \approx A e^{\frac{\tilde{\varphi}}{2}}$$

is applicable to most semiconductors (A is a multiplying factor of the order of unity). The validity of this approximation is quite obvious in view of the earlier remark concerning $\rho_1(\tilde{\varphi})$, at least for cases in which $\tilde{\varphi} > 5$. Since this last would require that $\varphi > 0.13$ V at room temperature and the "smooth" contact potential difference, V_c^0 is about 0.5-1 V, it becomes clear that the case $\varphi < 0.1$ V has no special interest and our approximation can be retained even in this region.

It is also clear that most of the potential drop will be localized in the gap if

$$\tilde{\varphi}_c \ll d \left[\frac{A \varepsilon_{sc}}{\varepsilon_g} L^{-1} e^{\frac{\tilde{\varphi}}{2}} + \frac{4\pi e}{kT \varepsilon_g} \sigma(\tilde{\varphi}) \right].$$

(10)

The first member of the right hand side of Eq. (10) has essentially the same form as in [1], but certain, quite different, variants of $\sigma(\tilde{\varphi})$ are now possible.

It is clear, however, that the maximum screening effect can be expected when $\sigma(\tilde{\varphi})$ is a rapidly increasing function of $\tilde{\varphi}$, that takes on large values even when $\tilde{\varphi} \approx 5$. Numerical calculations based on (1) show that the inequality of (10) is almost never applicable when the first term is predominant.

It should be noted that $\sigma(\tilde{\varphi})$ will, in general, be sensitive to variations in $\tilde{\varphi}$ if the donor (or acceptor) levels on the free surface are unsymmetrically located with respect to the Fermi level. This is indication that the original number of ionized states of each type is the determining factor in fixing the screening [1]. Analysis of the distribution function for the almost-degenerate Fermi gas shows the alteration of $\sigma(\widetilde{\varphi})$ with increasing $\widetilde{\varphi}$ to be most marked when the grouping of each type of level is highly asymmetrical in the neighborhood of the Fermi level.

These qualitative conclusions can be confirmed for various models, but there seems no necessity to do so since they are rather obvious.

B. Charged Free Surface, the Potential Energy of the Surface Electron ($\widetilde{\varphi}^f$) Having the Same Sign as V_C^0 (see figure, Variant a).

We emphasize that one must not confuse V_C^0 with the experimentally determined V_C when the surface is charged.

The potential of the free surface is obviously fixed by the relation

$$\varepsilon_{sc}\, L^{-1} \sqrt{ 2\int_0^{\widetilde{\varphi}} \rho_1(\widetilde{\varphi})\,d\widetilde{\varphi} + \frac{4\pi e}{kT}\,\sigma(\widetilde{\varphi}) } = 0.$$

(11)

It is clear that a smaller part of the potential drop is now localized in the gap, the second member of (9) being equal to zero even when $\widetilde{\varphi} = \varphi^f$. Thus the reference point for the calculations is not at $\widetilde{\varphi} = 0$ but at $\widetilde{\varphi} = \varphi^f$; if $\widetilde{\varphi}^f \approx V_C^0$, there cannot, in general, be any screening of the interior and both the gap field and the force of adhesion will be small. Any treatment of the surface that gives $\widetilde{\varphi}^f$ and V_C^0 values of the same sign for the semiconductor-metal pair therefore leads to reduction in the electrostatic component.

C. Surface Charged, φ^f and V_C^0 Being of Opposite Sign (see figure, Variant c). This case corresponds to maximum gap field for fixed value of V_C^0.

This case actually arises only when the second term of (9) is positive and of larger modulous than $\widetilde{\varphi}$. This term is equal to $\widetilde{\varphi}$ in absolute magnitude when $V_C^0 = 0$; it increases with increasing V_C^0, while $\widetilde{\varphi}$ diminishes. It is seen from (8) and (9) that the gap field satisfies the inequality $eE_g/kT > \widetilde{V}_C^0/d$. For a certain value of V_C^0, φ becomes equal to zero. This point corresponds to surface supercharging. Here the gap field is such that $E_g = (kT/e) \cdot (V_C^0/d)$. On the other hand, $\widetilde{\varphi}$ takes on positive values and remains considerably less than the second term. This is the case of maximum adhesion.

The earlier remarks of the analysis of Case A still apply to the behavior of the function $\sigma(\widetilde{\varphi})$ in the last two cases.

Conclusions

1. Study has been made of the gap electric field and the associated electrostatic component of the force of adhesion between a metal and a semiconductor with surface states.

2. Equations have been derived which can be used to calculate the electrostatic field in a gap of given width as soon as the spectrum of surface levels is known.

3. A complete classification has been set up, and a qualitative analysis of the various situations carried out, with estimates of actual gap fields.

4. The results indicate that surface states are essential if high values of the force of adhesion are to be reached.

Literature

1. V. P. Smilga and B. V. Deryagin, Dokl. Akad. Nauk SSSR 122:1049 (1958).
2. V. P. Smilga, Kolloidn. Zh. 22:615 (1960).
3. W. H. Bratten and J. Bardeen, "Surface properties of germanium," In collection: Problems of Semiconductor Physics [Russian translation], Moscow, IL (1957).
4. B. V. Deryagin and V. P. Smilga, "Electronic theory of adhesion," J. Appl. Phys. (in press).

THE EFFECT OF THE DOUBLE ELECTRIC LAYER
ON ROLLING FRICTION
(THE ELECTRICAL COMPONENT OF ROLLING FRICTION)

B. V. Deryagin and V. P. Smilga

Institute of Physical Chemistry, Acad. Sci., USSR
Laboratory for Surface Phenomena

The present work deals with a special mechanism designed to account for the existence of an electrostatic component of rolling friction. This mechanism is based on the breakdown of the double electric layer which is at rest in the interface between the contacted bodies. It will be shown that the rolling of a cylinder causes the charge distribution in the double electric layer to become unsymmetrical with respect to the mean contact point. This asymmetry gives rise to an angular momentum in the force of rolling friction. Rather unexpectedly, the mechanism predicts angular momenta of considerable magnitude in certain cases.

In addition, calculation will be made of the moment of the frictional force associated with the rolling of a cylinder without slip over the plane surface of another body physically different from the cylinder. From what was just said, it is clear that only that part of the angular momentum which arises from the double electric layer is to be taken into account here. It is seen from Fig. 1 that the boundary of the contact region is determined by angle θ_0, $h(\theta)$ being the distance from a point on the cylinder surface to the base surface.

Discussion will be limited to the case in which the base is metallic, and the cylinder either a dielectric or semiconductor of high specific resistance.

When at rest, the outer part of the cylinder carries a surface charge such that the potential difference between the metal and an interior point in the cylinder is equal to the difference of contact potentials (φ_c).

Naturally, these electrical charges can be considered concentrated in the "true" surface (i.e., in a region of approximately monolayer depth) only in the case of a metal.

The charged region extends from the surface into the interior of a dielectric or semiconductor, the depth of penetration being approximately equal to the Debye length.

The charge can also be considered to lie on the surface in the macroscopic problem of interest here. The Debye length, and associated depth of double layer penetration, will usually fall in the range from 10^{-6} to 10^{-2} cm. Although the Debye length can be of the order of several centimeters, or more, in high quality insulators, materials which have not been subjected to special treatment are appreciably contaminated near the surface and therefore show a sharp rise in the number of carriers and surface layer conductivity. The screening distance is reduced accordingly. In practice, one can therefore always assume the double layer to be concentrated in the region no more than 10^{-2} cm deep. (We note in passing, that Debye lengths of this magnitude are met in good insulators with 10^{10} carriers per cm^3.)

The surface charge density in the dielectric or semiconductor is determined from the relation

$$\sigma = \int\limits_{0}^{\infty} \rho(x)\,dx. \tag{1}$$

Here x is the distance from the contact boundary and $\rho(x)$ the bulk charge density.

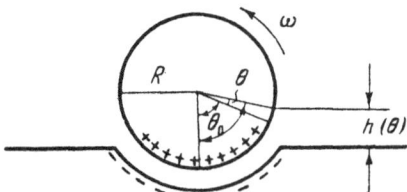

Fig. 1. Schematic representation of a mechanism to account for an electrostatic component of the force of friction in the rolling of a cylinder over a plane surface without slip.

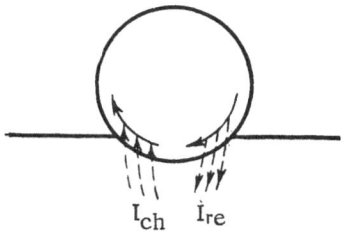

I_{ch} $\dot{I}_{re}$

Fig. 2. Schematic representation of the charging current, I_{ch}, and recombination discharge current, I_{re}, over the cylinder surface.

Since, as pointed out above, the conduction in the subsurface layer of insulator or high resistance semiconductor is usually much higher than the conduction within the sample, the bulk conductivity of the cylinder can be neglected in the calculations and the cylinder itself treated as a perfect insulator.

It will be convenient for the subsequent work to introduce a reduced double layer depth, defining the latter by the relation

$$d = \frac{\varphi_c}{4\pi\sigma} .$$

(2)

We will now find the surface charge density distribution on the resting cylinder. Generally speaking, it is not too difficult to obtain an exact solution to this problem (see, for example, [1]). In view of further developments, we will, however, draw on the approximation method suggested in [2]. There it was shown that "the surface of the cylinder can be treated as a system of plane condensers connected in parallel" over that region where the slope of the tangent to the cylinder surface relative to the base is much less than unity ($\theta \ll 1$). Taking account of the fact that the surface of the cylinder and the base are at the same potential, one then has

$$\sigma(\theta) = \frac{d}{h(\theta) + d} \sigma(0).$$

(3)

It is at once obvious from [3] that the surface charge will fall off to practically zero in the region where $\theta \ll 1$. As pointed out above, the double layer depth usually falls in the 10^{-6}-10^{-2} cm interval. Thus, with $h \approx 10^{-1}$ cm, one will have

$$\sigma(\theta) \leqslant 10^{-1}\sigma(0).$$

If the cylinder radius is assumed to be 10^1-10^2 cm, it is easily seen that $\theta \ll 1$ when $h(\theta) \approx 10^{-1}$ cm. This result is readily justified since $h(\theta)$ is defined through the relation $h(\theta) = R(\theta^2 - \theta_0^2)$.

Rolling of the cylinder over the plane breaks down the condition of equality of potentials and currents are set up in both the cylinder surface and the base. The surfaces of cylinder and base will be considered as systems of parallel connected condensers, just as for the condition of equilibrium. It will be assumed, once more, that the charge falls away to essentially zero in the region where $\theta \ll 1$.

It is obvious that this last assumption is not always valid, by any means, and criteria marking out range of its applicability will be given below. In attempting to develop the surface charge distribution over the cylinder, it proves most convenient to have recourse to a system of coordinates rigidly attached to the mean point of contact between cylinder and base (see Fig. 1).

Here we have, as the figure shows, considered for the sake of definiteness that the cylinder carries a positive, and the base a negative, charge. The area of direct contact between cylinder and base is determined by the cylinder radius and the angle θ_0; it varies with the normal pressure exerted by the cylinder on the base and can be evaluated by solving the contact problem in the theory of elasticity (see, for example, [3]).

Before passing to a mathematical formulation of the problem, we will explain the physics of the situation. Rolling carries those points of the cylinder surface which were originally close to the region of contact and highly charged to the right, thus separating them from the plane surface.

These sectors which have been separated from the point of contact are now more highly charged than they would be in a state of equilibrium, and the excess charge passes into the contact region where it recombines with the base charge. The situation to the left of the point of contact is, so to speak, the reverse of that just described.

Sections of the cylinder surface which are far removed from the region of direct contact carry a low charge; rotation brings these into the contact region where the equilibrium condition requires a higher charge than they are actually carrying. Thus there is a continuous surface charging to the left of the contact region.

The situation regarding current flow in the surface is shown schematically in Fig. 2. Since the total charge of the cylinder must remain constant in the steady state, it follows that the recombination discharge current, I_{re}, must be equal to the charging current, I_{ch}. Thus rotation gives rise to a current in the cylinder surface, recombination of cylinder and base charges occurring to the right of the mean contact point, and charges from the base continually passing into the cylinder to the left of this point. Further treatment will be essentially limited to the case in which charging and recombination occur at the point of immediate contact and in a narrow region surrounding this point. In other words, no account will be taken of the possibility of gaseous discharge between the surfaces.

A high potential difference between the condenser plates would be required for gaseous discharge to develop and this, in turn, would call for either high rotation rates or very low surface conductivity. Supercharging does not occur in other cases, both recombination and charging taking place either in the region of direct contact or by the tunnel effect in a region where the separation of cylinder and base is quite low. In a system of coordinates rigidly connected with the point of rupture, the current density in the cylinder surface will be given by the expression

$$j_{sur}(\theta) = - \lambda \{\sigma(\theta)\} \frac{d\varphi(\theta)}{Rd\theta} + \omega R \sigma(\theta).$$

(4)

Here $\lambda\{\sigma(\theta)\}$ is the surface conductivity, a quantity which depends on the concentration of the excess charges on the cylinder surface; the angle is defined by Fig. 1; $\varphi(\theta)$ is the potential; $\sigma(\theta)$ is the surface charge density per unit of surface area; and ω is the angular-rate of rotation.

The second term of (4) characterizes the convection current in the reference system of coordinates:

$$\lambda\{\sigma(\theta)\} = \mu_n [N_0 + n(\theta)] + \mu_p [P_0 + p(\theta)].$$

(5)

Here μ_n and μ_p are the electron and hole mobilities, respectively. N_0 and P_0 are the numbers of electrons and holes in the surface layer prior to contact, and $n(\theta)$ and $p(\theta)$ are the numbers of additional electrons and holes introduced into the surface layer.

If the cylinder is positively charged, it follows that

$$\sigma(\theta) = e[p(\theta) - n(\theta)] > 0.$$

Substituting

$$\lambda_0 = \mu_n N_0 + \mu_p P_0$$

and

$$\lambda_1(\theta) = \mu_n n(\theta) + \mu_p p(\theta),$$

one finds

$$\lambda\{\sigma(\theta)\} = \lambda_0 + \lambda_1(\theta).$$

(6)

We will now consider certain limiting cases.

I. $\lambda_0 \gg \lambda_1(\theta)$.

In other words, the alteration in the conductivity resulting from contacts in the surface layer is negligibly small in comparison with the intrinsic conductivity of the cylinder itself. Here Eq. (4) takes the form

$$j_{\text{sur}}(\theta) = -\lambda_0 \frac{d\varphi(\theta)}{R\, d\theta} + \omega R \sigma(\theta).$$

(4')

II. $\lambda_1(\theta) \gg \lambda_0$, and at the same time

$$p(\theta) \gg n(\theta)_n \quad \mu_p p(\theta) \gg \mu_n n(\theta).$$

Equation (4) then passes over to

$$j_{\text{sur}}(\theta) = -\frac{\mu_p \sigma(\theta)}{eR} \frac{d\varphi(\theta)}{d\theta} + \omega R \sigma(\theta).$$

(4")

We now recall that

$$j_{\text{sur}}(\theta) \equiv \text{const}$$

outside the recombination region.

Since the charge distribution is stationary in the reference system of coordinates, $\partial\sigma/\partial t = 0$, the equation of continuity

$$\frac{\partial\sigma}{\partial t} + \text{div } j = 0$$

(7)

can be applied outside the recombination region where

$$j(\theta) \equiv j_{\text{sur}}(\theta) = 0$$

to show that

$$\text{div } j_{\text{sur}}(\theta) = 0,$$

(7')

and therefore

$$j_{\text{sur}} = \text{const}.$$

However, if Eq. (4") is correct up to the region of the surface in which $\sigma(\theta)$ can be considered negligible, it then follows that

$$j_{\text{sur}}(\theta) = 0.$$

This indicates that one can set

$$j_{\text{sur}}(\theta) \ll \omega R \sigma(\theta)$$

over all portions of the cylinder surface where $\sigma(\theta)$ is negligibly small. We note that it is an essential feature of the problem under discussion here that $\sigma(\theta)$ can be considered negligibly small at any point, θ_{lim}, where the contribution to the moment of the friction force arising from the charge in $\theta > \theta_{\text{lim}}$ regions is low in comparison with the contribution from the charge in $\theta < \theta_{\text{lim}}$ regions. At the same time, our approximation is permissible only if θ_{lim} lies in the $\theta \ll 1$ region.

257

We will now formulate mathematically the conditions to be satisfied in order that $\sigma(\theta)$ fulfill these requirements. For Case II, (4") is to be replaced by

$$\frac{d\varphi(\theta)}{d\theta} = \frac{e\omega R^2}{\mu_p}.$$

(8)

Since the boundary points for the recombination region are essentially fixed by the angle θ_0, it follows that integration will lead to

$$\varphi(\theta) = \varphi(\theta_0) \pm \frac{\omega R^2}{e\mu_p} |\theta - \theta_0|.$$

(9)

The plus and minus signs of Eq. (9) are associated with surface regions to the right and left, respectively, of the mean point. The charge distribution, $\sigma(\theta)$, on the cylinder must be known before the moment of the force of friction can be calculated. In view of what has been said, and the fact that the potential of the metallic surface is practically constant (the high conductivity of the metal assures that the potential drop along the surface be quite small), one has

$$\varphi(\theta) = 4\pi y(\theta) \sigma(\theta).$$

(10)

Here $y(\theta) = d + h(\theta)$ is the distance between the double layer faces, and the potential of the base has been set equal to zero.

In the region under study $\theta \ll 1$, and $y(\theta)$ will approximate the second power parabola

$$y(\theta) = d + R\theta^2 - R\theta_0^2$$

(11)

when $\theta > \theta_0$. This expression passes over to $y(\theta) = d$ when $\theta < \theta_0$. Equation (10) is obviously quite general and applicable to both cases, I and II. Knowing the relation between $\sigma(\theta)$ and $\varphi(\theta)$, it is an easy matter to obtain the force of interaction between elements of the cylinder surface and the base surface in our approximation

$$dF = 2\pi\sigma^2(\theta) R \, d\theta.$$

(12)

We emphasize once more that all subsequent calculations are based on Eqs. (10) and (12), and applicable only to the region in which $\theta \ll 1$. Thus the validity of the entire analysis depends on the surface charge falling away practically to zero in this region.

In Case I, the differential equation fixing $\varphi(\theta)$ outside the recombination region is linear and readily integrable to give

$$\varphi = \varphi(\theta_0) e^{\frac{\omega R^2}{4\pi\lambda_0} \int_{\theta_0}^{\theta} \frac{dt}{y(t)}} - \frac{jR}{\lambda_0} \int_{\theta_0}^{\theta} e^{\frac{\omega R^2}{4\pi\lambda_0} \int_{u}^{\theta} \frac{dt}{y(t)}} \cdot du.$$

(13)

Equation (10) covering the relation between $\sigma(\theta)$ and $\varphi(\theta)$ was drawn on in integrating (4'). It must be remembered that both θ and θ_0 are negative when (13) is applied to the left of the mean point.

In determining $\varphi(\theta)$, account must be taken of the fact that the characteristic rates of charging and recombination are much higher than the rate of charge transfer along the cylinder surface. This is essentially equivalent to the assumption, natural in the present case, that the specific resistance of the contact is much less than the subsurface layer resistance of the cylinder.

Charge flow takes place in two steps: 1) transfer along the surface; and, 2) transfer from the cylinder to the base in the neighborhood of the direct contact.

The second of these two steps is rapid and departures from equilibrium in the charging and recombination regions not great, so that $\varphi(\theta_0)$ can be assumed independent of the velocity of rotation of the cylinder at its equilibrium value $\varphi(\theta_0) \approx \varphi_c$.

Thus one obtains the following expressions for the potential distribution over the cylinder surface:

for Case (I) $\lambda_0 \gg \lambda_1(\theta)$,

$$\varphi(\theta) = \varphi_c e^{\frac{\omega R^2}{4\pi\lambda_0}\int_{\theta_0}^{\theta}\frac{dt}{y(t)}} - \frac{jR}{\lambda}\int_{\theta_0}^{\theta} e^{\frac{\omega R^2}{4\pi\lambda_0}\int_{u}^{\theta}\frac{dt}{y(t)}} \cdot du;$$

(13')

for Case (II), $\lambda_1(\theta) \gg \lambda_0$, $p(\theta) \gg n(\theta)$ and $\mu_p p(\theta) \gg \mu_n n(\theta)$,

$$\varphi = \varphi_c \pm \frac{\omega R^2}{e\mu_p}|\theta - \theta_0|.$$

(9')

Equation (13'') must be simplified before the moment of the force of friction can be calculated. The current, j, which figures in the equation is actually an unknown parameter of the problem, being fixed by the values of λ_0, R, and ω. The determination of j is a very complex problem; it will, however, be subsequently shown that the second member of (13) can often be neglected in comparison with the first. This is indication that the condition

$$j_{\text{sur}} \ll \omega R \sigma(\theta)$$

(14)

must be fulfilled if regions in which $\sigma(\theta)$ is still rather high are to make any contribution to the moment of the force of friction in the original Eq. (4). In this case

$$\varphi(\theta) = \varphi_c e^{\frac{\omega R^2}{4\pi\lambda_0}\int_{\theta_0}^{\theta}\frac{dt}{y(t)}}.$$

(15)

At the same time, j_{sur} can be determined in the following manner. Let δ_1 and δ_2 be arc lengths measured right and left of the mean point to points on the cylinder surface where a condition reverse to (14) is fulfilled, namely

$$j_{\text{sur}} \gg \omega R \sigma(\theta).$$

Here, $\sigma(\theta) \approx 0$. Let $\varphi_1(x_1)$ and $\varphi_2(x_2)$ be the surface potentials for the cylinder at these points. Then

$$j_{\text{sur}} = \lambda_0 \frac{\varphi_1(\delta_1) - \varphi_2(\delta_2)}{2\pi R - \delta_1 - \delta_2}.$$

(16)

Since all of the calculations assume δ_1 and δ_2 to be much less than $2\pi R$, it follows that

$$j_{\text{sur}} \approx \lambda_0 \frac{\varphi_1(\delta_1) - \varphi_2(\delta_2)}{2\pi R}.$$

(16')

Equation (14) then takes the form

$$\lambda_0 \frac{\varphi_1(\delta_1) - \varphi_1(\delta_2)}{2\pi R} \ll \omega R \sigma(\theta).$$

(17)

In the classification of A. F. Ioffe [4], the specific resistance of semiconductors varies within wide limits, ranging from 10^{-5} to 10^{10} ohm·cm, while the specific resistance of insulators falls in the 10^{10}-10^{15} ohm·cm range.

Assuming $d \approx 10^{-4}$ cm, the λ_0 interval will be

$$10^{-19} \text{ohm}^{-1} < \lambda_0 < 10^{-1} \text{ ohm}^{-1}. \tag{18}$$

In the CGS system, this will be

$$10^{-7} \text{ cm/sec} < \lambda_0 < 10^{13} \text{ cm/sec}. \tag{18'}$$

Naturally, the conditions of (14) and (17) can be fulfilled under this almost unlimited variation of the value of λ_0, but values can also be found under which these conditions will no longer be satisfied. It is therefore useful to assume the validity of (14), and determine the range of applicability of the calculations by analysis of the results obtained.

Drawing on (10) and (15), one obtains the following expressions for $\sigma(\theta)$ in the right-hand and left-hand regions, respectively:

$$\sigma_1(\theta) = \frac{\varphi_C}{4\pi y(\theta)} e^{\frac{\omega R^2}{4\pi \lambda_0} \int_{\theta_0}^{\theta} \frac{dt}{y(t)}}, \tag{19'}$$

$$\sigma_2(\theta) = \frac{\varphi_C}{4\pi y(\theta)} e^{-\frac{\omega R^2}{4\pi \lambda_0} \int_{\theta_0}^{\theta} \frac{dt}{y(t)}}. \tag{19''}$$

The moment of the force of friction per unit length of cylinder is fixed by

$$M \int_{\theta_0}^{\pi/2} 2\pi \sigma_1^2(\theta) R^2 \theta \, d\theta - \int_{\theta_0}^{\pi/2} 2\pi \sigma_2^2(\theta) R^2 \theta \, d\theta. \tag{20}$$

The choice of upper limit in the integrals of (20) was naturally arbitrary. The exact value of this limit is of no significance, however, when $\sigma(\theta)$ falls off so rapidly with increasing θ that the integration region in which $\theta > \theta_{lim}$ (if $\theta_{lim} \ll 1$) makes no contribution to the moment of the friction force.

It follows from (20) that this condition can only be fulfilled if $|\theta| \lesssim 0.1$ and the functions $\sigma_1(\theta)$ and $\sigma_2(\theta)$ fall off more rapidly than $1/\theta$.

Since it is seen from (19') and (19'') that $\sigma_2(\theta)$ diminishes more rapidly with increasing θ than does $\sigma_1(\theta)$, it is clearly enough to formulate the conditions under which $\sigma_1(\theta)$ rapidly diminishes to zero in the region $\theta \ll 1$. It is in order to point out here that $\sigma_2(\theta)$ can be a monotonically decreasing function of θ, without the same being true of $\sigma_1(\theta)$, the latter passing through a maximum. In fact, (19') shows that the exponential factor increases with rising θ.

The integral in the exponential of (19') can be readily evaluated by elementary means, to give the following explicit expression for $\sigma_1(\theta)$:

$$\sigma_1(\theta) = \frac{\varphi_C}{4\pi y(\theta)} e^{\frac{\omega R^2}{4\pi \lambda_0} \frac{1}{\sqrt{dR}} \int_{u_0}^{u} \frac{du}{1-u_0^2+u^2}}. \tag{21}$$

Here

$$u = \sqrt{\frac{R}{d}} \, \theta, \quad u_0 = \sqrt{\frac{R}{d}} \, \theta_0.$$

With the relation

$$F(uu_0) = \int_{u_0}^{u} \frac{du}{1 - u_0^2 + u^2} \, ,$$

one has, depending on the value of u_0,

$$u_0 < 1, \quad F(uu_0) = \frac{1}{\sqrt{1 - u_0^2}} \left[\text{arc tg} \, \frac{u}{\sqrt{1 - u_0^2}} - \text{arc tg} \, \frac{u_0}{\sqrt{1 - u_0^2}} \right];$$

(22')

$$u_0 = 1, \quad F(uu_0) = \left(\frac{1}{u_0} - \frac{1}{u} \right) = \left(1 - \sqrt{\frac{d}{R}} \cdot \frac{1}{\theta} \right);$$

(22")

$$u_0 > 1, \quad F(uu_0) = \frac{1}{2\sqrt{u_0^2 - 1}} \ln \frac{\left(u - \sqrt{u_0^2 - 1} \right)\left(u_0 + \sqrt{u_0^2 - 1} \right)}{\left(u_0 - \sqrt{u_0^2 - 1} \right)\left(u + \sqrt{u_0^2 - 1} \right)}.$$

(22‴)

The exact value of these integrals is, in general, of no significance, however.

In fact, in view of the approximations involved, it should not be anticipated that the theory can do more than give an understanding of the descriptive aspects of the relations in question and an estimate of the order of magnitude of the moment of the force of friction. This information can be obtained through a study of the integral of (21).

In order to fix the conditions for sufficiently rapid fall-off of $\sigma_1(\theta)$ to zero, we first note that the upper limit on the integral of (21) tends to infinity when $\theta \approx 0.1\text{-}1.0$ (since $\sqrt{R/d} \gg 1$), and that the entire integral is a monotonically decreasing function of u_0. This last point can be readily checked by making the change in variable:

$$v = u - u_0.$$

In fact

$$F(u_0) = \int_{u_0}^{\infty} \frac{du}{1 + u^2 - u_0^2} = \int_0^{\infty} \frac{dv}{1 + v(v + 2u_0)} \, ,$$

(23)

from which it is seen that F increases monotonically with falling u_0, passing through a maximum value of $\pi/2$ when $u_0 = 0$.

From this it follows that the maximum possible value for $\sigma_1(\theta)$ at large θ is given by the expression

$$\sigma_1(\theta) = \frac{\varphi_C}{4\pi \left[d + R(\theta^2 - \theta_0^2) \right]} e^{\frac{\omega R^{3/2}}{8\lambda_0 d^{1/2}}}.$$

(24)

In order that $\sigma_1(\theta)$ be small, even in the region where $\theta \ll 1$, it is necessary that the exponent be not too large. Actually the exponent varies from zero to $\omega R^{3/2}/8\lambda_0 d^{1/2}$ as θ is changed from θ_0 to values at which $\sqrt{R/d} \, \theta \gg 1$ (or $u \to \infty$).

For example, the exponential term in the expression for $\sigma_1(\theta)$ increases by no more than several factors when $\omega R^{3/2}/8\lambda_0 d^{1/2} \approx 1$, and the alteration in the function is principally determined by the preexponential term. Thus $\sigma_1(\theta)$ will fall off to zero rapidly enough if the value of λ_0 is such that

$$\frac{\omega R^{3/2}}{8\lambda_0 d^{1/2}} \leqslant 1.$$
(25)

It is clear from (19') that $\sigma(\theta)$ will certainly tend toward zero more rapidly than $1/\theta$ if the exponent is a slowly changing function of θ, so that the region of high θ values need not be taken into account in calculating the moment of the force of friction.

If, for example, $\omega = 30$ radian/sec, R = 10 cm, and d = 10^{-4} cm, the interval of permissible λ values fixed by (25) is

$$10^4 \text{ cm/sec} < \lambda < 10^{13} \text{ cm/sec.}$$
(26)

It is seen that our theory will not be applicable to good insulators ($\lambda < 10^4$ cm/sec). This is quite natural since the charge will not be able to pass into the recombination region, but accumulate on the cylinder surface, if the surface conductivity is low. A second limitation is imposed on λ_0 by the requirement that condition (17) be fulfilled.

The terms $\varphi(\delta_1)$ and $\varphi(\delta_2)$ of (17) can be evaluated with the aid of (15). It is readily seen that there is an interval of λ values (approximately 10^4-10^7 cm/sec) in which (17) and (25) are satisfied simultaneously, and the solution of (15), (19') and (19") is self-consistent.

The self-consistency of the solution breaks down at high λ values, condition (17) cutting off the interval of permissible λ's on the upper side. This fact has definite physical significance, since the current in the outer surface of the cylinder will be high, and condition (17) no longer satisfied, when the surface conductivity is high.

Values of σ_1 and σ_2 from (19') and (19") substituted into (20) lead to the moment of the frictional force

$$M = \frac{\varphi_C^2 R^2}{4\pi^2} \int_{\theta_0}^{\pi/2} \frac{\theta}{y^2(\theta)} \operatorname{sh}\left[\frac{\omega R^2}{4\pi\lambda_0} \int_{\theta_0}^{\theta} \frac{dt}{y(t)} d\theta\right]$$
(27)

or, in view of (21),

$$M = \frac{\varphi_C^2 R}{4\pi^2 d} \int_{u_0}^{\infty} \frac{u\,du}{[1 - u_0^2 + u^2]^2} \operatorname{sh}\left[\frac{\omega R^{3/2}}{4\pi\lambda_0 d^{1/2}} \int_{u_0}^{u} \frac{dt}{1 - u_0^2 + t^2}\right].$$
(28)

All of these calculations apply to the case in which the argument of the hyperbolic sine is either less than, or equal to, unity and an approximation expression for M when this argument is much less than unity is therefore useful:

$$M = \frac{\varphi_C^2 R\omega}{16\pi^3\lambda_0} (R/d)^{3/2} \int_{u_0}^{\infty} \frac{u\,du}{(1 - u_0^2 + u^2)^2} \int_{u_0}^{u} \frac{dt}{1 - u_0^2 + t^2}.$$
(29)

Estimates show the moment of the frictional force per unit length to be as high as ten or twenty dynes in cases of this kind. It should be noted that a marked rise in the moment tends to drive the argument of the hyperbolic sine toward unity. Equation (29) relates the moment of the force to the various physical parameters of the problem and is, therefore, open to experimental check.

It is natural to assume that $\lambda_0 \ll \lambda_1(\theta)$ when the surface of the cylinder is a good insulator of very low intrinsic conductivity, thus passing to a study of Case II. It has already been shown that the potential distribution over the cylinder surface is covered by Eq. (9) when (4") is applicable right up to the regions where $\sigma(\theta) = 0$.

It is, on the other hand, obvious that (4") will apply only over that portion of the surface where $\sigma(\theta)$ is large enough for validity of the inequality $\lambda_1(\theta) \gg \lambda_0$. It is to be recalled that $\lambda_1(\theta) = (\mu_p/e)\sigma(\theta)$. In addition, the general relation must naturally be fulfilled, which is to say that $\sigma(\theta)$ must fall off rapidly to zero in order that the region $\theta > \theta_{lim}$ make no significant contribution to the moment of the frictional force when $\theta_{lim} \ll 1$.

It has already been pointed out that this requires that $\sigma(\theta)$ tend to zero more rapidly than $1/\theta$ when the value of θ is high (see Eq. (20)). In the contrary case, the integral involved in the calculation of the moment of the frictional force will diverge as the upper limit tends to infinity. It is seen from (9) and (10), that $\sigma_1(\theta)$ falls off with $1/\theta$ in Case II, so that the upper limit cannot be chosen arbitrarily in calculating the moment. This is indication that Eq. (4") does not give a correct description of the region in which $\sigma_1(\theta) = 0$.

For an exact solution it is therefore necessary to use the exact (4) in place of (4") at a certain point on the surface and again pass over to (4') at other, still more distant, points. This makes the entire problem quite complex. Equation (9) can be used only for a quite accurate evaluation of the order of magnitude of the moment of the force of friction, the upper limit of integration, θ_{lim}, being so selected in each case that Eq. (4") will be valid and the condition $j_{sur}(\theta) \ll \omega R\sigma(\theta)$ fulfilled over the region in which $\theta_0 < \theta < \theta_{lim}$.

The moment of the force of friction for good insulators can be obtained by neglecting all surface conduction and assuming that removal of charge from the cylinder surface to be by gaseous discharge alone.

The calculations then become quite simple. The charge at surface points directly to the left of the region of immediate contact is identically equal to zero while points to the right of this same region carry fixed charge of equilibrium value, this situation persisting up to θ_{lim} where gaseous discharge reduces the charge essentially to zero.

It is then seen from Eq. (20) that

$$M = \pi\sigma^2_{eq} R^2 (\theta^2_{lim} - \theta^2_0).$$ (30)

In view of (11), it is possible to write

$$M = \pi\sigma^2_{eq} R [y(\theta_{lim}) - d]$$ (31)

or

$$M = \frac{R\varphi^2_C}{16\pi d} \left[\frac{y(\theta_{lim})}{d} - 1 \right].$$ (32)

Determination of the value of θ_{lim} requires recourse to the Paschen law of gaseous discharge (see [2]) which relates the discharge potential, V_{lim}, and the reduced gap width, ph (p is the gas pressure in mm Hg), through an equation of the form

$$V_{lim} = \frac{Bph}{C + \ln(ph)},$$ (33)

in which B and C are constants depending on the nature of the gas and the electrode metal. Relaxation of the gaseous discharge has been completely neglected here, discharge being assumed to begin as soon as the discharge potential, V_{lim}, has been established. It is clear from (32) and (33) that the moment of the force of friction in the insulator will be independent of the rotation rate in our approximation. Since the discharge gap is usually such that $h_{lim} \gg d$, h_{lim} can be identified with $y_{lim}(\theta)$ and the potential drop within the dielectric neglected in comparison with the potential drop in the gap between the two surfaces.

Equation (31) can then be written as

$$M = \pi R\sigma^2_{eq} \, y(\theta_{lim}) = \frac{R\sigma_{eq}}{4} V_{lim}$$ (34)

or

$$M = \frac{R\sigma_{eq}}{4} \frac{Bpy(\theta_{lim})}{C + \ln[y(\theta_{lim})]}.$$ (34a)

263

The critical gap depth, $y(\theta_{lim})$, can be obtained by applying the obvious relation

$$V = 4\pi y(\theta)\,\sigma_{eq}$$

(35)

and solving Eqs. (33) and (35) graphically.

A scheme of this kind for the determination of V_{lim} has been presented in [2]. It is clear that both $y(\theta_{lim})$ and V_{lim} are completely fixed by the value of σ_{eq}. Thus it follows from (34) that the moment of the frictional force is entirely independent of θ_0 (i.e., independent of the contact area) in the case under discussion here. It should be noted that this result traces back to the geometry of contact assumed in the figure at the beginning of this paper.

For air at $p = 760$ mm Hg and $\sigma_{eq} = 1.3 \cdot 10^4$, calculation shows that $y(\theta_{lim}) \approx 10^{-4}$ cm and $V_{lim} \approx 5 \cdot 10^3$ V. From this it follows that the moment of the frictional force per unit length for a cylinder of 20 cm radius is approximately 10^6 dynes.

Reduction in the pressure essentially reduces the discharge potential and therefore leads to an increase in the moment of the force of friction.

Conclusions

Rolling of a semiconducting cylinder, or a cylinder of dielectric, over a metallic surface gives rise to an electric component of rolling friction.

This component results from an asymmetrical distribution of double layer charge with respect to the point of contact. This asymmetry arises, in turn, from delayed recombination of the faces of the double layer through surface conduction, or gaseous discharge at high rotation rates.

The calculations developed here make it possible to express the moment of the frictional force as a function of rotation rate, contact area, and surface conduction. Similar equations apply to the rolling of a metallic cylinder over a semiconducting surface.

Under certain conditions, the electric component of the rolling moment can reach a value of 1 kg per 1 cm length of cylinder.

Literature

1. L. D. Landau and E. M. Lifshits. The Electrodynamics of Continuous Media, Moscow, Gostekhizdat (1957).
2. B. V. Deryagin and N. A. Krotova. Adhesion, Moscow, Izd. Akad. Nauk SSSR (1949).
3. L. D. Landau and E. M. Lifshits. The Mechanics of Continuous Media, Moscow, Gostekhizdat (1953).
4. A. F. Ioffe. The Physics of Semiconductors, Moscow, Izd. Akad. Nauk SSSR (1957).

ALTERATION OF THE SURFACE STATE OF A SEMICONDUCTOR IN ADHESIONAL BONDING WITH A POLYMER

N. A. Krotova

Institute of Physical Chemistry, Acad. Sci., USSR
Laboratory for Surface Phenomena

Introduction

The concept of the electrical character of the adhesional bond has recently attracted more and more interest among specialists in this field. The electrical theory of adhesion [1] views the adhesional bond as resulting from a double electric layer formed between the contacting solid surfaces.

Here the work required for rapid surface separation and the electrical charge density in the double electric layer are related by the equation

$$A = 2\pi\sigma^2 h = \frac{\sigma v}{2},$$

(1)

in which v is the potential difference in the double layer, and h is the maximum distance of face separation prior to the onset of gaseous discharge. The value of σ is constant for a given solid-solid couple, depending on the chemical constitution of the components, and on the nature of the functional group if one of the components of the couple is a polymer [2].

The most direct proof of the electrical nature of adhesive forces is the emission of high-speed electrons during rapid stripping of polymer from metal or glass, an effect discovered by V. V. Karasev, N. A. Krotova, and B. V. Deryagin [3]. It has been shown by L. P. Morozova [4] that the velocities of the electrons emitted by a polymer in such separation can be correlated with the strength of adhesion of this polymer to the substrate. The high velocity of the emitted electrons is indication of the existence of a high discharge potential in the gap between the separating surfaces. The role of electrical forces in adhesional phenomena has also been pointed out by the Americans Skinner, Savage, and Rutzler [5], who have proposed band models for the treatment of the adhesional bond.

B. V. Deryagin and V. P. Smilga have recently developed an electronic theory of the double electric layer responsible for the mechanical strength of the adhesional bond.

One of the papers of these authors [6] has used the concepts of solid state band theory in treating the role played by surface states in the semiconductor. Here it was suggested that strong adhesional forces would be developed in semiconductors with a large number of surface states. The system used by Bardeen and Brattain to explain the rectifying action of germanium [7] was drawn on to introduce surface states into the discussion of adhesional phenomena.

The problem reduces to that of developing a functional relation between the contact potential and the concentration of ionizing centers on the free semiconductor surface. Using statistical methods to treat the alteration in the contact potential, V_c, accompanying juncture of metal and semiconductor, these authors were led to the expression

$$V_C = 2d \left(\frac{\varkappa k T_\varepsilon}{e} sh \frac{\varphi}{2} + C\pi e sh\varphi \right) + \frac{kT}{e} \overline{\varphi}.$$ (2)

Here d is the gap width, $\varkappa$ the inverse Debye length

$$\overline{\varphi} = \frac{e\varphi}{kT}$$

and $e\varphi$ represents the bending of the semiconduction band resulting from contact with the metal.

The calculations of these authors indicate that the field, E, in the gap between the two separating surfaces increases markedly as the contact potential rises and the number of surface states, C, increases. The force of adhesion, F, can be calculated through the equation

$$F = \frac{E^2}{8\pi}.$$ (3)

The case of contact between a semiconductor and a dielectric (polymer) was not considered in [7] and requires a different method of approach [8]. Study of this case has, however, considerable practical interest in connection with the protection of semiconductors and p − n junctions so as to assure stable operation of semiconducting devices. The question of the mechanical strength of the adhesional bond is of undoubted interest for semiconductor-polymer systems. The strength of the adhesional bond is not, however, the only factor involved in the selection of a protective covering for the semiconducting device. The polymer chosen must permit the desired surface state to be maintained in the semiconductor and must not, therefore, alter the properties of the latter in the wrong direction [9].

Experiments in the neighboring fields of adsorption and catalysis on semiconductors can be drawn on in solving these problems.

The electronic theory of adsorption on semiconductors [10] considers the chemisorbed particle as a surface defect which charges the surface, or alters the charge which the surface originally carried, thereby changing the surface states and bending the energy bands. Chemisorption can change both the magnitude and the sign of the surface charge.

The bending of the energy bands can lead to alteration of the work function and the surface conduction. A positive surface charge enriches the surface layer with electrons and decreases the number of holes. A negative surface charge gives rise to the opposite effects.

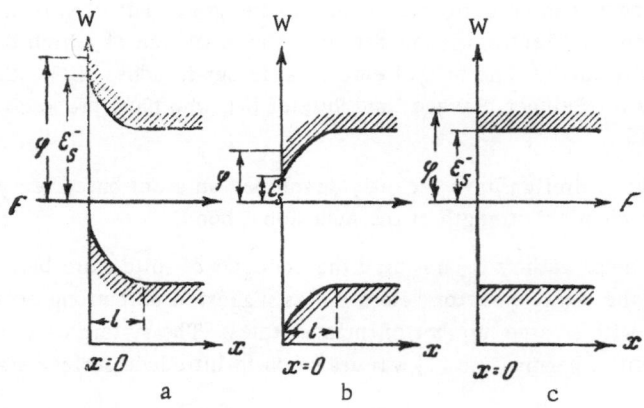

Fig. 1. The change in the work function, φ, of the semiconductor under alteration of the surface potential, ε. a) Negative surface charge; b) positive surface charge; c) electrically neutral surface; l is the screening length.

Thus the chemisorbed particle, affecting both the magnitude and sign of the surface charge, can function as either a donor or an acceptor.

The magnitude of the surface charge is usually determined from the work function. This procedure is entirely acceptable for the system gas-semiconductor and the method of its application is well developed.

The band diagram represents the work function as the distance between the Fermi level and the level corresponding to the potential outside of the semiconductor. When the surface charging is such that $\Delta q > 0$, the work function, $\Delta \varphi$, will be diminished, whereas a charging such that $\Delta q < 0$ entails an increase in $\Delta \varphi$ (Fig. 1).

The usual methods of determining the contact potential and work function are not applicable to semiconductor-polymer systems. Here the band bending and the surface charge can be determined by observing the affect of a field on the surface conduction. Data from such experiments makes it possible to follow the change induced in the surface state of a semiconductor through adhesional bonding with a polymer.

A second important problem for study in this field is the effect of treatment of the semiconducting surface prior to deposition of the polymer film. Various chemical and physical treatments can markedly alter the adhesion. This effect is obviously related to surface activation with the production of centers which are effective in both adsorption and adhesion.

Adsorption Centers

Formation of a polymer film on a solid surface is usually carried out by bringing the surface into contact with a liquid phase containing the monomer, and then polymerizing the latter in late stages of the technological process. In other cases, a solution of the polymer is brought into contact with the solid surface. Thus regular adsorptional interaction must play a predominant role in the initial stages of adhesional bonding. It is possible, in principle, to apply the theory of semiconductor adsorption to adhesion when the solid in question is a semiconductor [6]. The problem here is largely one of kinetics and catalysis. The concept of "adsorption center" is of especial interest to the theories which have been proposed. Consideration is usually limited to simply interpreted cases of the adsorption of gaseous molecules on a semiconducting surface. The rate of adsorption is covered by the expression

$$\frac{dN}{dt} = N^* S \alpha \varkappa p. \tag{4}$$

Here N is the number of adsorbed molecules of given type, N^* is the total number of adsorption centers per unit surface area, S is the effective surface area of the adsorbed molecule, p is the gas pressure, t is the time, $\varkappa$ is the probability that a molecule impinging onto an adsorption center will be captured by the latter, and

$$\alpha = \frac{1}{2\pi MkT},$$

M being the mass of the adsorbed molecule, T the absolute temperature and k the Boltzmann constant. The following remarks apply to the general case. For activated adsorption, it is generally assumed that

$$N^* = \text{const and } \varkappa = e^{-\frac{E}{kT}}, \tag{5}$$

E being the activation energy.

Figure 2 shows an adsorption curve with a potential barrier. This barrier functions in such manner as to bar energy-deficient molecules from the surface. The only gas molecules capable of adsorbing are those which approach the surface with kinetic energies in excess of the barrier height.

There is no reason to suppose that the number of adsorption centers, N^*, is constant. The activation energy, E, may be determined solely by the nature of the adsorption centers and then represents the

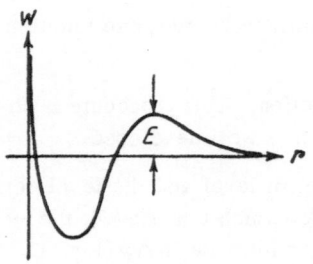

Fig. 2. Adsorption curve. W) system energy; r) distance between the particle and the adsorbent surface; E) activation barrier.

energy of center formation. Here one can, according to F. F. Vol'kenshtein, write

$$N^* = e^{-\frac{E}{kT}} \text{ and } \varkappa = \text{const.}$$

The activation energy will depend on both the adsorbent and the adsorbing molecules when several kinds of molecules are adsorbed on different centers. It can be assumed in the general case that

$$N^* = e^{-\frac{E_1}{kT}}, \quad \varkappa = e^{-\frac{E_2}{kT}}. \tag{6}$$

Here the activation energy is additively composed of two terms,

$$E = E_1 + E_2. \tag{7}$$

The valence electron of an atom adsorbed on a lattice[10] of singly charged positive and negative ions will no longer be exclusively associated with this atom but will be pulled into the adsorbent, the strength of the resultant bond being determined by the degree of attraction. The electron cloud surrounding the nucleus of the atom will be deformed from the spherical symmetry which it possessed in the isolated atom. This is an instance of a one-electron bond; it can be rather arbitrarily designated as "weak" in order to distinguish it from the "strong" bond resulting from donor-acceptor interaction. The chemisorbed atom can trap either a free electron or a free hole of the lattice in such way that the latter will participate directly in the chemisorptional bond which results. This chemisorbed particle can be considered as a structural defect. It can function as a localization center for the free electrons of the lattice, acting as a trap and acceptor. The result is the formation of a strong two-electron bond through the valence electron of the foreign atom and a free electron from the conduction band of the crystal lattice itself. In other words, this bond results from exchange between the electrons of lattice and foreign atom. Depending on its nature, the chemisorbed particle can also function as a center for the hole localization and then serves as a donor.

The adsorption centers are not localized on the surface. The number of these centers is not fixed but varies with the temperature and the presence of impurities in the adsorbent crystal lattice; it is also affected by various external factors such as ultra violet irradiation. Introduction of donor additives increases the concentration of adsorption centers, while addition of acceptor additives reduces this concentration, at least in cases where the adsorption centers are free electrons. Acceptor additives diminish the concentration of the electron gas, and increase the concentration of the hole gas, on the surface. Donor additives act in the opposite direction.

Additional adsorption centers can be formed by irradiation of the surface with light of such frequency that electrons are carried from the normal band into the conduction band. This illumination increases the concentration of adsorption centers, just as does an additive, and thereby affects both the adsorption equilibrium and the adsorption kinetics. Experiments [11] showing increased adsorption capacity of the semiconductor under illumination are well-known. Here it is found that the increase in adsorption capacity occurs at exactly those frequencies which are photoelectrically active [12]. Maximum sensitivity with respect to adsorptional properties coincides with maximum internal photoelectric effect.

Alteration of the Semiconductor Surface Through External Physical Effects

Subjecting the surface of the solid body to external illumination, or irradiating it with ionizing radiation, gives rise to surface defects which are adsorptionally active [13].

It is a well-known fact that defects of this kind can, in turn, give rise to surface centers which are effective in recombination and capture. Various methods are available for the controlled introduction of defects in the surface. A. V. Rzhanov and his coworkers [14] have proposed a model for those surface defects which give rise to recombination and capture centers on the surface of monocrystalline germanium.

A great deal of work, theoretical and experimental, has been recently published on the properties of recombination centers, traps, and adhesion centers, and the changes resulting in these from ion bombardment in glow discharge and various kinds of irradiation. We mention only those among these studies which seem to us to be of most interest.

It has been experimentally shown that two types of traps are formed when n-germanium is bombarded with positive argon ions in glow discharge and then annealed, namely: 1) centers located at the approximate center of the forbidden band, and 2) centers located in the neighborhood of the valence band.

Germanium has a large number of acceptor levels and its surface proves to be n-type as a result [15]. The work of A. V. Rzhanov and his colleagues has shown surface recombination in germanium which has been thermally treated to be determined by the joint action of several centers, each giving a comparable contribution to the overall recombination rate. These centers arise in the course of oxygen atom fixation at the germanium-oxide film interface [16].

The data of Lofersky and Rappoport [17] indicate the lifetime of minority carriers to be markedly affected by bombardment of the germanium surface with 1 mev electrons. These workers conclude that the lifetime is controlled by a single-center (recombination) in both n-type and p-type germanium. The energy of this center is 0.21 eV, measurement being from the bottom of the conduction band. The capture cross section proves to be 18 times larger for minority holes than for minority electrons. The absolute value of the cross section for hole capture is $4.5 \cdot 10^{-15}$ cm.

Lorlie and Curtis [18] have studied those defects which arise from irradiation of the germanium surface. The nature and concentration of the capture centers was experimentally studied through data on the lifetime of minority carriers. These authors conclude that this characteristic time is a very sensitive to the presence of defects arising from irradiation.

The irradiation of n-type and p-type germanium in this work was with the γ-radiation from Co^{60}, with fission neutrons, and with 14 mev monochromatic neutrons. The characteristics of the recombination centers resulting from irradiation, i.e., the positions of the energy levels, the capture cross sections, and the probability of capture for n-type germanium, were determined here.

Rapid and Slow States on the Semiconductor Surface

The term surface state is widely employed in discussing the physics of the semiconductor surface. This term was first proposed by Tamm [19] who pointed out that surface states corresponding to levels in the forbidden band could arise even in the ideal case of a lattice whose surface was free of foreign atoms and defects in the form of vacancies and dislocations.

Others have called attention to the fact [20] that these levels can merge to form a surface band partially overlapping the bulk bands, thus making it possible for electrons from this band to wander freely over the surface. In addition to Tamm levels, surface states can also arise from the adsorption of foreign atoms and from surface defects such as vacancies and dislocations. While the density of the Tamm levels is determined by the number of surface atoms, the density of defect and impurity levels can be varied at will. These levels can also merge to a band at high density. We will distinguish two types of states on the semiconductor surface.

R a p i d s t a t e s are characterized by the fact that the state charge alters rapidly enough to keep pace with the space charge (in an external field, for instance), the time required for alteration being of the order of a microsecond or less. The density of rapid states is of the order of $(10^{11} - 10^{12})/cm^2$.

S l o w s t a t e s have relaxation times of the order of seconds or minutes at room temperature. These states are supposed to be localized on the outer surface of the oxide film on the semiconductor, the rapid states being concentrated on the interface between the semiconductor and the film. Relaxation times of this order are presumed to result from the fact that bulk — surface interaction occurs by charge seepage through the oxide layer, a process requiring considerable time. Changes in the slow states can be brought about externally by illumination, or the application of an electric field, with gradual return to the equilibrium state. The density of slow states is $10^{15} cm^{-2}$.

Two models have been proposed to account for the experimental observations on slow states (time dependence of the characteristics of the semiconductor). Kingston and McWhorter [21] have suggested that surface variability will lead to a spectrum of relaxation times, the relaxation time varying in passing from one portion of the surface to another. Another model due to Morrison [22] explains the experimental decay curve, even for a uniform surface, by considering electron exchange through a surface barrier to be the rate limiting step. Departure of the surface state charge from equilibrium alters the rate of capture, directly, as a result of the change in the number of vacant surface traps, and indirectly, as a result of the variation of barrier height with surface charge. Both these suggestions recall the model proposed for chemisorption [23] where an experimental adsorption curve similar to the decay curve for the semiconductor characteristics is explained through surface variability and electron transitions.

F. F. Vol'kenshtein [10] has pointed out that chemisorption on the semiconductor surface can be valence-saturated, radical, or ion-radical. These forms, and the transitions between them, depend on adsorption center localization and delocalization of electrons and holes.

Illumination of the semiconductor alters the adsorption capacity by changing the concentrations of electron or hole gas on the surface. This leads to an alteration in the various forms of chemisorption. Application of an external electric field brings about the same result.

This author has not touched on the question as to which surface states of the semiconductor are, in the first instance, involved in adsorption. It can, however, be assumed that the establishment of equilibrium and equilization of Fermi levels will proceed at different rates over the various portions of the nonuniform surface, and that the relaxation time for the slow states will vary accordingly. Electron transfers corresponding to passage from one type of chemisorption to another may require more or less extended periods of time. Although the experimental material available at the present would indicate that it is the slow states which are involved in adsorption, the data is contradictory and lacking in precision.

The structure of the solid surface is so complex, and the electron transfers occuring on it so numerous, that it may well be that both the slow and the rapid states participate in adsorption.

Investigation of the Surface State of the Semiconductor by Determination of the Surface Recombination and Conduction in the Presence of a Field

Semiconductors are particularly interesting for studies on adhesional phenomena since the magnitude of the adhesion depends on the number of charged surface states [6]. The presence of these states accounts, according to Bardeen, for the fact that the surface is associated with a potential barrier, whose height (φ_0) can be varied by the application of a transverse electrical field. Alteration of the potential barrier affects the experimentally determined surface characteristics, e.g., the conduction and recombination.

The theory of bulk recombination has been developed by Schockley and Read [24] from the statistics of electron-hole recombinations within the semiconductor. A similar analyses of surface recombination has been carried out by Brattain and Bardeen (see [25]). Here recombination ties up with the presence of adsorbed particles and other surface defects. It follows that surface recombination and conduction are structurally sensitive properties. The statistical equation of Brattain and Bardeen for the rate of surface recombination contains various parameters characterizing the surface state of the semiconductor. This equation has the form

$$ S = \frac{N_t c_p (p_b + n_b)}{2n_i \exp(q\varphi_0/kT) \{\mathrm{ch}\,[(E_t - E_i - q\varphi_0)/kT] + \mathrm{ch}\,[q(\varphi_S - \varphi_0)/kT]\}}. \tag{8} $$

Here N_t is the density, per unit surface area, of those surface centers which participate in the recombination; E_t is the discrete energy level corresponding to these recombination centers; E_i is the Fermi level for intrinsic semiconduction (center of the forbidden band); n_i is the concentration of electrons (or holes) in intrinsic semiconduction; n_b and p_b are the equilibrium carrier concentrations in the bulk; c_n and c_p are the respective probabilities of capture of electrons and holes; q is the elementary charge; φ_0 is the height of the potential barrier; $q\varphi_0 = kT/2 \ln(c_p/c_n)$; and $(\varphi_S - \varphi_0)$ is the difference in barrier heights, the value with the external field applied minus the original value.

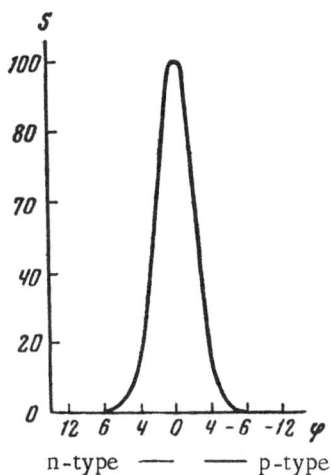

Fig. 3. The relation between the rate of surface recombination and the surface potential (Eq. (8)).

The potential barrier height is frequently expressed in terms of kT, and one then has

$$\psi = \frac{\varphi}{kT} .$$

(9)

Equation (8) is represented graphically in Fig. 3 which shows the rate of surface recombination to be symmetrical around the point $\varphi_s = \varphi_0$.

The effective lifetime of the minority carriers is usually used to characterize recombination processes. This time can be directly measured by the use of suitable appratus. The decay of optically excited excess minority carriers is balanced against either the change of potential in an RC circuit, or some other known standard, and directly read from the scale of a cathode oscillograph.

The rate of surface recombination can be obtained from the carrier lifetime if the dimensions of the sample, the diffusion constant of the minority carriers, and the carrier lifetime in the bulk are all known. A nomogram can be used to calculate the rate of surface recombination when these parameters are available.

The relation between the rate of surface recombination, S, and the potential can be developed experimentally by applying a constant, or alternating, electrical field at right angles to the surface, thereby altering the energy barrier and band bending. Alteration of the barrier height continues until equilibrium is established. From this it follows that relaxation effects must be taken into account in applying the constant field. This is the method employed when it is a matter of slow states with relaxation periods measured in seconds or minutes. By use of an alternating field of correctly chosen frequency, the rapid surface states can be held in continual equilibrium and the slow states prevented from screening the field [26]. In this way it is possible to study rapid states with relaxation times of the order of microseconds.

The surfaces conduction is a second important parameter characterizing the surface state of the semiconductor. For fixed specific resistance and temperature, the surface conduction per unit surface area of space charge layer is directly determined by the height of the potential barrier, $\Psi = \varphi/kT$. Theoretical calculation

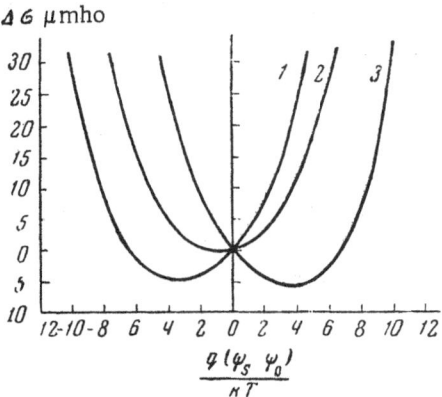

Fig. 4. The relation between the surface conduction and the surface potential. 1) Number of donors, $Nd = 10^{14}$ cm^{-3}; 2) number of donors equal to the number of acceptors; 3) number of acceptors, $Na = 3.16 \cdot 10^{14}$ cm^{-3}.

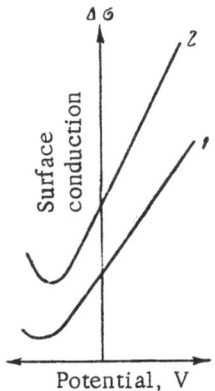

Fig. 5. Oscillographic dark (1) and light (2) curves developed for a germanium surface modified by treatment with an alkyl chlorosilane.

by Garret and Schriffer [27] have considered the reduction in carrier mobility on the semiconductor surface. The curve obtained by Schriffer is shown in Fig. 4. The characteristic parameters of the semiconductor surface (barrier height or band bending, charge, etc.) can be calculated by the method developed by Brown [28] when there is zero applied field. This method is based on the assumption that there is a single point of intersection of the theoretical and experimental curves, this corresponding to the minimum on the conduction vs field strength curve. Here the number of carriers of each sign is the same in the subsurface layer.

Slow states with extended relaxation times are studied under a fixed field. Information concerning the rapid states can be obtained by combining the field effect with stationary photoconduction [29]. Experiments of this kind are carried out in the following manner. A high amplitude sinusoidal field is applied at right angles to the surface of the sample. A curve showing conduction as a function of field strength can then be observed on the screen of the oscillograph, this curve passing through a characteristic minimum. By applying the Brown theory to this curve, one can obtain the band bending, surface charge, and charge captured by rapid surface states, parameters characterizing the surface state of the semiconductor. If the sample is illuminated with intermittent light at an interruption frequency at least as high as the field frequency, two curves, the dark and the light, will be simultaneously traced out on the screen as potential functions. Oscillographic curves of this kind for an n-germanium surface modified by treatment with an alkyl chlorosilane are shown in Fig. 5. The vertical separation between curve points corresponding to the same field strength represents the photoconduction. The latter is proportional to the effective lifetime of the minority current carriers. In this way, one can develop a curve showing surface recombination as a function of field strength.

Data obtained in this manner to show the alteration of the lifetime of the minority carriers (in μsec) resulting from modification of the n-germanium surface are presented below:

Prior to silane treatment 100   102   72
Following silane treatment . . . 600   400   460

The method developed by Rzhanov and his coworkers [30] has many advantages and permits rapid determinations; it was first applied to the semiconductor-polymer interface by G. A. Sokolina [31].

Alteration of the Semiconductor (Germanium) Surface Through Formation of the Adhesional Bond

The physics of the semiconductor surface is still a new field in which there has recently been intense activity in connection with problems in the manufacture of semiconducting devices.

The problem of interest to us is a more narrow one, namely, that of developing a mechanism for the processes involved in the formation of an adhesional bond between semiconductor and metal, or dielectric (polymer). Both cases have considerable practical significance.

The first case (semiconductor-metal) is important for the development of new technological methods in the production of semiconducting devices, while the second, has significance for the problems which arise in attempting to improve the functioning of these devices by protecting them with lacquer coatings.

The original papers of the present author and G. A. Sokolina (with Yu. A. Khrustalev) [31] described the alteration of slow and rapid surface states of germanium resulting from surface treatment by glow discharge or

TABLE 1

n-Germanium, etched in SR-4		n-Germanium, treated in glow discharge	
Adhesion, g /mm^2	Effective lifetime, μsec	Adhesion, g/mm^2	Effective lifetime, μsec
75	600	740	100

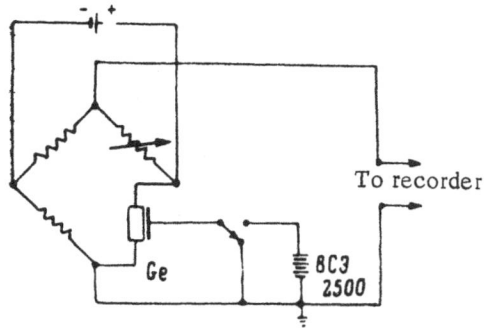

Fig. 6. Circuit for the measurement of $\Delta\sigma$ under an applied field.

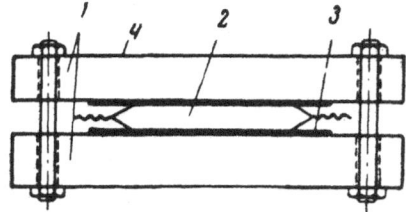

Fig. 7. Schematic representation of a germanium sample clamped in the condenser. 1) Metallic sheets; 2) sample with leads to bridge; 3) mica; 4) to high potential supply.

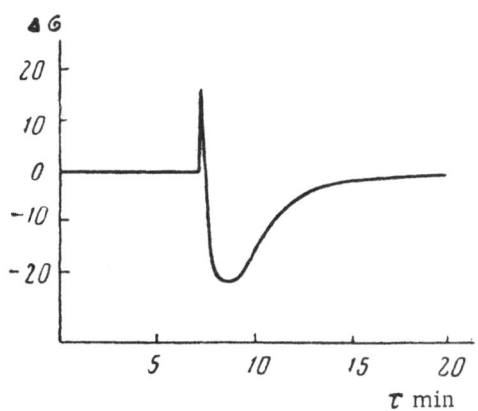

Fig. 8. Time (τ) dependence of the relaxation of conduction ($\Delta\sigma$) in slow state studies.

ultraviolet irradiation. Experiment proved that such treatment could be helpful in improving the adhesion of the polymer to a solid base such as glass [32]. The rapid surface states were here characterized by the effective lifetime of the minority carriers, as determined from the fall-off in the photoconduction compared with the voltage drop in an RC circuit.

Treatment in the glow discharge increases the number of surface states of the semiconductor. The considerations of [6] indicate that such treatment should enhance the surface adhesability. We have actually observed this effect in the case of the adhesion of indium to n-type germanium (Table 1). The adhesion was determined on an AZS-1 adhesiometer, using the method which we proposed in [31]. Statistical analysis of the results obtained from measurements of the adhesion of indium to a germanium surface which had been etched in SR-4 and then treated in the glow discharge indicates that most of the untreated samples had zero adhesion while the adhesion of most of the treated samples fell in the 500-600 g/mm² interval. The adhesability of the germanium was diminished by vacuum heating at 500°C and 10 mm Hg for 1-2 hours.

Table 1 shows the adhesion of indium to the untreated n-germanium surface and to the n-germanium surface which had been treated in the glow discharge, with a comparison of the effective minority carrier lifetimes.

These figures are consistent with the theoretical calculations by V. P. Smilga and B. V. Deryagin [6] which show that the field arising through semiconductor-metal contact increases rapidly with the number of ionized centers originally present on the semiconductor surface, all other conditions being constant. These authors define the adhesional strength by the expression

$$F = \frac{E^2}{8\pi},$$ (10)

so that it is proportional to the second power of the field strength. The experiments in question therefore establish a relation between the adhesion and the number of surface states associated with ionized centers on the free semiconductor surface.

The alteration of the rapid states on the germanium surface resulting from monomer (methacrylic acid and acrylonitrile) adsorption was studied in a second series of experiments. A germanium sample previously etched with SR-4 was evacuated to 10^{-4} mm Hg in a glass vacuum system where the monomer was stored in a side arm. A ground joint was rotated at the end of evacuation to allow the gaseous monomer to come into contact with the semiconductor surface. Polymerization was subsequently brought about by heating the sample. Our experimental data are presented below; they show an alteration in the effective lifetime of the minority

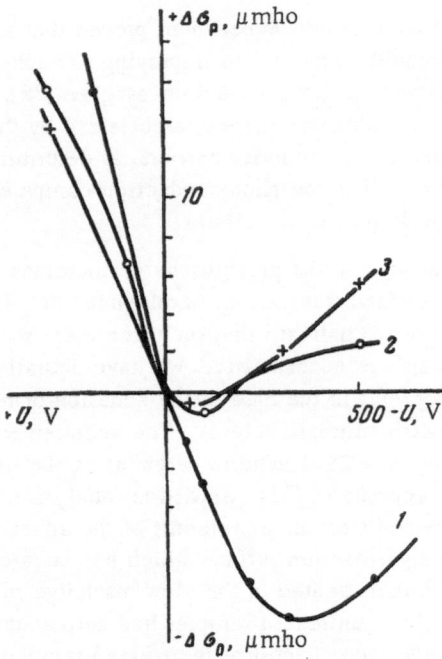

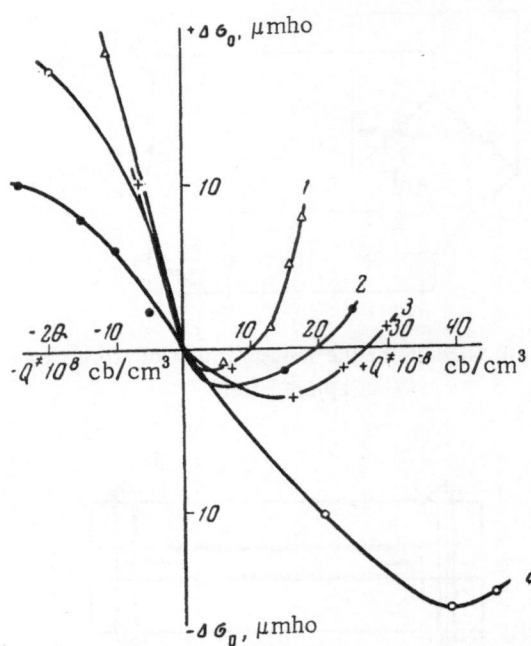

Fig. 9. The effect of the external medium on the form of the $\Delta\sigma$ vs field strength curve for germanium. 1) Under atmospheric conditions; 2) in vacuum, 10^{-4} mm Hg; 3) vacuum heated at 800°C for 10 minutes.

Fig. 10. Field induced $\Delta\sigma$ curves for germanium surfaces covered with vinyl series polymer films. 1) Polystyrene; 2) polyvinyl alcohol; 3) polyvinyl acetate; 4) perchlorvinyl.

carriers of n-type germanium, and, in turn, reflect the change in the number of rapid surface states resulting from monomer (acrylonitrile, in this case) adsorption on the surface.

After H_2O_2 treatment.	800	730	600
After evacuation and admission of acrylonitrile vapors.	600	560	520

The data of the literature indicate a relation between slow surface states localized on the outer surface of the oxide film in contact with the medium surrounding this surface, and the foreign atoms adsorbed on it. The slow states were characterized by the field effect in surface conduction in our own experiments. It is a well-known fact that surface adsorption of foreign substances can alter the band bending and surface charge. By applying a transverse electric field to the sample and tracing out the conduction curve, one can decide how the curve has been affected by adhesional bonding between polymer and semiconductor, and thereby be led to conclusions concerning the surface changes induced by bond formation. The apparatus used here is shown schematically in Fig. 6.

The alteration of the conduction was determined with a Wheatstone bridge, one arm of which contained the germanium sample. This sample had been held in the air for two months prior to experiment so that its surface had acquired an equilibrated oxide film. Alteration of the conduction of the sample was reflected in a change in the potential, which was recorded, in turn, on a ÉPP-09 recorder with a transit time of 1 sec.

The sample of germanium in the one arm of the Wheatstone bridge served as a capacitor plate of a condenser to which a high voltage was applied from a direct VSÉ-2500 supply. The construction of this condenser is shown schematically in Fig. 7.

These experiments were carried out with continuous registration of the time variation of those parameters which are sensitive to alteration in the surface conductivity of the germanium in an externally applied transverse electrical field. An example of a curve showing the time dependence of the surface conduction is presented in Fig. 8.

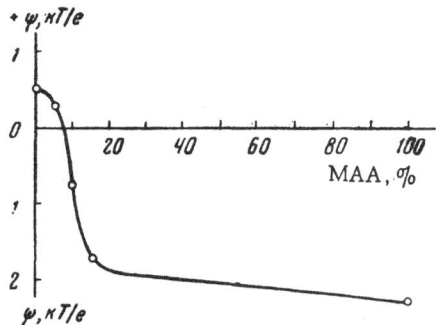

Fig. 11. The effect of the concentration of methacrylic acid (MAA) in copolymers of methylmethacrylate and methacrylic acid on the band bending.

It is assumed that damping results from relaxation effects associated with carrier capture by those slow surface levels which correspond to centers on the external face of the oxide film.

Typical curves showing the conductivity of high-resistance n-germanium under an applied field are presented in Fig. 9. These curves can be represented either as $\Delta\sigma = f(U)$ or as $\Delta\sigma = \varphi(Q^{\neq})$, $Q^{\neq}$ being the induced charge on the condenser, as determined from the capacity of the latter, $\Delta\sigma$ the change in surface conduction, and U the potential. The minimum point corresponds to intrinsic conduction where the number of electrons is equal to the number of holes. For a semiconductor with known bulk characteristics, the values of $\Delta\sigma$ at the minima on the experimental and theoretical curves must be identical since this common point on the two curves corresponds to intrinsic conduction.

The $\Delta\sigma$ minimum is deeper on the curve developed in air than it is on the vacuum curve. This difference is due to atmospheric moisture (see Fig. 9). Long evacuation, followed by vacuum heating to 200-300°C, has no effect on the position of the minimum.

We have carried out systematic studies on the effect of the nature and concentration of the functional groups in the copolymer chain on $\Delta\sigma = f(U) = \varphi(Q)$ curves developed in vacuum.

Vinyl series polymers were used in these studies (Fig. 10). Copolymers of methylmethacrylate and methacrylic acid were used in studying the effect of the carboxylic group concentration in the copolymer chain. There is a single carboxylic group in the molecule of methacrylic acid, $CH_2 = C - CH_3$. Increasing the content of this
$$\overset{\displaystyle |}{COOH}$$
acid in the polymerizing mixture leads to an increase in the number of carboxylic groups in the chain of the resulting polymer. $\Delta\sigma = f(U) = \varphi(Q)$ curves for copolymers of methylmethacrylate and methacrylic acid (MAA) are shown in Fig. 11.

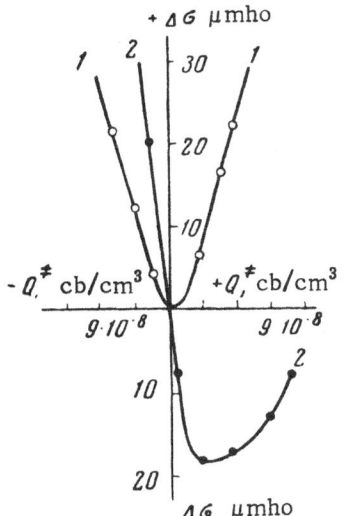

Fig. 12. Field induced $\Delta\sigma$ curves for the free germanium surface (1), and for the germanium surface carrying a gelatin film (2).

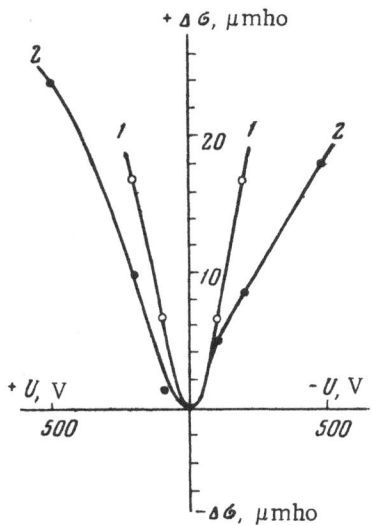

Fig. 13. Field induced $\Delta\sigma$ curves for the free germanium surface (1), and for the germanium surface carrying a polyoxyethylene film (2).

Figure 12 shows curves covering the field induced conduction at the germanium-gelatin interface. Gelatin contains various polar functional groups (NH_2, OH, and COOH). The presence of a film of this polymer on the germanium surface leads to a marked increase in the value of $\Delta\sigma$, comparison being with the value for the free surface.

We have also studied the effect of a polyoxyethylene polymer which was free of polar functional groups but contained singly bound oxygen in the chain: $(-CH_2-CH_2-O-CH_2-CH_2-O-)_x$. Deposition of this polymer on the germanium surface did not lead to any alteration in the value of $\Delta\sigma_{min}$ (Fig. 13).

Comparison of these curves leads to the conclusion that polar functional groups are capable of bringing about a marked alteration in the position of $\Delta\sigma_{min}$, while polymers free of such groups (polystyrene, methylmethacrylate, polyoxethylene) have comparatively little effect on it.

Conclusions

The application of modern methods to the study of the semiconductor-polymer interface makes it possible to obtain very significant information concerning the alteration of the surface characteristics of the semiconductor resulting from formation of the adhesional bond. The principal experimentally determined parameters characterizing the surface properties of the semiconductor are the surface conduction and the rate of surface recombination. The theory of the so-called field effect in these parameters has been rather fully developed. By comparing the experimentally obtained data with the results of theoretical calculations, and drawing on special numerical methods [34], it is possible to determine the band bending, surface potential, charge captured by surface states, and other important parameters of the surface of contact between semiconductor and polymer film. Study of the field effect in surface recombination and conduction gives a highly sensitive method of following surface changes resulting from irradiation with visible and ultraviolet light and other types of radiation, treatment in the glow discharge, and chemical modification. Thus, a study of the rate of surface recombination and surface conduction can be combined with investigations on the strength of the adhesional bond to obtain a clue to the mechanism of the action of these various factors on the adhesional properties of the semiconductor surface. In principle, it is possible to obtain a detailed analysis of those changes which result in the slow and rapid states at the polymer-semiconductor interface from formation of the adhesional bond.

Our experiments have shown the possibility of measuring the rate of surface recombination and conduction resulting from the formation of a polymer film on the surface of the semiconductor.

They have also indicated that deposition of a polymer film on the germanium surface can alter the magnitude and sign of the band bending, ψ, and the surface charge, g, in the absence of an external electric field (Table 2).

Application of the field leads to an alteration of the induced charge, $Q^{\neq}$, and the band bending, ψ. The induced charge is given by

$$Q^+ = Q_{ss} + Q_{sc}, \tag{11}$$

where Q_{SC} is the space charge in the subsurface layer, and Q_{SS} is the charge captured by the surface states (traps).

TABLE 2

Working material	Band bending, ψ		Surface charge, q coulombs/cm^2
	kT/e	mV	
Pure germanium	+2	+50	$-2.5 \cdot 10^{-9}$
Polyoxyethylene	+2	+50	$-2.6 \cdot 10^{-9}$
Polystrene	+0.6	+ 1.5	$-0.6 \cdot 10^{-9}$
Polyvinyl alcohol	+0.15	+ 3.75	$-0.1 \cdot 10^{-9}$
Polyvinylacetate	−0.25	− 6.26	$+0.15 \cdot 10^{-9}$
Perchlorvinyl	−2.6	−65	$+3.8 \cdot 10^{-9}$
Gelatin	2.75	−69	$+4.2 \cdot 10^{-9}$

TABLE 3

Polymer	Band bending, ψ		Surface charge, q, coulombs/cm^2
	kT/e	mV	
Copolymer, methacrylate with 6.5% MAA	+0.25	+ 6.25	$- 1 \cdot 10^{-9}$
The same, with 10.8% MAA	−0.75	−18.8	$+ 1 \cdot 10^{-9}$
The same, with 17.2% MAA	−1.75	−44.0	$+ 2 \cdot 10^{-9}$
Methacrylic acid (MAA)	−2.25	−55.7	$+3.5 \cdot 10^{-9}$

Knowing the relation of $\Delta\sigma$ to ψ, the value of Q_{ss} in the absence of an external field can be calculated.

Comparison of the band bending, ψ, and the surface charge, q, parameters affected by the presence of the functional groups in the polymer on the germanium surface, leads to the following conclusions.

The polymers which have been studied here contained various functional groups in the chain; most of these proved capable of altering both the sign of the band bending and the surface charge of germanium which had been in contact with the atmosphere. The surface charge proved to be negative, and the space charge in the subsurface layer was therefore positive (Tables 2 and 3).

The data of Tables 2 and 3 indicate that the absolute magnitude of the surface charge resulting from contact with a polymer depends on the polarity of the functional groups which the polymer contains. Polyoxyethylene induces practically no alteration in the surface charge, whereas gelatin, with various strongly polar groups, alters its sign and gives the surface a pronounced positive charge.

Increasing the concentration of the polar functional groups in the copolymer chain leads to an increase in the surface charge. Here it is also observed that supercharging takes place when the methacrylic acid content in the copolymer reaches a certain definite value. This occurs when the concentration of methacrylic acid in the copolymer falls in the interval from 6.5% to 10.8%. Pure methacrylic acid gives a positive charge of considerable magnitude to the surface. From this it can be concluded that the presence of polar functional groups in the polymer can markedly affect both the magnitude of band bending, ψ, and the surface charge, q. In view of these facts, it would be natural to chose a protective lacquer whose chain was free of polar functional groups of high dipole moment. Our experiments have shown the semiconductor surface to be quite sensitive to the chemical structure of the protecting polymer film.

The experimental part of this work, and the calculations, were carried out by G. A. Sokolina with the aid of Yu. A. Khrustalev.

I would like to express my thanks for the valuable advice offered by B. V. Deryagin and F. F. Vol'kenshtein.

Literature

1. B. V. Deryagin and N. A. Krotova, Usp. Fiz. Nauk 36:387 (1948).
   B. V. Deryagin and N. A. Krotova. Adhesion. Moscow, Izd. Akad. Nauk SSSR (1949).
   N. A. Krotova and L. P. Morozova. Dokl. Akad. Nauk SSSR 127(1):141 (1959).
   L. P. Morozova and N. A. Krotova. Kolloidn. Zh. 20(1):59 (1958); Dokl. Akad. Nauk SSSR 115:747 (1957).
   N. A. Krotova, L. P. Morozova, G. A. Sokolina, and B. V. Deryagin. Dokl. Akad. Nauk SSSR 129:149 (1959).
   N. A. Krotova. Concerning Gluing and Adhesion, Moscow, Izd. Akad. Nauk SSSR (1950).
2. N. A. Krotova, Yu. M. Kirillova, and B. V. Deryagin. Zh. Fiz. Khim. 30:1921 (1956); Dokl. Akad. Nauk SSSR 97:475 (1954).
3. V. V. Karasev, N. A. Krotova, and B. V. Deryagin. Dokl. Akad. Nauk SSSR 88:777 (1953); Dokl. Akad. Nauk SSSR 89:108 (1953);
   V. V. Karasev and N. A. Krotova. Dokl. Akad. Nauk SSSR 99:715 (1954).
4. N. A. Krotova, L. P. Morozova, and B. V. Deryagin. Dokl. Akad. Nauk SSSR 129:149 (1959).

5. S. M. Skinner, R. L. Savage, and J. E. Rutzler. J. Appl. Phys. 24(4):434 (1953).
   B. V. Deryagin and V. P. Smilga. Dokl. Akad. Nauk SSSR 121:877 (1958).
   V. P. Smilga, Zh. Fiz. Tverd. Tela 1:307 (1959).

6. V. P. Smilga and B. V. Deryagin. Dokl. Akad. Nauk SSSR 122:1049 (1958).

7. W. H. Brattain and J. Bardeen. "Surface properties of germanium," In collection: Problems of Semiconductor Physics [Russian translation] Moscow, IL (1957).
   J. Bardeen. Phys. Rev. 71(10):717 (1947).

8. V. P. Smilga. In collection: Research in Surface Forces, Moscow, Izd. Akad. Nauk SSSR (1961), p. 76. [English translation: Consultants Bureau, New York (1963).]

9. V. A. Presnov and V. F. Syporov. Zh. Fiz. Tvered. Tela 2(3) (1960).

10. F. F. Vol'kenshtein. Usp. Fiz. Nauk 50:253 (1953); The Electronic Theory of Catalysis on Semiconductors, Moscow, Gosfizmatizdat (1960).

11. A. N. Terenin. Zh. Fiz. Khim. 6:186 (1936).
    L. N. Kurbatov. Zh. Fiz. Khim. 11:1049 (1940).

12. J. A. Hedvall and S. Nord. Z. Elektrochem. 49:467 (1943).
    A. Kabayashia and S. Kawaji. J. Phys. Soc. Japan 10:270 (1955); 11:369 (1958).

13. A. N. Terenin. In collection: Problems of Kinetics and Catalysis (1955), pp. 8, 17.
    Yu. P. Solonitsyn. Zh. Fiz. Khim. 32:1241 (1958).
    V. F. Kiselev, K. G. Krasil'nikov, and E. A. Sysoev. Dokl. Akad. Nauk SSSR 116:990 (1957).
    A. V. Rzhanov. In collection: The Surface Properties of Semiconductors, Moscow, Izd. Akad. Nauk SSSR (1962), p. 101.
    V. A. Presnov and L. L. Lyuze. In collection: The Surface Properties of Semiconductors, Moscow, Izd. Akad. Nauk SSSR (1962), p. 217.

14. A. V. Rzhanov, Yu. F. Novototskii-Vlasov, I. G. Neizvestnyi, S. V. Pokrovskaya, and T. I. Galkina. Zh. Fiz. Tverd. Tela 3(3):822 (1961).

15. Shyh Wang and C. Wallis. J. Appl. Phys. 30:285 (1959).

16. A. V. Rzhanov, N. M. Pavlov, and M. A. Selezneva. Zh. Tekhn. Fiz. 27:1707 (1957).

17. J. G. Loferesky and A. Rappoport. J. Appl. Phys. 30:1181 (1959).
    B. A. Krasyuk and L. I. Gibov. Semiconductors of Germanium and Silicon, The Scientific—Theoretical Institute for Light and Heavy Metals (1961).
    V. G. Litovchenko and V. I. Lyashchenko. Zh. Fiz. Tverd. Tela 2(7):1592 (1960).

18. A. Lorlie and J. Curtis. J. Appl. Phys. 30:1174 (1959).
    C. H. Werthum. J. Appl. Phys. 30:1166 (1959).
    M. V. Chukichev and V. S. Vavilov. Zh. Fiz. Tverd. Tela 3(3):935 (1961).
    N. A. Vitovskii, E. V. Meshovets, and S. M. Ryvkin. Zh. Fiz. Tverd. Tela 1:1381 (1959).

19. I. E. Tamm. Zh. Éksperim. i Teor. Fiz. 3(1):34 (1933).

20. E. Goodwin. Proc. Cambridge Phys. Soc. 35:205, 20 (1939).
    L. V. Azaroff. Introduction to Solids, New York (1960)

21. R. H. Kingston and A. McWhorter. J. Phys. Rev. 103:534 (1956); 95:1491 (1955).

22. S. R. Morrison. In collection: Semiconductor Surface Physics, R. H. Kinston (ed.), University of Pennsylvania Press, Philadelphia, Pennsylvania (1957), p. 169; In collection: Problems of Semiconductor Physics [Russian translation], Moscow, IL (1959), p. 323.

23. S. R. Morrison. In collection: Semiconductor Surface Physics [Russian translation], Moscow, IL (1959), p. 186.
    D. J. Pratt and I. K. Kolm. In collection: Semiconductor Surface Physics [Russian translation], Moscow, IL (1959), p. 217.

24. W. S. Shockley and W. H. Read. Phys. Rev. 87:835 (1952), in collection: Semiconductor Electronic Devices [Russian translation], Moscow, IL (1953), p. 121.
    G. Goudet and C. Moulean. Les Semicouducteurs, Paris (1958).
    G. Doplep. Introduction to the Physics of Semiconductors [Russian translation], Moscow, IL (1959).

25. A. Mzni and E. Kharnik. In collection: Semiconductor Surface Physics [Russian translation], Moscow, IL (1959).

W. H. Brattain and J. Bardeen. Bell System Tech. J. 82:1 (1953); In collection: Problems of Semiconductor Physics [Russian translation], Moscow, IL (1953), p. 237; Current Carrier Recombination in Semiconductors [Russian translation], Moscow, IL (1959).

D. T. Stevenson and J. P. Keyes. Physica 20:1041 (1954); [Russian translation appeared in the collection: The Electrophysical Properties of Germanium and Silicon; and in: Soviet Radio (1956), p. 367].

26. A. E. Yunovich. In collection: The Surface Properties of Semiconductors, Moscow, Izd. Akad. Nauk SSSR (1962), p. 127.

Yu. F. Novototskii-Vlasov and N. G. Neizvestnyi. Pribory i Tekhn. Éksperim., No. 1:127 (1961).

V. G. Litovchenko and V. I. Lyashchenko. Zh. Fiz. Tverd. Tela 2:1592 (1960).

V. G. Litovchenko, V. I. Lyashchenko and O. S. Frolov. In collection: The Surface Properties of Semiconductors, Moscow, Izd. Akad. Nauk (1962), p. 147.

27. J. R. Schriffer. Phys. Rev. 97(3):641 (1955); Problems of Semiconductor Physics [Russian translation], Moscow, IL (1957).

28. W. L. Brown. Phys. Rev. 590:100 (1955).

29. Yu. F. Novototskii-Vlasov and N. G. Neizvestnyi. Zh. Prib. i Tekhn. Éksperim., No. 1:127 (1961).

V. G. Litovchenko and O. V. Snitko. Zh. Fiz. Tverd. Tela 2(5) (1960).

E. Lasser, K. Vysotskii, and B. Bernshtein. In collection: Semiconductor Surface Physics [Russian translation], Moscow, IL (1959), p. 247.

30. A. V. Rzhanov. Zh. Fiz. Tverd. Tela 1:522 (1959).

A. V. Rzhanov, Yu. F. Novototskii-Vlasov, and N. G. Neizvestnyi. Zh. Fiz. Tverd. Tela 1:1959.

31. G. A. Sokolina, N. A. Krotova, and Yu. A. Khrustalev. Dokl. Akad. Nauk SSSR 147:1409 (1962).

32. N. A. Krotova, L. P. Morozova, and G. A. Sokolina. The mechanical and electrical characteristics of adhesion, In collection: Proceedings, Third International Congress on Surface Phenomena, Cologne (1960).

33. N. A. Krotova, G. A. Sokolina, and L. P. Morozova. Dokl. Akad. Nauk SSSR 302:1959 (1947).

34. Semiconductor Surface Physics, [Russian translation under the editorship of G. E. Pikus] Moscow, IL (1959).

A STUDY OF THE ELECTRICAL PROPERTIES OF THE FRESHLY FORMED SURFACE THROUGH ITS EMISSION OF HIGH SPEED ELECTRONS

A. M. Polyakov and N. A. Krotova

Institute of Physical Chemistry, Acad. Sci., USSR
Laboratory for Surface Phenomena

It is a well-known fact that the freshly formed surface possesses special properties. For example, sputtered metallic films adhere much more firmly to a fresh glass surface than to a surface which has been aged. Sand blasting a metallic surface prior to use will considerably increase its adhesibility. It is usually considered that this increase in adhesion is due to an increase in surface area and removal of oxide films. It should be noted, however, that mechanical bombardment with sand serves to expose fresh layers of the metal and leads and, at the same time, to imbedding of quartz in the metal. These processes can be attended by unexpected effects, such as the emission of exoelectrons. It has been found that the quartz particles are intense emitters of these electrons [1].

Processes leading to inocculation and polymerization proceed intensively on a freshly formed quartz surface, or on the fresh surface of a polymer [2]. Cases have been reported in which the fresh surface functioned as an initiator of polymerization. It is considered at the present time that the nature and concentration of electronic defects on the surface, or in the subsurface layer, determine the catalytic activity of the solid body. The active centers are positions particularly favorable for electron exchange, rather than the edges and corners on the catalyst surface, as assumed by the Taylor theory [3]. It has recently been possible to extend these studies into the new field of mechanochemistry, especially with polymeric substances [4]. It has been shown that mechanical breakdown — dispersion, plastification (rolling), and, in general, all processes involving polymer destruction — markedly increase the reactivity of the polymer and favor reaction of the latter with other components. It is supposed that free radicals are produced in the course of mechanical destruction, and that these then undergo interaction with the surrounding medium. This opens the possibility for mechanochemical syntheses. Thus the interaction of these radicals with the reactive centers of the polymer chain can lead to the establishment of three-dimensional structures. Mechanical destruction of the polymer in the presence of various monomers can result in the production of highly diversified structures.

The formation of free radicals in mechanical destruction can be detected by various methods, infrared spectroscopy and paramagnetic resonance, for example [5].

In addition to this work on the enhanced reactivity of pulverized and repeatedly deformed bodies, mention should also be made of a special effect which accompanies mechanical treatment and breakdown of the solid, namely, electron emission.

The emission of slow exoelectrons was first observed by Kramer (Kramer effect) [6]. A rather extensive literature has been built up around exoelectronic emission and the laws applying to it. Exoelectrons have energies of the order of an electron-volt, and intensities of the order of ten or a hundred counts/(sec·cm^2). It is characteristic that this type of emission usually arises as the result of subjecting the sample to simultaneous mechanical treatment and interaction of some other type, optical excitation, for example. Exoelectronic emission is assumed to be closely related to luminescence, and is generally approached through solid state band

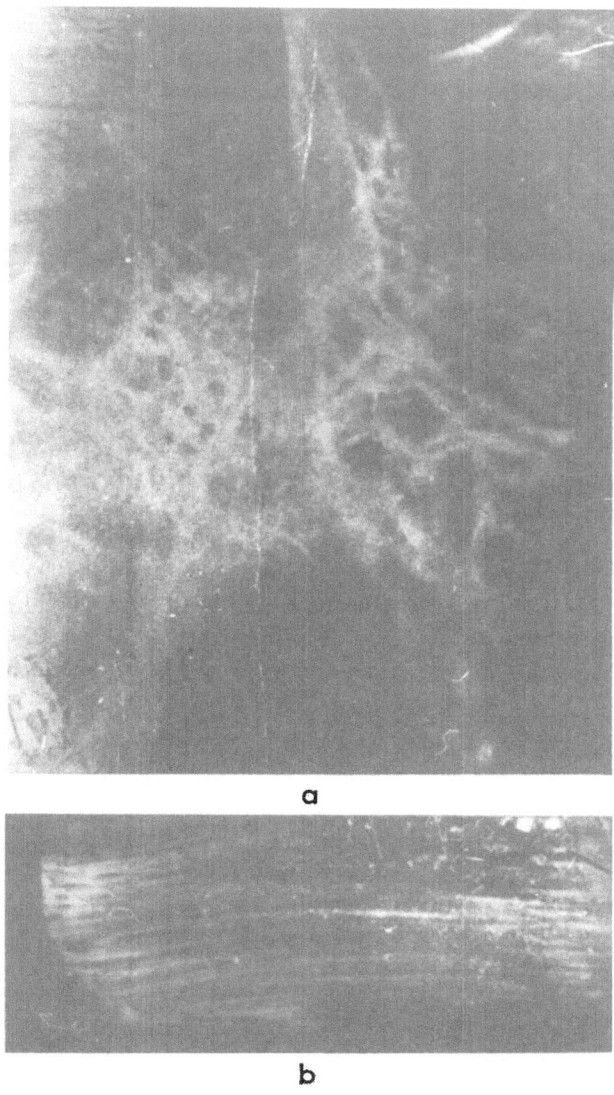

Fig. 1. Photographs of freshly formed surfaces, obtained
through emission of mechanoelectrons. a) Fractured zinc
blend; b) polymer (nitrile rubber) surface after stripping.

theory. In both cases, prior treatment must have led to the formation of donor centers capable of supplying electrons to the conduction band under excitation. Luminescence also requires that there be activator centers in the neighborhood of the valence band to combine with electrons from the conduction band with the emission of light. The donor centers themselves are sufficient in the case of exoelectronic emission, dissociation of these centers supplying electrons to the conduction band, which are then emitted beyond the confines of the solid body [7]. This makes it clear that the exoelectrons cannot have high energies.

Another type of emission of high speed electrons was observed by B. V. Deryagin, V. V. Karasev, and N. A. Krotova in vaccum stripping of a polymer film from a solid base [8]. Emission during crystal cleavage was observed by N. A. Krotova and V. V. Karasev somewhat later [9]. The velocities of the emitted electrons were here determined by: 1) beam deflection in the magnetic field, and 2) barrier penetration. The velocity of the electrons emitted in breakdown of the adhesional bond is of the order of 10 keV. Crystal breakdown (quartz) gives rise to electrons of still higher velocities, possibly of the order of several hundred keV. L. P. Morozova has shown that the electron velocity depends on the strength of the adhesional bond [10]; it also varies with the crystal lattice type. It has been found that emission does not accompany breakdown of the amorphous solid [11].

281

These processes are related to the luminescence which is observed in breakdown of the solid body, an effect designated by the none too happy term "triboluminescence" [12]. Luminescence and emission are here completely different from the Kramer effect. The effects associated with breakdown of the adhesional bond are naturally explained by the electronic theory of adhesion which has been developed by B. V. Deryagin and N. A. Krotova [13], and by B. V. Deryagin [14]. From this point of view, the luminescence in air, and medium vacuum, is a mere electrical discharge, while the high vacuum effect is spontaneous emission resulting from the presence of a strong electrical field in the gap between the separating surfaces. These effects have nothing in common with exoelectronic emission which should be logically referred to as "mechanoemission," the electrons involved being "mechanoelectrons."

Study of the mechanoemission from freshly formed surfaces opens the possibility of obtaining new information concerning surface states, and gives an approach to the problems of enhanced surface activity.

The possibility of obtaining an image on a sensitive film by the use of mechanoelectrons is very interesting. Vacuum experiments with simple apparatus which were carried out by V. V. Karasev [11] showed the possibility of obtaining a geometrical representation of a rupture surface (Fig. 1a) and the surface of a stripped film (Fig. 1b). The fact that the various portions of the surface (peaks, ridges, valleys) emit at different intensities makes it possible to obtain the rupture image. Figure 1a shows the convoluted fracture surface of zinc blend. The high intensity of radiation from the ridges serves to clearly bring out the surface structure here. A photograph of the surface of a nitrile rubber film which had been stripped from glass is shown in Fig. 1b; here the cellular structure is brought out by the emission of mechanoelectrons from the deformed net elements. Such experiments could have considerable interest for the study of supermolecular structures in various polymers.

The apparatus used here is somewhat similar to the early models of the emission microscope; these, it will be remembered, were not equipped with an electron-optical system. The source of electrons in the emission microscope is a cathode which is heated, and sometimes subjected to a supplementary illumination, as well, so that both thermoelectrons and photoelectrons are emitted. Image formation in the cases under discussion results from exposure of the film to the mechanoelectrons. The latter are torn loose under the action of the intense fields set up in the cracks and fissures, even without the application of a supplementary external field to induce autoelectronic emission. The photograph obtained under the action of the mechanoelectrons directly indicates the structure of the freshly prepared surface, and the electron density in the various portions of this surface.

Study of the solid surface relief, i.e., of the mechanically produced streaks and other defects, can be carried out by a variant of the replica method. A polymer film is deposited on the surface under study. A fresh defect surface is exposed when this film is stripped away [15]. It should be noted that the base surface always functions as an electron donor in the formation of a contact between polymer and metal or glass; as a result, the polymer film is always electron enriched and the freshly formed polymer surface invariably emits electrons once the contact is broken.

We have recently constructed a complicated adhesiometer which makes it possible to follow the mechanical parameters and intensity of electron emission resulting from rupture of the adhesional bond between polymer and base (glass or metal).

An apparatus designed for study of mechanoemission at the instant of rupture of bonded specimens, or at the instant of fracture of crystals and polymeric materials, is shown in Fig. 2. The emitted electrons are here registered on a sensitive x-ray film set in the cylindrical plate holder, 2. The triggering mechanism, 3, is so constructed that the lever 4 first opens this holder and the conical prism 5 then throws on the stripping load. The shock wave associated with the rapid application of the force is smoothed out by the special damper 6 [18]. The stripping force (rupture force) is recorded by a system of strain gauge sensing elements which are connected in a bridge circuit (Fig. 3). The signal from these sensing elements is fed into a special UD-3 amplifier and then recorded on a N-700 loop oscillograph. The appratus can also be equipped with a plane film holder and a diaphragm system for defining a narrow bundle of electrons. A fixed magnetic field can be set up between this system and the holder, and the mean energy of the electrons determined from the beam deflection. When the electrons are film recorded, it is impossible to determine the electron emission intensity and the time variation of this intensity in the course of the experiment, so that the laws applying to the emission from the freshly formed surface cannot be established; moreover, the low sensitivity of the film eliminates the possibility of re-

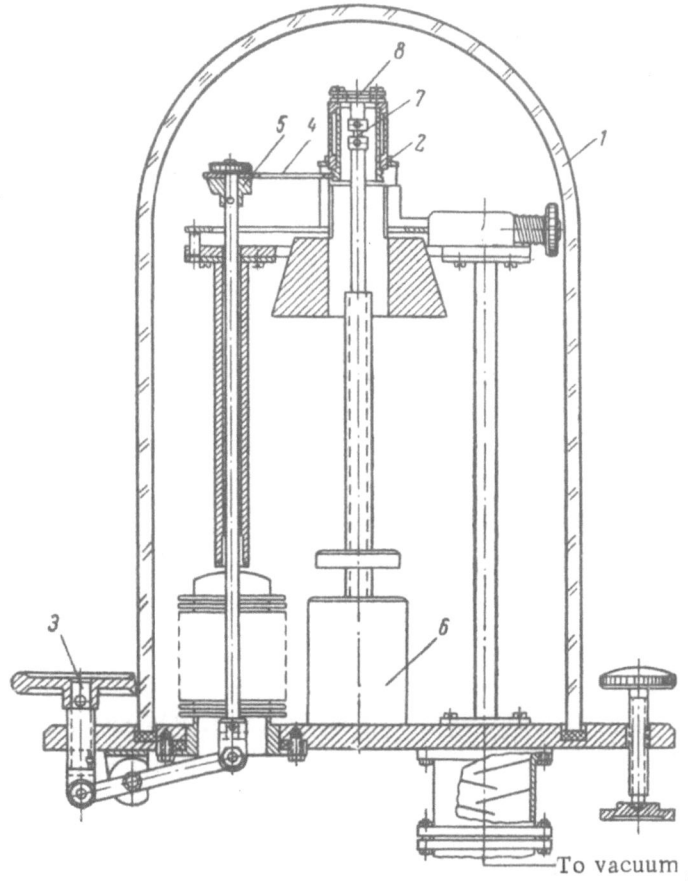

Fig. 2. Apparatus for the simultaneous determination of adhesion and mechanoemission in stripping. 1) Glass bell; 2) film holder; 3) bellows (triggering mechanism); 4) mechanism for opening film holder; 5) conical prism for applying load; 6) damper; 7) sample; 8) strain gauge sensing elements.

To vacuum pump

cording low energy electrons. For these reasons, a special VÉU secondary electron multiplier tube constructed in the State Optical Institute [16] was used for recording the electron emission.

These open type multipliers have a high amplification coefficient (10^8-10^9), excellent working stability, and low sensitivity to the introduction of air at the end of the experiment. With their aid, it was possible to register electrons of energies ranging from several eV to hundreds of keV.

Figure 3 shows the block diagram for the system used in registering the electron emission which accompanies the stripping of a polymer film from a glass or metal roller. The mechanical characteristics of the adhesion were recorded by the indicated circuit. The emission intensity was measured with the multiplier tube, using a uniform voltage divider. The cathode of the multiplier was grounded, and the anode charged to a high positive voltage (up to 5 kV) by an Orekh voltage stabilizer. The primary emission current could be measured by switching a microameter into the anode circuit; this current was of the order of 10^{-14} amp at the instant of stripping gutta percha from glass. When in the counting mode, the multiplier signal was fed into the electronic counting system. The total number of counts registered during the stripping process, or in the course of a definite time interval following stripping, was read from a Floks PS-10,000 pulse counter. The intensity of stripping emission, its time dependence, and the variation of emissivity over the freshly formed polymer film surface, were recorded on the Tyul'pan ISS counting rate meter and transcribed on an ÉPP-09 electronic potentiometer with a carriage period of 1 sec (see Fig. 3). All of the experiments were carried out at 10^{-5} mm Hg. The vacuum section of the system is included within the dotted lines of Fig. 3.

The working materials were polymer films which had been formed on the surface of a glass roller. The surface of this roller was cleaned with chromic acid solution, washed with distilled water, and then dried in an oven, prior to film deposition. The film was deposited by rotating the roller in a vessel containing a solution of the polymer, removing this vessel, and then vaporizing the solvent by rotating the roller in the air. The test polymers were elastomers of α- and β-gutta percha, copolymers of butadiene rubber and methacrylic acid

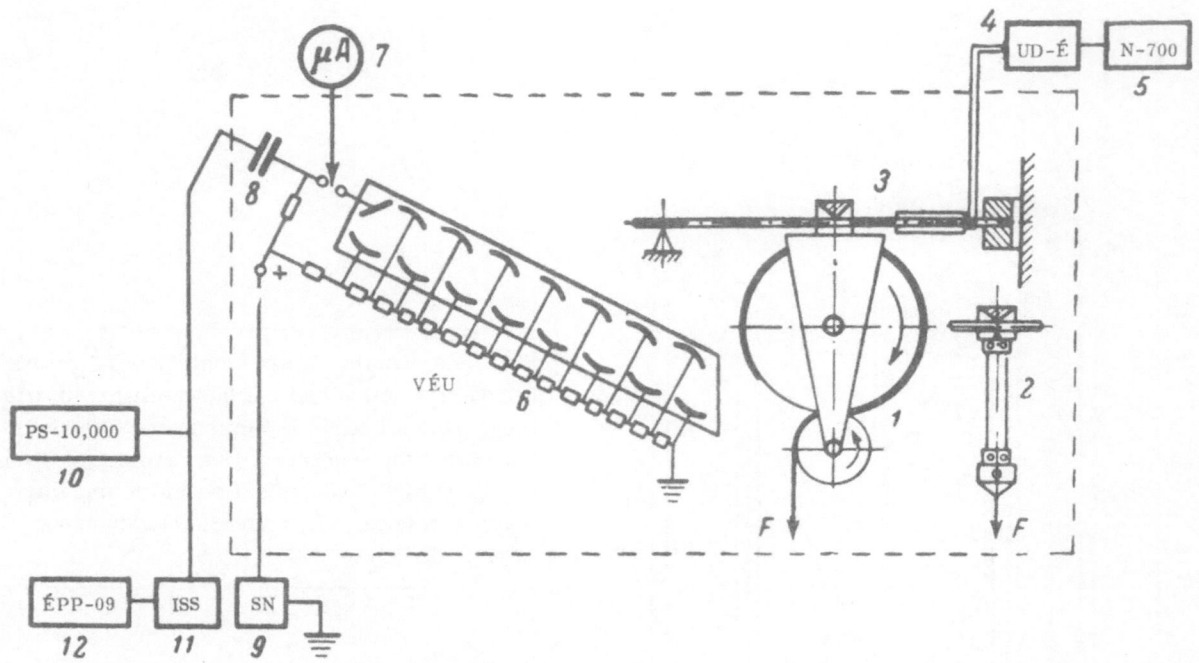

Fig. 3. Block diagram of the apparatus for studying mechanoemission. 1) Roller with film; 2) sample of stretched film; 3) strain gauge sensing elements; 4) signal amplifier for the sensing elements; 5) loop oscillograph; 6) electronic multiplier; 7) microameter; 8) condenser; 9) voltage stabilizer; 10) counter; 11) counting rate meter; 12) electronic potentiometer.

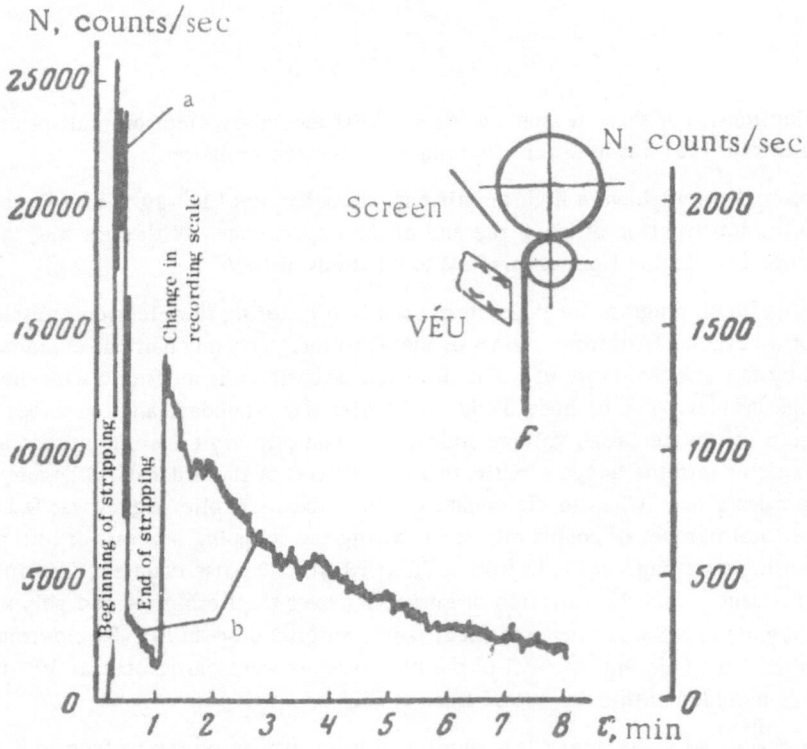

Fig. 4. Recording of the time variation of the after-emission from a β-gutta percha film freshly stripped from glass. a) Initial after-emission; b) after-emission following completion of stripping.

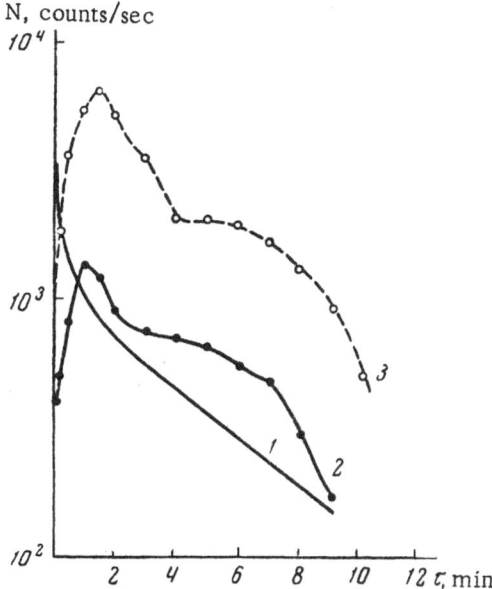

N, counts/sec

Fig. 5. The effect of the holding time, τ', on the time (τ) variation of the after-emission from a stripped β-gutta percha film. Stripping rate, 0.5 cm/sec. Origin at the end of stripping. 1) Experiment performed 2 hours after film formation; 2) the same, but after 48 hours; 3) the same, but after 96 hours.

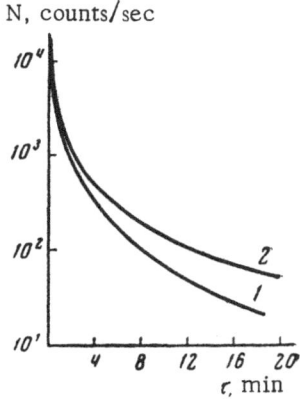

N, counts/sec

Fig. 6. The fall-off of the emission from nitrile rubber. 1) Time of holding the sample, 1 day; 2) the same, but 5 days.

(carboxylate rubbers), and divinyl and acrylonitrile copolymers (nitrile rubbers). It is a well-known fact that the valuable technical properties and high adhesibility of these copolymers are due to the presence of functional carboxyl (—COOH) and nitrile (—CN) groups in the molecule. Since the nitrile groups are acceptors, the electronic theory of adhesion [17] indicates that their presence should lead to the establishment of strong adhesional bonding with surfaces of donor materials such as glass which contains donor hydroxyl groups. This has been observed to be the case in our earlier experiments.

Theory also indicates that strong adhesion is to be expected in systems with high double layer charge density at the polymer — base interface [13, 14]. It is clear that the emission intensity in stripping should increase with the initial charge density. This assumption is confirmed by the observation that the emission intensity falls off markedly if the film is discharged as the intensity is being measured.

What has been said suggests that it would be of interest to study the intensity of the mechanoelectronic emission resulting from breakdown of adhesional bonds involving polymers of various chemical structures and adhesional properties.

It was first necessary, however, to establish some of the laws applying to mechanoelectronic emission; this was done, largely through experiments with gutta percha films.

Figure 4 shows a typical recording of the electron emission from a freshly formed β-gutta percha film (as a function of the time, τ). These experiments were carried out in such manner that the gap between the separating surfaces could be specially screened, and the electrons emitted at the instant of stripping prevented from entering the multiplier tube (the experimental arrangement is indicated in the upper right-hand corner of Fig. 4). It is clear from the figure that the highest intensity of after-emission was that observed when a film freshly stripped several seconds earlier passed before the multiplier (a). A freshly formed section of the film was held in front of the multiplier at the end of stripping. The intensity of emission from this section fell off rather rapidly (b), but was still of considerable magnitude (as high as 100 counts/sec) even after 1-1.5 hours. It is interesting to note that the emission intensity could be increased by a supplementary illumination with visible light.

Curve 1 of Fig. 5 is a typical semilogarithmic plot of the fall-off emission. With samples which had been held for long periods after preparation (several days), the initial fall-off was found to be followed by a sudden rise in intensity with a subsequent second diminution (curves 2 and 3). It is obvious that processes occurring on the freshly formed polymer surface tend to restore the surface layers disrupted in stripping to a state of equilibrium, and that these processes are accompanied by enhancement of emission, possibly as a result of chemical

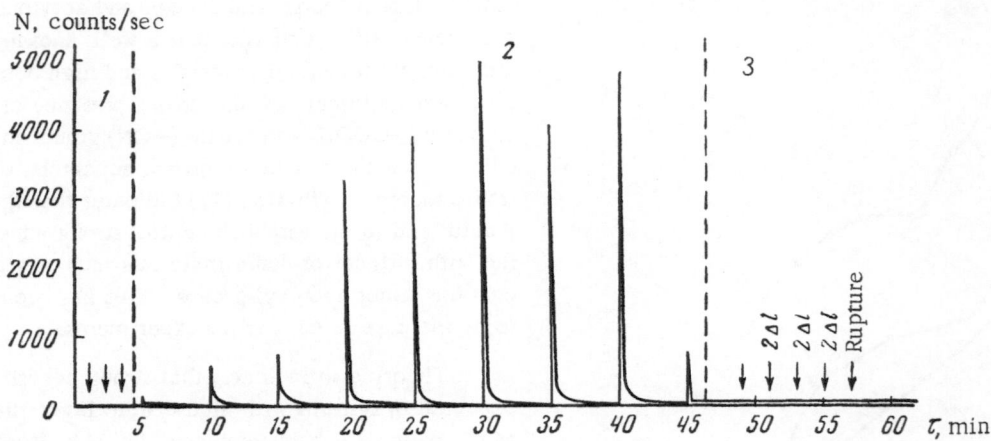

Fig. 7. Time variation of the mechanoemission from deformation of the polymer film. 1) Elastic deformation; 2) plastic deformation; 3) hardening; Δl is the elongation.

reactions which are not yet understood. These processes must be different for old surfaces and for surfaces which have been freshly prepared. It is seen from Fig. 6 that the time of contact of film and base (holding time for the sample prior to experiment) also affects the time (τ) variation of the emission. The fall-off in intensity after stripping is more rapid for the freshly prepared surface than it is for the aged sample.

The initial intensity of emission, χ, for polymers of different structures was compared in terms of the number of counts recorded in the first 10 minutes, including here the counts associated with what was considered to be instantaneous stripping. The fact that the time of holding prior to experiment (τ') had no effect on this parameter is easily seen from the following data for the nitrile rubber (SNK-40) – glass system:

$$\tau', \text{days} \dots\dots\dots\dots \quad 1 \qquad 2 \qquad 3 \qquad 7$$
$$\chi \cdot 10^{-3}, \text{counts/10 sec} \dots \, 352 \quad 565 \quad 516 \quad 406$$

The initial emission intensity, χ, varied from polymer to polymer, as is seen from the results of our experiments on the stripping of nitrile rubber (SNK-40), β-gutta percha, SKS-30+15% methacrylic acid (MAA), and acetylcellulose from glass:

Polymer	SKN-40	β-gutta percha	SKS-30+15% MAA	Acetylcellulose
$\chi \cdot 10^{-3}$, counts/10 sec . .	500	300	25	0.2

The highest initial emission was observed with the nitrile rubber and the lowest with the acetylcellulose, just as would be anticipated. Thus a definite correlation exists between the strength of the polymer to glass adhesion and the intensity of emission in stripping.

Emission of mechanoelectrons is observed not only in stripping, but also in the mechanical deformation and disruption of the polymer film. Experiments here were carried out in the following manner. The roller in front of the VÉU was replaced by a clamp into which the film was set, (see Fig. 3, 2), the arrangement being such that the film could be periodically elongated by 10% of its original length. Figure 7 is a recording of the electron emission resulting from deformation of a β-gutta percha film (sample dimensions, 7 x 5 x 0.5 cm). Stretching in the region of elastic deformation (indicated by arrows) did not give rise to electron emission. The first emission was noted at the instant of appearance of a slip band on the sample. Further deformation led to the propagation of this band along the sample (Fig. 7, 2), each stretching being accompanied by a sharp emission peak. Emission ceased and the region of mechanical hardening was entered when the slip band covered the entire sample (Fig. 7, 3). Further stretching led to no detectable emission, even if continued to rupture. It is interesting to note that the intensity of emission under deformation falls off rapidly with the passage of time, thereby differing from the emission in breakdown of the adhesional bond.

Conclusions

1. A complex AK-1 adhesiometer has been constructed for studying the emission of mechanoelectrons in the stripping, mechanical deformation, and breakdown of polymer films.

This adhesiometer permits simultaneous determination of the mechanical parameters and the intensity of electron emission.

2. A special secondary electron multiplier with an amplification factor of 10^8-10^9 was used to record the electron emission. The signal from the multiplier was fed into an electronic counter and recorded with an electronic potentiometer.

3. The emission current resulting from stripping was recorded on a microameter in the anode circuit of the multiplier. The emission current proved to be of the order of 10^{-14} amp/cm^2 in the β-gutta percha – glass system.

4. It was found that mechanoelectrons are emitted not only from the gap between the separating surfaces but also from the fresh polymer surface formed by stripping, this emission continuing for some time after stripping has been completed. This effect has been arbitrarily designated as after-emission, in order to distinguish it from gap emission.

5. Gap emission resulting from breakdown of the adhesional bond has been shown to be an auto-electronic effect originating in the strong gap field which is set up by separating the faces of the double electric layer at the polymer – base interface.

6. The fall off of after-emission from the freshly formed surface has been found to follow a law of the form $N = a\tau^{-b}$, τ being the time, and a and b constants. Cases are observed, however, in which the emission from the freshly formed surface at first falls and then rises again.

7. It has been suggested that after-emission is associated with the breakdown of the centers responsible for the levels which are set up in the forbidden band, and near the conduction band, when the adhesional bond is formed. Stripping sets up a high intensity field and excites these centers so that they accumulate the energy required for subsequent electron emission.

8. It has been shown that the initial intensity of emission in the stripping of a polymer from glass depends on the chemical nature and structure of the polymer. Nitrile rubber gives the highest initial intensity, and acetyl-cellulose a very low intensity.

9. It is clear that the intensity of emission under stripping depends on the strength of adhesion of polymer to glass.

10. It has been found that plastic deformation of β-gutta percha leads to emission bursts with a maximum intensity of 5,000 counts/sec.

Literature

1. W. Kusterer and O. Bruna. Anz. Österr. Akad. Wiss. Math.-Naturw. Kl. 92(1):48 (1955).
2. V. A. Kargin and N. A. Platé. Vysokomolekul. Soedin. 1(2):330 (1959).
3. K. Hauffe, Angew. Chem. 67:189 (1955). G. M. Schwab. Angew. Chem. 67:433 (1955). H. Nassenstein and R. Menold. Acta. Phys. Austriaca 10(4):453 (1957).
4. N. K. Baramboim. The Mechanochemistry of Polymers, Moscow, Gostekhizdat (1959).
5. Collection: The Formation and Stabilization of Free Radicals [Russian translation under the editorship of A. Bass and G. Brair], Moscow, IL (1962).
6. I. Kramer. Der metallische Zustaud, Göttingen (1950).
7. W. Hanle and G. Gourge. Phys. Blätten 14:499 (1958).
   G. Gourge. Z. Physik, 153:186 (1958).
8. N. A. Krotova, V. V. Karasev, and B. V. Deryagin. Dokl. Akad. Nauk SSSR 109:728 (1956).
9. N. A. Krotova and V. V. Karasev. Dokl. Akad. Nauk SSSR 92:607 (1953).

10. L. P. Morozova, N. A. Krotova. Kolloidn. Zh. 20:59 (1958).

11. V. V. Karasev and N. A. Krotova. Dokl. Akad. Nauk SSSR 99:715 (1954).

12. V. V. Karasev, N. A. Krotova, and B. V. Deryagin. Dokl. Akad. Nauk SSSR 89:108 (1953).

13. B. V. Deryagin and N. A. Krotova. Adhesion, Moscow, Izd. Akad. Nauk SSSR (1949).

14. B. V. Deryagin. "Problems of Adhesion," Vestn. Akad. Nauk SSSR, No. 7 (1954).

15. L. P. Morozova and N. A. Krotova. Dokl. Akad. Nauk SSSR 115:747 (1957).

16. A. M. Tyutikov and A. I. Efremov. Dokl. Akad. Nauk SSSR 118:286 (1958).

     A. M. Tyutikov and Y. A. Shuba. Opt. i Spektroskopiya 9, 5:631 (1960).

     A. M. Tyutikov, Zh. Prib. i Tekh. Éksperim., Vol. 1 (1962).

17. V. P. Smilga and B. V. Deryagin. Dokl. Akad. Nauk SSSR 122, 6:1049 (1958).

     V. P. Smilga. Dissertation, Moscow (1962).

18. L. A. Laius. Peredovoi Nauchio-Tekhnicheskii i Proizvodstvennyi Opyt, Izd. VINITI, No. 11-58-135 (1958).

A STUDY OF THE ADHESIONAL INTERACTION BETWEEN POLYMER AND MODIFIED SURFACE BY INFRARED SPECTROSCOPY

N. A. Krotova and L. P. Morozova

Institute of Physical Chemistry, Acad. Sci., USSR
Laboratory for Surface Phenomena

The possibility of regulating the strength of adhesion so as to obtain a desired type of bonding is a problem of considerable practical significance. The most effective method of regulation is by chemical modification of the base surface, with the introduction of groups which are active, or inactive, in adhesion.

Chemisorptional processes in adhesion have recently acquired especial interest in connection with the theoretical work of B. V. Deryagin and V. P. Smilga [1] on donor — acceptor interaction at the polymer — base interface and the adhesional bonding between substances of radically different molecular structures. This last possibility has been opened up by the development of a wide variety of methods for the chemical modification of surfaces of glasses, polymers, and the mineral fillers in glues and lacquers, as well as other materials [2]. Modification can give to the surface either hydrophobic or hydrophilic properties, and thereby markedly alter its ability to enter into adhesional bonding [3]. It should be noted that these same ends were achieved earlier by the introduction of various surface active substances which adsorbed on the base surface [4]. The advantages of chemical modification as a means of regulating adhesional properties are obvious in view of the high stability of chemisorptional compounds.

The increase in the number of methods available for base surface modification and the extensive application of these methods has naturally suggested the study of the surface chemical compounds formed between the molecular groups of the modified surface and the functional groups of the monomer which is eventually polymerized into the adhesate film.

Infrared spectroscopy of the type used in the work of A. N. Terenin, N. G. Yaroslavskii, and A. N. Sidorov [5], and the studies of the present authors, proves to be most suitable for this purpose. The work described in the sequence is an extension of these earlier studies on the chemisorptional formed between polymer and modified glass surface [6].

The study of surface compounds should be carried out in such manner as to assure reflection of the interfacial processes in the infrared spectrum. This can be done either by using highly porous materials or by having to resort to the method of multiple reflection. We have made use of the first of these possibilities, our working materials being microporous glasses in which the pore diameter was of the order of 40 A [7]. The extensive surface development in these glasses permitted the chemical interaction at the polymer — glass interface to be studied through infrared spectroscopy. The absorption spectra were studied in an IKS-14 infrared spectrometer. Our earlier papers have described the cell in which the microporous glass was degassed by evacuation and heating, and then brought into contact with the working material [8].

Studies by infrared spectroscopy can serve as a basis for conclusions concerning the bonding of polymer to glass surface. We have been able to show that such interaction frequently proceeds through hydrogen bond formation. Thus interaction between gaseous methacrylic acid and a calcined glass involves the establishment of hydrogen bonding between the carboxyl groups of the acid and the hydroxyl groups of the glass. This is indicated by the displacement of the skeletal hydroxyl peak toward lower frequencies. Calculation gives a value of

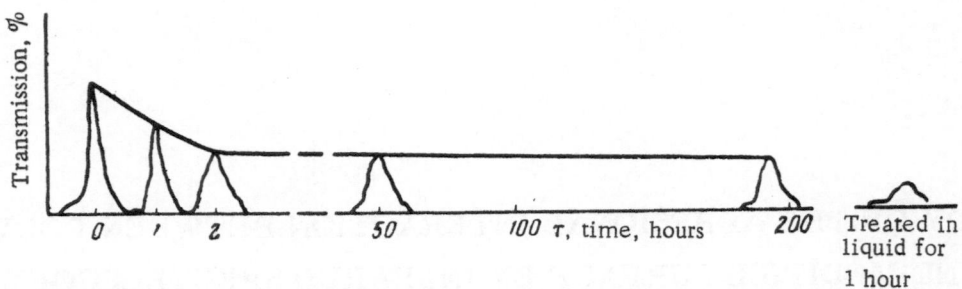

Fig. 1. The effect of the time of treatment of glass with dimethylchlorosilane on the degree of substitution of the free hydroxyl groups on the surface.

2.9 A for the length of the hydrogen linkages formed in bonding the COOH groups of the methacrylic acid to the OH groups of the glass. Linkages of this kind are said to be long. This type of hydrogen bonding is principally the result of electrostatic interaction [9].

Experiment shows that the molecular configuration can affect the type of surface compound formed. Thus ortho-carbethoxyphenylmethacrylamide (o-CEPhMA) shows intramolecular hydrogen bonding while the corresponding para-compound has intermolecular hydrogen bonding. The configurational difference in these two compounds leads to differences in the O...OH distances in the hydrogen bonds established between monomer and glass. The displacement of the skeletal hydroxyl peak is 20 cm^{-1} for the ortho-compound and 40 cm^{-1} for the para-compound.

We have attempted to obtain an understanding of the chemistry of adhesional phenomena by studying the chemisorptional bonding set up between the polymer and a glass surface modified by an alkyl chlorosilane. The glass was modified by treatment with methyl, dimethyl, and trimethyl chlorosilane. It is a well-known fact that treatment with an alkyl chlorosilane hydrophobizes the glass surface, forming on it a firmly bound film of silicoorganic compound which can be removed only by hydrofluoric acid or mechanical treatment. Attack by an alkyl chlorosilane involves chemical interaction of the active hydrogen of the glass and the halogen of the silane, with the formation of Si—O—Si bonds. The result is that the glass surface is methylated and acquires the ability to repel water [10].

The fact that the height of the skeletal hydroxyl peak diminishes as the time of treatment is increased is indication of a diminution of the number of these groups on the surface (Fig. 1). This process is limited in time. The hydroxyl peak does not completely disappear, even after extended interaction, but the groups themselves must be excited since the peak displacement persists.

The reactivity of the modified glass is much lower than that of the original. This is reflected in the chemisorption of methacrylic acid and styrene by the original glass and the modified glass. Polymers of methacrylic acid and styrene form hydrogen bonds in reacting with the original glass surface. Figure 2a shows the principal region of the infrared spectrum of the calcined microporous glass.

The absorption band at 3720 cm^{-1} corresponds to the fundamental vibration frequency of the free OH groups. Figure 2b shows the infrared spectrum of styrene. The absorption bands at 3085, 3060, and 3030 cm^{-1} correspond to C—H stretching vibrations in the benzene ring. Figure 2c shows the infrared spectrum of a microporous glass which had been brought into contact with styrene. The OH group absorption bands have been displaced toward lower frequencies, and considerably widened. This is indication of hydroxyl group excitation. An absorption band corresponding to stretching vibrations in the styrene also puts in its appearance.

Figure 3a shows the infrared spectrum of a microporous glass which had been modified by trimethylchlorosilane. The absorption band at 2972 cm^{-1} is associated with the stretching vibrations in the silane methyl groups. A comparison with Fig. 3b showing the infrared spectrum of the modified glass after contact with styrene, indicates practically no interaction between styrene and the surface.

These results are in good agreement with our own data on the adhesion of polystyrene with the original glass and the modified glass. The work of adhesion at a stripping rate of 1 cm/sec was 4460 erg/cm^2 for the

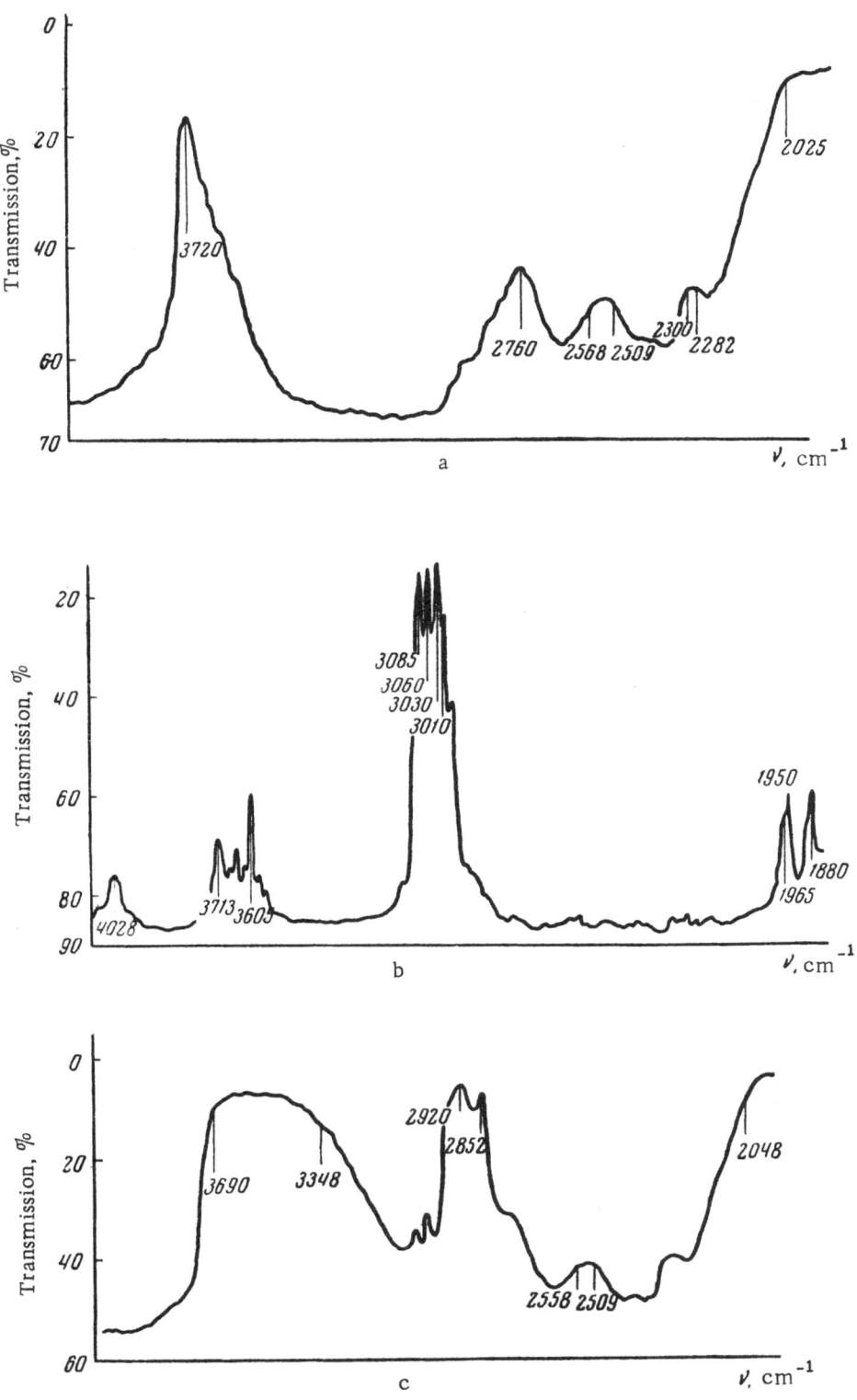

Fig. 2. The interaction of styrene with a microporous glass. a) IF spectrum of the microporous glass; b) IF spectrum of styrene; c) IF spectrum of the glass treated with styrene.

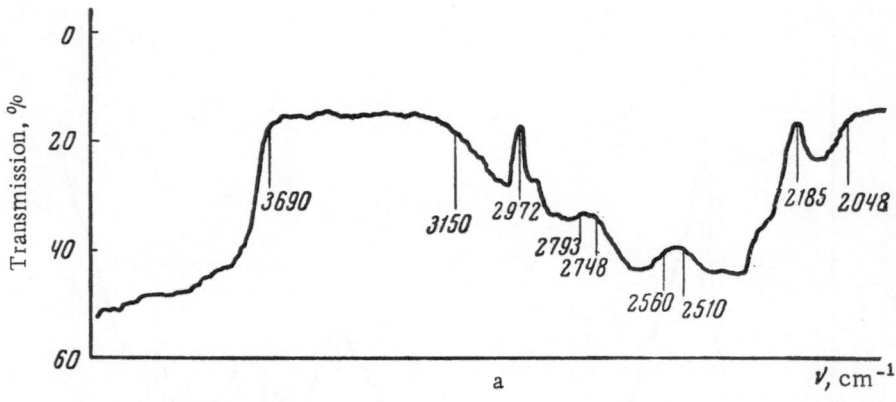

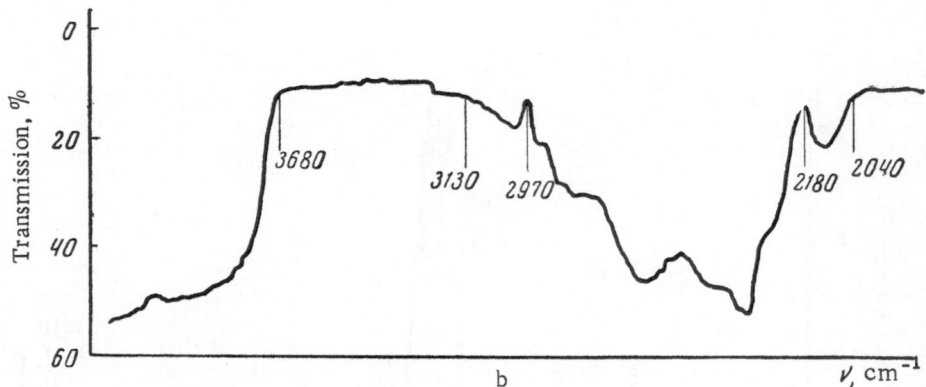

Fig. 3. The interaction of styrene with a microporous glass modified by treatment with trimethylchlorosilane. a) IF spectrum of the microporous glass modified with trimethylchlorosilane; b) the same, but after supplementary treatment with styrene.

original glass and 88.5 erg/cm^2 for the trimethylchlorosilane treated glass, the adhesion being two orders lower in the second case. Supercharging of the contact surfaces is observed to result from the formation and subsequent breakdown of the junction between polymer and modified surface. Stripping polystyrene from the original glass charges the polymer film negatively and the glass surface positively. Here the glass surface is largely composed of hydroxyl groups and serves as a donor. Change in the charge sign occurs after stripping is completed, the glass being negative and the polystyrene film positive, so that the glass surface has acquired acceptor properties. Treatment of the glass with an alkyl chlorosilane sharply diminishes the concentration of surface hydroxyl groups, replacing these by methyl groups in which the donor properties are less pronounced. The methyl groups and the benzene ring are nonpolar, and only slightly different in donor — acceptor properties [11]. Thus there should be practically no donor — acceptor interaction between these groups, an experimentally confirmed conclusion.

We have so far been concerned with cases in which the adhesion was diminished through modification of the glass by treatment with an alkyl chlorosilane. The problem of increasing the adhesion can prove to be equally important for practical purposes. The industrial gluing of rubbers by glues containing polar functional groups (perchlorvinyl and nitrocellulose glues, for instance) follows the method of F. A. Sapegin in which the rubber surface is first given a preliminary treatment with sulfuric acid [12]. N. N. Stefanovich and N. A. Krotova have recently recommended an analogous method for the industrial treatment of the polyethylene surface with a view to increasing the adhesability. The aim of the remainder of the present work is to obtain an understanding

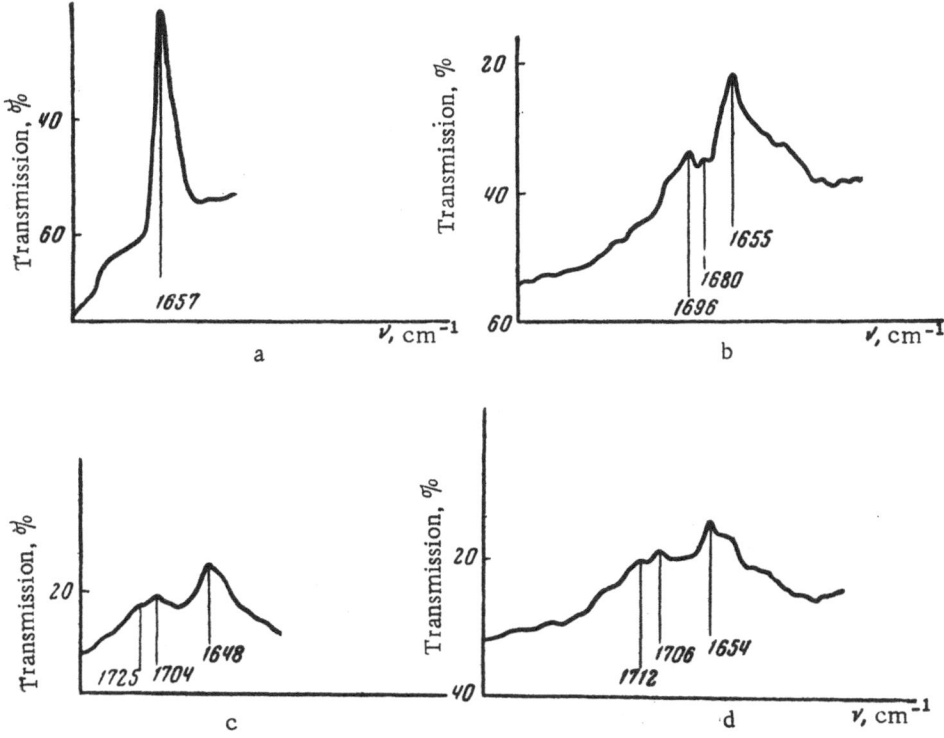

Fig. 4. Alteration of a gutta percha film surface resulting from treatment with sulfuric acid. a) IF spectrum of the original gutta percha; b) IF spectrum of the gutta percha after 5 minute treatment; c) the same, but after 10 minute treatment; d) the same, but after 15 minute treatment.

of the surface chemical compounds resulting from such treatment. The working material here was a gutta percha film 20 μ thick. Figure 4a gives the infrared spectrum of the gutta percha. The surface of this film was treated with concentrated sulfuric for 5, 10, and 15 minutes, at room temperature. The film was then carefully washed to neutrality and dried. Comparison of the infrared spectra of the gutta percha before and after treatment (Figs. 4a, b, c) leads to the conclusion that the surface has been oxidized, this being indicated by the appearance of characteristic peaks for the carbonyl groups in the spectrum. A five minute treatment of the gutta percha led to the appearance of an infrared absorption band at 1700 cm^{-1} which corresponds to the C = O carbonyl group stretching vibrations. Increasing the time of treatment increases the intensity of this band, thus indicating an increase in the number of surface carbonyl groups. The absorption band at 1655 cm^{-1} (double bond stretching) is also altered by the treatment, its intensity gradually diminishing because of reduction in the number of these bonds.

The adhesive properties of the gutta percha are considerably enhanced by oxidation. This is clearly shown by the following data applying to perchlorvinyl as the adhesive:

Base — gutta percha	Original	Oxidized
Work of adhesion, erg/cm^2	1260	39,000

It follows from these figures, that surfuric acid treatment of the gutta percha surface increases its perchlorvinyl adhesion approximately 30-fold. This indicates that carbonyl groups are more active in adhesion than are double bonds. Surface supercharging is not observed after modification. This fact is easily understood since substituent chlorine stands above the carbonyl group in the donor — acceptor series, and functions as a donor with respect to both CO groups and double bonds.

Conclusions

1. It has been shown that stable adhesional compounds can be obtained by chemical alteration (modification) of the base surface. This is a very convenient and satisfactory method since it changes the chemical structure and adhesional properties of the surface in the desired direction but leaves the body of the component materials unaltered.

2. Formation of a polymer film on a modified surface is frequently accompanied by supercharging, obviously because of alteration of the properties of the contacted donor — acceptor pair.

3. Infrared spectroscopy has been used to study the structure of the modified surface and the character of its bonding with the functional groups of the monomer (methacrylic acid, styrene, etc.).

4. Treatment of the glass surface with an alkyl chlorosilane increases the concentration of methyl groups and decreases the concentration of hydroxyl groups. The change here is particularly marked in the first few seconds, after which the surface reaction tends to slow down. The degree of modification (concentration of donor hydroxyl groups) and the adhesional properties of the surface can be controlled by altering the time of treatment.

5. The modified glass surface reacts differently to monomers of styrene, methacrylic acid, and so on. In many cases, hydrogen bonding is set up between the hydroxyl groups of the glass and the functional groups of the polymer or monomer. Calculations of the hydrogen cross link lengths point to the electrostatic nature of this bonding. On the other hand, there is weak interaction, or no interaction at all, between the monomer and a glass modified by alkyl chlorosilane treatment.

6. The adhesion of polystyrene to glass is weakened by surface modification. Reduction in adhesability and surface supercharging are observed after even brief modification (several seconds).

7. Increased polymer adhesion can frequently be obtained by treating the surface with an oxidizing agent. Infrared studies of surface oxidized gutta percha point to the rupture of double bonds and oxidation of carbonyl groups. The adhesion of gutta percha to perchlorvinyl is markedly increased by oxidation.

The authors would like to conclude by expressing their thanks to A. A. Babushkin for his valuable advice and help with this work.

Literature

1. B. V. Deryagin and V. P. Smilga. In collection: Proceedings, Third International Congress on Surface Phenomena, Vol. II, Cologne (1960).
   V. P. Smilga. The Electronic Theory of Adhesion, Dissertation, Moscow (1961).
2. A. V. Kiselev, N. V. Kovaleva, A. Ya. Korolev, and K. D. Shcherbakova. Dokl. Akad. Nauk SSSR 124:617 (1959).
   A. V. Kiselev and V. I. Lygin. Usp. Khim. 31(3):351 (1962).
   A. B. Taubman, S. N. Tolstaya, V. N. Borodina, and S. S. Mikhailova. Dokl. Akad. Nauk SSSR 142:407 (1962).
3. N. A. Krotova, L. P. Morozova, and B. V. Deryagin. Dokl. Akad. Nauk SSSR 129:149 (1959).
4. A. B. Taubman. Khim. Nauka i Promy. No. 5:566 (1959).
5. A. N. Terenin, N. G. Yaroslavskii. Acta Physicochim. 17:240 (1940).
   N. G. Yaroslavskii. Zh. Obshch. Khim. 24(1):68 (1950).
   A. N. Sidorov. Zh. Fiz. Khim. 30:995 (1956).
6. N. A. Krotova, L. P. Morozova, and G. A. Sokolina. Collection: Proceedings, Third International Congress on Surface Phenomena, Vol. II, Cologne (1961), p. 368; Zh. Fiz. Tverd. Tela 3(7):1999 (1961).
   N. A. Krotova and L. P. Morozova. Kolloidn. Zh. 24(4):473 (1962).
7. O. S. Molchanova and I. V. Grebenshchikov. Zh. Obshch. Khim. 12:587 (1942).
8. N. A. Krotova and L. P. Morozova. In collection: Research in Surface Forces, Moscow, Izd. Akad. Nauk SSSR (1961), p. 83. [English translation: Consultants Bureau, New York (1963).]

9. N. Stüart. Die Struktur des freien Molekuls, Berlin (1952).

   A. A. Ketelaar. Chem. Constitution, Amsterdam (1957).

10. V. Bazhant, V. Khvalovski, and I. Rotouski. Silicones, Moscow, Goskhimizdat (1960).

11. A. Remik. Electronic Concepts in Organic Chemistry [Russian translation], Moscow, IL (1950), p. 45.

12. F. A. Sapegin and E. A. Nisnevich. Patent No. 98076 (1952).

DETERMINATION OF THE CHARGE ON DUST PARTICLES DUE TO DEPOSITION ON, AND SEPARATION FROM, SOLID SURFACES

A. D. Zimon, N. I. Dovnar, G. A. Belkina, and G. V. Nozdrina

Institute of Physical Chemistry, Acad. Sci., USSR
Laboratory for Surface Phenomena

Introduction

The present communication will outline an experimental technique for determining the charge acquired by dust particles during deposition on, and separation from, a solid surface, and the results obtained with it.

Electrical effects play a fundamental role in adhesional processes. There is every justification for speaking of an electrostatic component of the adhesion force [1]. The effect of this electrostatic component on the adhesion of dust particles has been pointed out theoretically [2], and evaluated experimentally [3, 4].

The nature of these electrostatic forces is not yet well understood; however, and the electrical effects occurring in the region of contact between dust particle and solid surface have not been studied.

The earlier studies of these effects [4] involved visual measurement of the electrical charge with the aid of an electrometer.

The system which we now recommend for this purpose incorporates an electrometric amplification circuit. The use of electronic and loop oscillographs eliminates the necessity of visual observation, and opens the possibility of photographing the time variation of the electrical processes occurring in the zone of contact between particle and base.

With this system it is possible to: a) determine the sign and absolute magnitude of the charge on the dust particle resulting from contact with, or separation from, various surfaces; b) fix the time variation of the dust particle charge, both visually and photographically; and, c) establish a correlation between the dust particle adhesion and the charge developed in separation, the latter being determined through the variation of the separating force.

The smallest charge measured here was $5 \cdot 10^{-13}$ coulombs.

Block Diagram for the System

The system for studying the electrical component of the adhesional force is composed of three blocks (Fig. 1): the vibrator for the dust-covered plate (I), the amplifier (II), and, the photographing section (III).

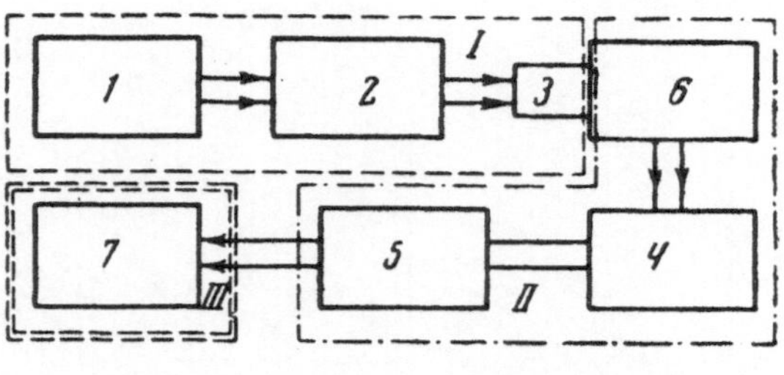

Fig. 1. Block diagram for the system.

I. The vibrator block consists of a ZG-10 audio generator, a U-50 power amplifier (2), and a vibrator (3).

The vibrator is a R-10 electrodynamic loudspeaker with a light weight coupling of organic glass attached to its diaphragm. A detachable test plate holder is clamped to this coupling. This plate can be oriented by rotating the holder in an articulated joint.

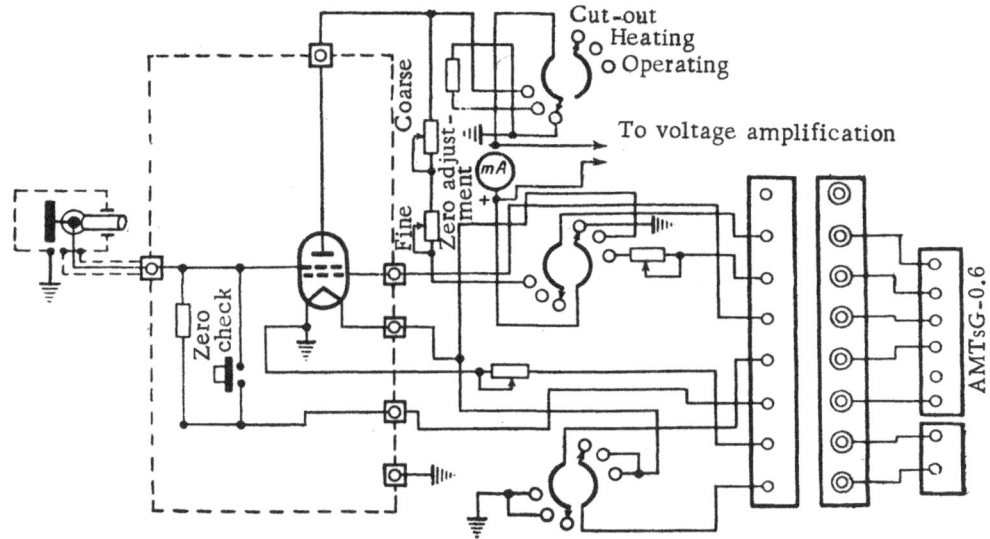

Fig. 2. Electrometric current amplifier for the 1É1P tube.

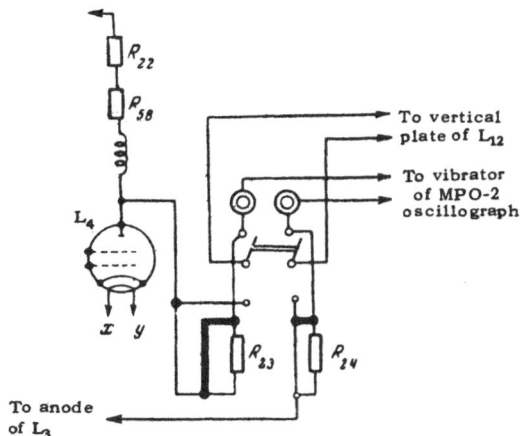

Fig. 3. Changes in the circuit of the ÉO-7 oscillograph amplifier.

The plate and holder are set into a grounded covered aluminum housing which serves as an electrostatic shield.

The frequency and amplitude of vibration of the dus-covered plate can be varied by altering the input potential of the power amplifier through the audio generator, thereby altering the detaching force acting on the dust particles [3].

II. The amplification block consists of an electrometric current amplifier (4, currrent amplification factor, $k_i = 1.9 \cdot 10^6$) a voltage amplifier (5, the vertical amplifier of an ÉO-7 oscillograph, voltage amplification factor, $k = 1800$), and the shielding housing (6).

III. The photographing block consists of an MPO-2 loop oscillograph.

The amplification section amplifies the electrical impulses generated in depositing dust particles on the plate, or in removing particles from it.

The electrometric current amplifier was designed around a single 1É1P tube; the major components of its circuit are shown in Fig. 2.

The changes introduced into the ÉO-7 circuit are shown in Fig. 3. All of the electrical processes are observed on the screen of the oscillograph.

The vibrator (Class VIII) of the MPO-2 loop oscillograph is connected in parallel with the vertical plate of the electronic oscillograph through a 180 kohm resistance. This oscillograph forms the third block of the system and permits cinematographic recording of the impulses registered on the ÉO-7 screen. A time trace is simultaneously recorded from a Class I vibrator which supplies 50 cps alternating current. The rate of film feed is optimal for a sensitivity of 45 GOST-25 units (in mm/sec).

The system was calibrated by discharge of a low capacity (C = 3 pf) condenser across the input of the elec-trometric amplifier. The impulses resulting from this discharge were photographed by the loop oscillograph(Fig. 4). A calibration curve was constructed to show the relation between the area, S, under the curve representing the time dependence of the discharge current,and the charge, $Q^{\neq}$, on the condenser (Fig. 5). The charge sign was determined from the direction of the first impulse.

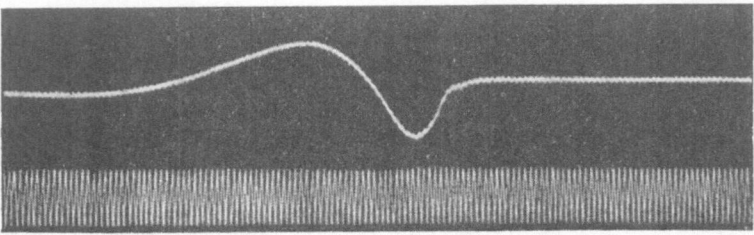

Fig. 4. Impulse resulting from condenser discharge (C = 3 pf).

Method of Determining the Charge

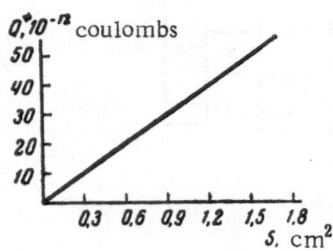

Fig. 5. Calibration curve showing the relation between the condenser charge and the area, S, under the current vs time curve.

The current was turned on and the system allowed to warm for 10-15 minutes. The test plate was set in its holder and weighed on the analytical balance. The holder and plate were then attached to the vibrator coupling in such manner that the plate was in the horizontal plane. The cover to the shielding chamber was then closed, and the loop oscillograph of the photographing section turned on. The test dust was deposited on the plate through an opening in the shielding cover. Photographing was stopped when the ray of the ÉO-7 oscillograph returned to its rest position.

A weighed cup was placed under the plate, and the latter then turned in its holder to the vertical position, a part of the dust falling off.

This cup was then removed and another put in its place. The MPO-2 oscillograph of the photographing section was once more turned on, and the vibrator put into circuit 1-2 second later. The cup and the holder with the plate were weighed after vibration and the dust originally present on the plate and dislodged by vibration determined by difference. The number of dust particles involved was determined from the density of the working material and the previously determined particle size distribution curve.

From the area under the curve traced out on the film by the initial impulse, and the calibration curve, it was possible to determine the total charge on the dust, either in deposition on the plate or in removal from it.

Weighing was replaced by an activity count [5] when working with a "tagged" dust.

This same system can be used for determining the sign and magnitude of the charge carried by weakly charged bodies, more especially, for the weak currents associated with the stripping of paints from base substances.

Experimental Results on the Determination of the Dust Charge

The adhesion of dust to a surface results from the action of various forces. Coulombic forces play a very considerable role here, their effect being manifest at the very instant of contact. A double electric layer is set up on contact, and the forces associated with this process are superposed on the coulombic forces.

Experimental results obtained with the system described above can be used to evaluate the force of coullombic interaction. The charges arising when dust particles of various diameters make contact with various surfaces are presented in the accompanying table, the sign of the charge being negative in every case.

We note that the charge initially carried by the dust particle in the air can be many times greater (hundreds of times greater in the electrofilter) than the total charge recorded by the system [6]. The coulombic interaction is then quite significant and exerts a more nearly determining influence in fixing the contact area. The coulombic force of interaction between dust particle and metallic surface is reduced by coating the latter.

Certain regularities have been brought out through measurement of the dust particle charge resulting from separation from the base. Thus we have been able to establish the existence of a direct proportionality between the dust adhesion (adhesion number) and the particle charge (q) resulting from separation from the base (Fig. 6).

298

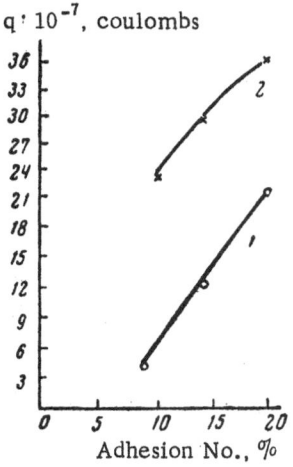

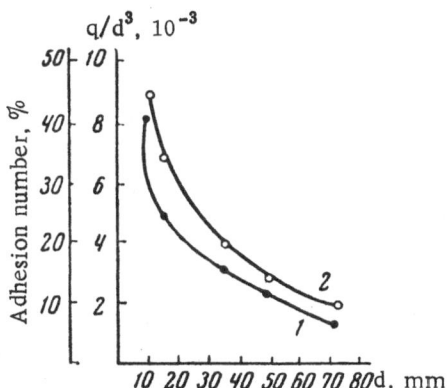

Fig. 6. The relation between the adhesion number and the dust particle charge d = 50 ± 5 μ (1), and 70 ± 5 μ (2); coated metallic base; particle separation by vibration at $3.6 \cdot 10^{-3}$ and $1.9 \cdot 10^{-2}$ dynes, respectively.

Fig. 7. The adhesion number (curve 1) and relative charge, q/d^3, of the dust (curve 2), and their dependence on the dust particle diameter (vibration, separating force $2.4 \cdot 10^{-4}$ - $1.9 \cdot 10^{-2}$ dynes).

The data cited, and Fig. 7, indicate that the relation between relative charge, q/d^3, and dust particle size is exactly the same as that existing between adhesion and particle diameter in similar cases:

Base material	Dust particle diameter, μ	Magnitude of charge, esu
Glass	20—30	$1.5 \cdot 10^{-8}$
	45—55	$8.1 \cdot 10^{-7}$
	65—75	$4.2 \cdot 10^{-6}$
	85—95	$3.9 \cdot 10^{-6}$
Uncoated metal (copper)	20—30	$2.1 \cdot 10^{-8}$
	45—55	$3.3 \cdot 10^{-7}$
	65—75	$3.1 \cdot 10^{-6}$
	85—95	$1.8 \cdot 10^{-7}$
Coated metal (steel)	20—30	$6.6 \cdot 10^{-8}$
	45—55	$5.7 \cdot 10^{-7}$
	65—75	$3.7 \cdot 10^{-6}$
	85—95	$6.3 \cdot 10^{-7}$

Here, q is the dust particle separation charge, just as before, and d is the particle diameter.

It is a well-known fact [1] that the electrostatic component of the adhesional force results from the formation of the double electric layer in the contact zone; its magnitude can be determined from the equation

$$F_{el} = 2\pi\sigma^2 S_c = \frac{2\pi q^2}{S_c},$$

(1)

in which σ is the charge density in the double electric layer, q is the charge acquired by the dust particle on separation from the base, as experimentally measured, and S_c is the area of contact between particle and base, a quantity that can be evaluated through the Hertz formula [7].

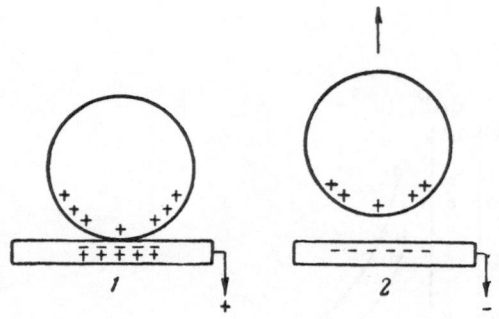

Fig. 8. Charges, measured when the dust particle makes contact with (1), and separates from (2), the base.

Thus the charge developed in separating dust particles $25 \pm 5 \ \mu$ in diameter from a coated metallic base is $5 \cdot 10^{-8}$ units, and the electrostatic component $5.3 \cdot 10^{-4}$ dynes, so that the latter is commensurate with the force of adhesion.

The electrostatic interaction can be understood in terms of the concepts advanced below.

In making contact with the base, the charged particle induces a charge of equal magnitude but opposite sign into the latter (Fig. 8). This, in turn, leads to the appearance of excess charges of the same sign which can then be measured in the proposed system. The residual charge on the base is measured when the dust particle separates, this charge being of the same magnitude, but opposite sign, as the charge carried by the particle itself. This description assumes the charges resulting from contact and separation of the dust particle to be of the same magnitude and opposite sign. The experimental data are not, however, consistent with this assumption. The charge carried by the particles on separation is not equivalent to the charge developed on contact (see cited table), and is even, on occasion, of the same sign.

Further discussion will be based on the concepts of atomic physics and the physics of semiconductors [8, 9].

Figure 9a shows the energy levels of the contacting bodies in schematic form. Additional impurity levels contribute donor or acceptor properties to the surface.

It is useful to consider two types of contact processes: semiconductor (dust particle) — metal (base), and semiconductor (dust particle) — semiconductor (glass, coated metallic base).

The Fermi levels are equalized, and the free and filled bands bent, in the contact zone A (Fig. 9b), while a contact potential, φ_c, is established on the surface.

Figure 9c shows the change in the semiconductor energy levels resulting from contact with a metal, the Fermi level of the metal being supposed to lie below the Fermi level of the dust particle material (glass). Donor conductivity develops in the metal under these conditions, the metal becoming an electron donor and the semiconductor an electron acceptor.

Contact between the dust particles and the coated surface can be considered as an instance of a limited-area contact between two semiconductors, one of which is already in contact with a metallic base (Fig. 9c). Here a contact potential is also established at the interface between the dust particles and the coated surface. It can be supposed that the magnitude of this potential, φ_c, will depend on the conduction band bending, and that the latter will, in turn, be determined by the depth of the coating.

Elimination of the electrostatic component of the adhesion force in particles which carry zero charge on deposition requires that $\varphi_c = 0$, i.e., that there be no bending of the conduction band.

Since experiment shows that the particle is always charged, elimination of the electrostatic component of adhesion requires that *

$$|U| = |\varphi_c|, \tag{2}$$

U being the contact potential arising from the initial charge of the contacting particle, and φ_c the potential difference resulting from differences in the electronic work functions, which is to say, from differences in the Fermi levels (Fig. 9).

Condition (2) is specific for dust adhesion.

─────────

*Equation (2) is valid when U is constant, which is to say, when the initial charge of the dust particle is fixed.

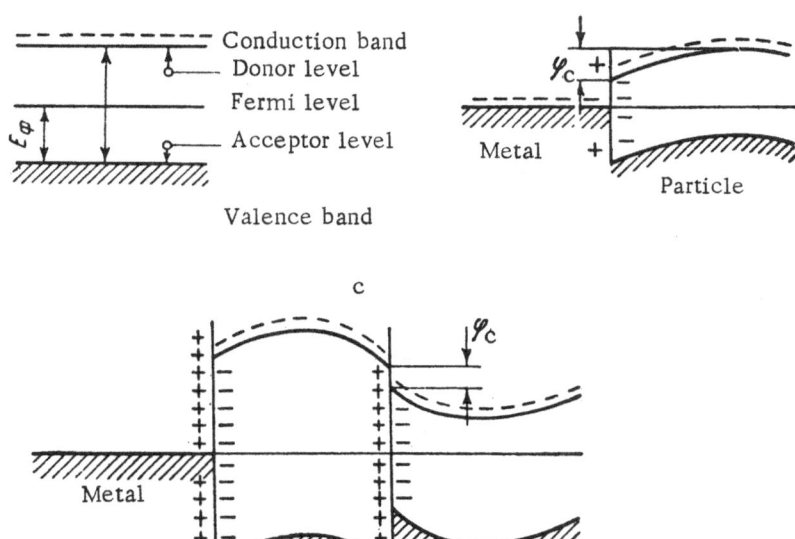

a

Conduction band
Donor level
Fermi level
Acceptor level

Valence band

b

Metal

Particle

c

Metal

Coating   Particle

Fig. 9. Energy levels. a) Level scheme; b) contact with metal; c) contact with coated metal.

When this condition is satisfied, the charge developed on the dust particle on separation is zero (isoelectric point), and the electrostatic component has no effect on the adhesion force.

The experimental data indicate that this condition is indeed fulfilled in certain cases. For example, zero charge is developed in the separation of glass dust particles, 20-30; 40-60; 70; 80-100 μ in diameter for a metallic base.

The electrostatic component can be altered to affect the adhesional force in two difference ways; first, by altering the Fermi level difference, of the two contacting bodies, and second, by acting on the impurity levels.

The data indicate that the value of φ_C for the coated surface is rather high and exceeds the original potential difference, i.e., $\varphi_C \gg U$. Thus the electrostatic component of the adhesional force can be quite significant in fixing the over-all force of adhesion to a coated surface, comparison being with the case of the uncoated surface.

In treating the second type alteration of φ_C (changes in the impurity levels) one uses the following expression for the surface charge:

$$\sigma = \overline{e}\,(p_d - h_a), \tag{3}$$

σ being the charge density in the double electric layer, e the charge on the electron, P_d the concentration of free donor centers on the surface, and h_a the concentration of occupied acceptor levels, also on the surface.

Alteration of the electrostatic component requires a change in the impurity levels to give still other donor (acceptor) relations. This result can be obtained by giving donor or acceptor properties to the surface groups, following the procedure proposed by N. A. Krotova [10] for reducing the electrostatic component in film adhesion

The authors conclude by expressing their sincere thanks to B. V. Deryagin for his continued attention and help in carrying out this work.

Literature

1. B. V. Deryagin and N. A. Krotova. Adhesion, Moscow, Izd. Akad. Nauk SSSR (1949).
2. V. P. Smilga. In collection: Research in Surface Forces, Moscow, Izd. Akad. Nauk SSSR (1961), p. 76. [English translation: Consultants Bureau, New York (1963).]
3. B. V. Deryagin and A. D. Zimon. Kolloidn. Zh. 23(5):544 (1961).
4. A. D. Zimon. Paint and Varnish Materials and Their Application, No. 2 (1963).
5. Zh. Tekenov. Dokl. Akad. Nauk SSSR, No. 2 (1962).
6. A. I. Kosenko. The Electrical Charge on Particles of Quartzite and Iron Ore Dust, Kiev, Izd. Akad. Nauk USSR (1954).
7. S. D. Ponomarev, et al.. Rigidity Calculations in Machine Construction, Moscow, Gostekhizdat (1958).
8. É. F. Shpol'skii. Atomic Physics, Moscow, Gostekhizdat (1951).
9. Introduction to the Physics of Semiconductors [Russian translation], Moscow, IL (1959).
10. N. A. Krotova, et al.. All-Union Conference on Colloidal Chemistry, Summaries of Reports, Moscow, Izd. Akad. Nauk SSSR (1962).

A STUDY OF THE ADHESIONAL STRENGTH OF POLYMERS

N. A. Krotova and Yu. P. Toporov

Institute of Physical Chemistry, Acad. Sci., USSR
Laboratory for Surface Phenomena

Adhesional phenomena are complex effects which can be studied through mechanical, electrical, and spectroscopical methods. With the aid of these methods it is possible to obtain more complete information concerning the various aspects of adhesion. For this reason, monographs on adhesion are generally divided into separate sections, each dealing with one more or less narrow problem [1-3]. The extensive practical application of adhesional phenomena requires, however, information on the mechanical strength of adhesive bonds and protective coatings. The fact that adhesion is a surface effect makes for difficulty in the experimental determination of mechanical strength here. For this reason, the mechanical strength must be related to a strictly defined area in the interface between the contacting bodies.

Adhesion is generally measured in terms of the force, or work, required for separating the materials joined by the adhesional bond. In practice, however, it is a rare case for the breakdown to occur at the adhesional contact. Mechanical tests can lead to rupture in the one or the other of the contacting bodies. It proves to be impossible to control the rupture conditions and the character of the breakdown. It should be kept in mind here that the rate of application of the rupturing force is often a determining factor in mechanical tests on the adhesional bond. This fact can be illustrated by the following examples. Low velocity stripping of a plasticized rubber film from a metal or glass surface frequently ruptures the film itself, and leaves a thin layer of rubber on the base surface. On the other hand, no visible traces of the rubber remain on the base if the stripping rate exceeds a certain critical value [1]. Gutta percha can be readily separated from the metal of a metal — gutta percha — leather system by low velocity stripping. Sudden application of a large force with almost instantaneous breakdown of the bond tears the leather fibers, but leaves intact the adhesion between the gutta percha and the metal and leather. Bond rupture in such systems has been studied by the present authors through high speed cinematography. The velocity dependence of the character of the breakdown of the adhesional bond can be explained in terms of a variation of the relaxation times of the strained bond elements with the rate of force application. This leads to a stress concentration redistribution within the bond, with the weak sectors proving to be in the interface. It is, in principle, possible to adjust the rate of application of the force so as to obtain adhesional breakdown of the bond. The character of the breakdown becomes particularly important when it is a matter of analyzing the dynamic strength of adhesive bonds subjected to high-speed loading.

The rate is also a factor of importance in connection with the electrical aspects of adhesional phenomena, the adhesional strength proving to be a function of the speed of separation in various systems in which adhesional breakdown of the bond occurs over a rather wide range of velocities. The electrical theory of adhesion ascribes the mechanical strength of the adhesional bond to the electrostatic attraction between the faces of a double electric layer established at the film — base interface during bond formation. This theory [1, 4] explains the form of the typical logarithmic adhesiogram showing the adhesional strength as a function of rate (Fig. 1) in the following way. The fact that the adhesional strength is low and practically independent of the velocity in Section I of the adhesiogram is due to more or less complete double-layer discharge during separation. The work of separation increases and becomes rate-dependent once a certain limiting velocity is reached and Section II of the adhesiogram entered; here charge leakage is something less than complete. High values of the adhesional strength are reached in Section III; these are explained in the theory as the effect of gaseous discharge and approach to maximum work of separation. Charge leakage cannot occur at these high velocities,

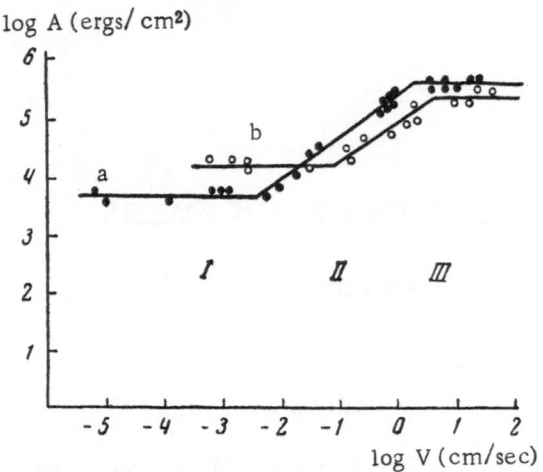

log A (ergs/cm²)

log V (cm/sec)

Fig. 1. Adhesiograms of the stripping of β-gutta percha from steel, obtained with: a) AR-1 adhesiometer; b) AZS adhesiometer.

and the value of the adhesional strength is high as a result. The work of separation depends on the density of surface electrification, which is fixed in any one system. Adhesiograms with three clearly developed sections are observed with films of cellulose esters, vinyl series polymers, gutta percha, and other materials on glass and metal surfaces.

Adhesion was measured by stripping the polymer film with the aid of a falling weight inclined plane adhesiometer or an AR-1 adhesiometer. It was not possible to observe the third section of the adhesiogram for each of the investigated systems, this being displaced toward velocities higher than those realized with the available adhesiometers. In this connection, it should be pointed out that the Institute of Physical Chemistry of the Academy of Sciences of the USSR has developed an AZS adhesiometer which can set up the separation ve-

locities required for adhesional bond rupture and at the same time measure the rupturing force and the work expended in rupture. The adhesiometer was intended for adhesion studies at constant stripping velocity, working under various velocity patterns. It has been used to determine the work of polymer film stripping under continuous alteration of the velocity (AZS-2 system) and for measuring the force required for separating two cylindrical samples butt-joined by a polymer film. Here, one of the samples serves as an anchor for the film, bonding to it being strengthened by careful selection of the materials and roughening the surface. The end of the other cylinder is polished to assure adhesional rupture on it. The samples are prepared in the following manner for the stripping tests. A roller of the base material is rotated in a vessel containing a solution of the experimental polymer, and the film formed by continued rotation of this roller in the air. An attaching strip is set in the film surface while still wet. One end of this strip is connected to the dynamometer and the roller with its adhering film set in a special chuck. This chuck can be lowered at fixed velocity with the aid of a guide screw. This same apparatus can be used to measure the stress variations of electrical origin which arise in the stripping of the polymer film [1]. The frequency and amplitude of these variations are determined by the nature of the film, the method of preliminary treatment, the separation rate, and the nature of the surrounding medium [5]. This adhesiometer also permits investigation of the mechanical properties of the free polymer film. The most recent version of the AZS-2* is shown schematically in Fig. 2; here a Pirozhkov system with a continuously variable frictional regulator is used to obtain a smooth alteration in the separation rate. The stripping force is registered by a system of wire resistance tensiometers attached to the console of the spring dynamometer and connected in an ordinary bridge circuit. Measurement is by bridge balancing, the working current being amplified and then recorded on a loop oscillograph. By analysis of the oscillograms it is possible to determine the adhesion characteristics and mechanical properties of the free films over a wide range of temperatures. For this purpose, heating and cooling jackets are provided which can be set at will over the moving part of the instrument. Figure 1 shows adhesiograms for the stripping β-gutta percha films from steel, these having been obtained with an AR-1 adhesiometer (a) and with an AZS adhesiometer (b). It is clear that these adhesiograms are of the same general character, despite the fact that they were developed at different times with different samples of the polymer.

With the AZS adhesiometer it is possible to extend the range of stripping velocities so as to obtain complete three-section adhesiograms for systems showing only two-section diagrams in the falling weight adhesiometer. The AZS adhesiometer is not, however, designed for studies on the velocity dependence of the break-

---

*Work with the AZS-2 indicated the need for improvement in the regulating system if the adhesiometer was to be used for studies at high velocities. A transmission box was used to affect change of velocity in the AZS-1 [Dokl. Akad. Nauk SSSR 124:302 (1959)].

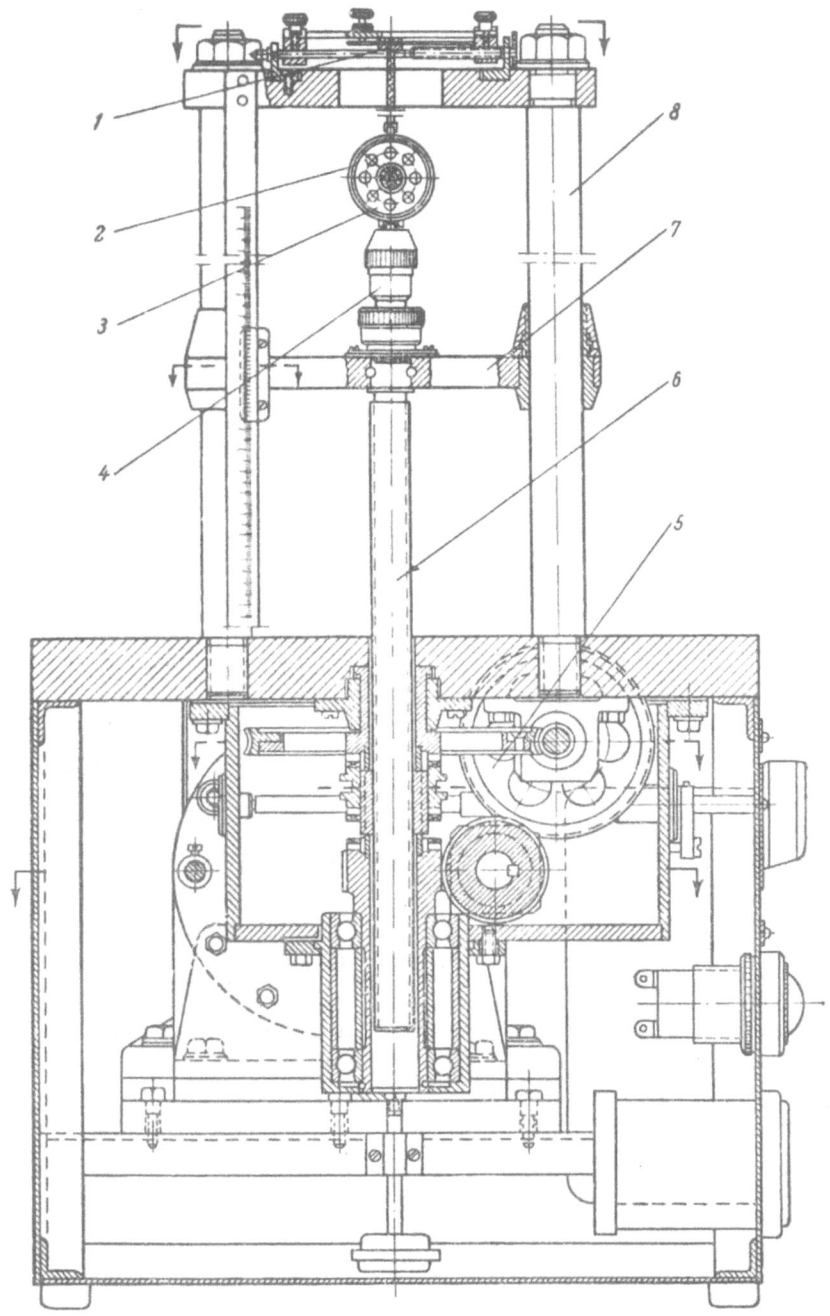

Fig. 2. AZS-2 adhesiometer. 1) Spring dynamometer; 2) test film; 3) working roller; 4) chuck; 5) velocity regulator; 6) guide screw; 7) adjustable platform; 8) frame with chuck.

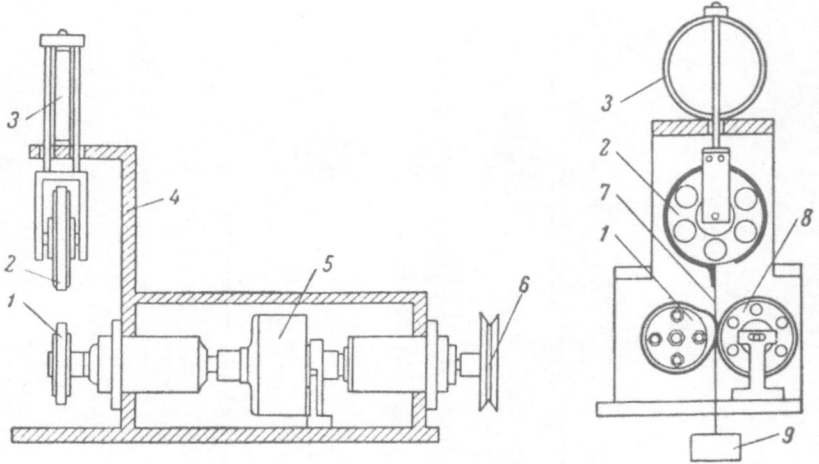

Fig. 3. Schematic representation of the APO-1 adhesiometer. 1) Eccentric; 2) ring with film; 3) dynamometer; 4) frame; 5) magnetic coupling; 6) drive pulley; 7) attaching strip; 8) pressure wheel; 9) load.

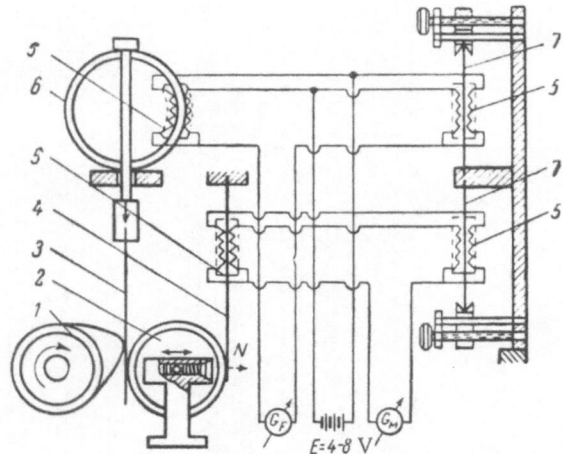

Fig. 4. Schematic representation of the attachment and inclusion of the tension sensors in the APO system (when used in mechanical and frictional experiments): 1) Eccentric; 2) pressure roller; 3) sample; 4) compression spring; 5) working tension sensors; 6) dyanamometer; 7) balancing spring.

down of the adhesional bond. As already remarked, experiment shows that separation at high rate of change of velocity will often lead to adhesional rupture, the surface being laid bare. It is, for instance, known that liquids behave like solid bodies under rapid loading [6]. Experiments on the stripping of liquid films and oils from solid surfaces [7] show rupture occurring within the film itself at low separation velocities. The surface is laid bare at high separation rates, and particles may even be detached from the base material (metal).

The behavior of elastomers and plasticized polymers under mechanical stress indicates that these systems are intermediate between true liquids and true solids. They are, accordingly, quite sensitive to the rate of mechanical deformation. Alteration of the mechanical properties of these materials is reflected in the type of breakdown of the adhesional bond. There has been little study of the behavior of the adhesional bond under various loading rates, especially in shock loading at high velocity, although there can be no doubt of the practical significance of such investigations.

An APO adhesiometer for periodic stripping was designed for parallel studies of the mechanical properties of free films and the behavior of adhesional bonding in shock loading. The construction here is such that the film can be periodically stripped from the base at a high, carefully controlled, velocity. The general form of the adhesiometer is shown in Fig. 3. The test film is deposited on the surface of the detachable ring, 2, which revolves freely in a bearing. The design permits the test ring to be easily replaced. Tie rods are used to brace the bearing support against an elastic steel ring, 3, which is firmly attached to the frame, 4. One end of the attaching tape, 7, which is deposited on the polymer film is passed through the gap between the eccentric, 1, and the rim of the freely revolving wheel, 8. The shaft of this wheel can be rotated in the horizontal plane, so as to be continually parallel to the shaft of the eccentric. Special springs acting on the ring shaft hold the

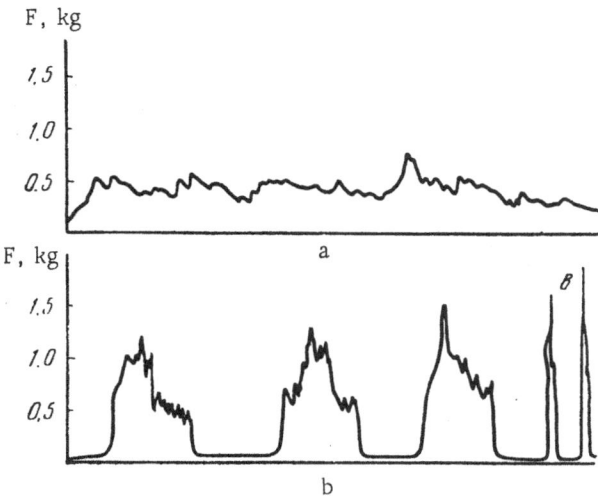

Fig. 5. Oscillograms for stripping β-gutta percha films from the surface of an aluminum ring. Stripping rate: a) 0.06 cm/sec; b) 6 cm/sec.

ring at a certain minimum distance from the eccentric surface. The rotating eccentric makes periodic contact with the rim of the wheel, 8, grabbing the tape in the gap and stripping a section of the test film from the base. The force of adhesion of the film is determined from the deformation of the dynamometer ring, 3, as indicated by the wire sensing elements. Figure 4 shows the method of attaching and the measuring bridge circuit, with the sensors. This bridge is fed from a storage battery and uses an N-700 oscillograph and a GB-111 galvanometer as measuring devices, thus permitting the adhesion to be determined without amplification. It is customary to use a standard tensiometer for measuring these forces. This same diagram shows in schematic form the APO-1 adhesiometer used for studying the mechanical and frictional properties of the free polymer film. The mean rate of stripping of the test film depends on the profile and rate of rotation of the eccentric, 1 (see Fig. 3). A magnetic coupling eliminates the possibility of irregularities in the rotation of the eccentric at the instant of switching on the motor. Stripping can be carried out at a constant rate, or at a rate varying according to a predetermined program, depending on the eccentric configuration. Continuous stripping at fixed velocity can be realized by replacing the eccentric by a disk of definite diameter, and the velocity itself altered over a wide interval by the use of a reductor. APO adhesiometer oscillograms recording the force required for stripping β-gutta percha films at various velocities are shown in Fig. 5 by way of illustration.

The stripping force (or work of stripping) can be developed as a function of the stripping rate from these oscillograms. The absolute value of the force is given by the maximum ordinate on the oscillogram, while the work is measured by the area under the curve. Periodic stripping with the eccentric can give both a general curve showing work as a function of velocity over a wide range of velocities, and special curves applying to narrow velocity intervals within the stripping cycle. Here it is necessary to split the oscillogram cycle area into elementary sections, and determine the area of each section and the mean velocity corresponding to it. Graphical integration then leads to a curve showing the work of adhesion as a function of velocity over the cycle. The velocity dependence of the specific stripping force can be developed in the same manner if pull-off is assumed to occur over the entire contact area in each elementary section. It has been shown that the APO-1 can be used to study the mechanical properties of the free polymer film, particularly when it is desired to compare adhesional characteristics under a fixed velocity pattern. Mechanical tests are carried out on the same system. Figure 4 shows that the upper part of the film, 3, is then set in a clamp attached to the dynamometer ring, 6. The lower part of the film is coated with some material to prevent slippage and inserted in the gap between the eccentric, 1, and the wheel, 2, which is held against the eccentric by a sheet spring, 4. The protuberance on the eccentric periodically engages the end of the film as the eccentric revolves, and stretches it. The film deformation is determined by the eccentric profile. Deformation can be carried out at fixed or varying velocity, depending on the eccentric design.

The advantage of this system is that it can be used for studying the frictional properties of various materials. Here one works with a thin sheet, either of the test material or some substance covered with this material. The upper end of this sample is attached to the dynamometer through an articulated clamp and the lower end set in the gap between the surface of the tension wheel and the eccentric. The eccentric can be made from any desired material; here, it functions as an indenter, the frictional forces resulting from slippage of the eccentric along the surface of the test sheet being measured on the dynamometer. The variation of normal pressure on the frictional contact during eccentric rotation is determined by the eccentric profile, the gap between the eccentric and wheel, the strength of the spring used for holding the wheel against the eccentric, and the initial

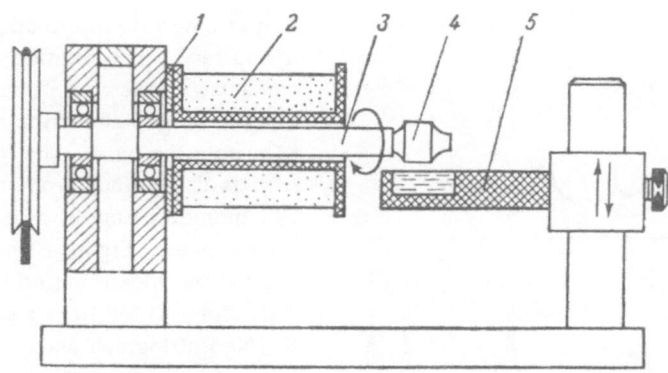

Fig. 6. Apparatus for depositing a polymer film on the ultracentrifuge rotor. 1) Frame; 2) electromagnet; 3) rotating shaft; 4) rotor; 5) cell with polymer solution.

pressure of the eccentric on the wheel. The normal contact force is measured by sensors attached to the spring and recorded on the same oscillograph tape as the frictional force. Thus it is possible to study the frictional properties under any desired variation of the normal load, even in the limiting case where the load is constant and the surfaces contact periodically. This gives a rather simple means of evaluating the abrasive resistance of protective films.

The problem of determining the strength of adhesional bonding is reasonably complex when it is a matter of polymerized resin lacquers. Here it is no longer desirable to use an attaching band for stripping the coating since a considerable error arises from the stresses thereby introduced into the film. It is, moreover, often impossible to strip the film without destroying it. The plant methods for testing lacquers are primitive (notch, wedge methods) and only rough. An effective quantitative method for determining the strength of lacquer coatings has been proposed by S. A. Shreiner and P. I. Zubov [8]; this consists in determining the critical stress required for spontaneous stripping at the film — base interface. This stress is considered to be the adhesional strength at critical film thickness.

Another group of mechanical methods makes use of high accelerations at the film — polymer interface. One such method [9] involves the rapid deceleration of a bullet carrying the test film on its surface, and another applies ultrasonics to determine the strength of the adhesional bond [10]. A method recently introduced applies the force generated in the rapid rotation of a special rotor to strip the film of test material from the rotor surface [11].

We have used an UTs-11-A ultracentrifuge constructed by Mikrotekhna in Prague for measuring adhesions by this method; here the coated rotor is driven at 30,000 rps, which is to say, at a linear surface velocity of about 800 m/sec (rotor diameter, 8 mm), or an acceleration of 15 million times that of gravity. The freely suspended ferromagnetic rotor of the ultracentrifuge is housed in a glass tube and maintained in position in the solenoid field through a servomechanism. A vacuum of $1-5 \cdot 10^{-6}$ mm Hg is set up in this tube. The current strength in the solenoid is regulated so that the magnetic field holds the rotor in position, without allowing it to move either up or down. The rotor is set in motion by generating a rotating electromagnetic field in two pairs of windings. These windings are bridge-fed through a drive generator based on a transition oscillator of 40 kc output frequency. The rotation rate of the rotor is determined photoelectrically. For this purpose, the rotor is illuminated by a beam of light, and the reflected beam focused onto a photoelement. The fact that a part of the rotor surface is covered with the test film gives rise to pulsations in the intensity of the reflected light, and the current generated on the surface of the photoelement varies at a frequency which is identical with the frequency of rotation of the rotor. One pair of deflecting plates of the electron ray tube of an oscillograph is fed by the alternating component of the current delivered by the photoelement, while the other pair is fed from an audio generator of regulated and carefully measured frequency. Synchronization is through the Lissajous figures on the screen, these becoming stable when the generator frequency is equal to, or a multiple of, the rotation frequency of the rotor. Stripping of the test film from the rotor surface is marked by a sharp reduction in the amplitude of the

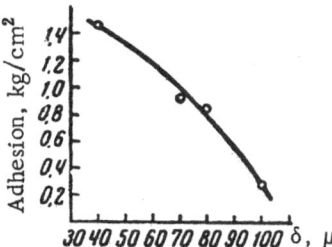

Fig. 7. The relation between the specific force of adhesion and the depth, δ, of the unplasticized perchlorvinyl film stripped in the ultracentrifuge.

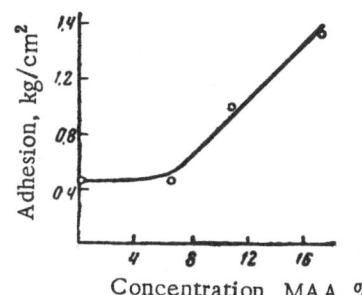

Fig. 8. The relation between the strength of adhesion of the polymethylmethacrylate film and the film content of methacrylic acid.

potential at the photoelement. The velocity at this instant is the principal characteristic of the film adhesion. The rim velocity and acceleration can be determined from the rotational velocity and the known diameter of the rotor. The weight of the test film can be used to calculate the centrifugal force at pull-off and this value then taken as a measure of the strength of adhesion.

With the ultracentrifuge, it is possible to strip the polymer film from the metal without having to make use of an attaching band. The method used for depositing the film is quite important when investigating the adhesional bond with the ultracentrifuge. It should be noted that the film must be colored in order to avoid difficulties in the photoelectric determination of the rotation rate. Nigrozine is the most suitable dye for polymers dissolved in nonpolar liquids since it gives an intense coloration at vanishingly low concentration. There are two methods which can be used for depositing the polymer film on the rotor. In the first, the rotor is set into a special holder and drops of a rather viscous solution of the polymer or lacquer deposited on its upper or lower surface. The resulting film is then allowed to solidify in a thermostat. The second method uses the special set-up of Fig. 6 to deposit a uniform layer over the entire rotor surface. Here a bearing supported horizontal shaft, 3, serves as the core for an electromagnet, 2. The strong field from the magnet holds the rotor, 4, against the end of the shaft. Rotation brings the surface of the rotor into contact with the solution in the cell, 5. This cell is supported in such manner that it can be moved up and down. Solidification of the film is brought about by continued rotation of the rotor after lowering the cell with the solution. A 3 x 3 mm section of this film is then marked out and the rest of the film peeled away from the rotor surface.

Experiment shows that the tightest adhesion is obtained with films deposited by the first of these two methods. Data on the strength of adhesion of polymethacrylate films deposited on the rotor of the ultracentrifuge by these methods are presented below, the values given being in kg/cm²:

Drop method	1.1	1.6	3.45
Cutting method . . .	0.46	1.0	1.55

The differences here trace back to the fact that the small film is formed under nonuniform conditions and experiences a pronounced edge effect. The drop method was used in all subsequent experiments.

It is a well-known fact that the bonding strength of a lacquer coat varies with its depth. This depth effect also appears in adhesion studies in the ultracentrifuge, being especially marked in unplasticized systems. Figure 7 showing the relation between the specific force of adhesion and the film depth, δ, in nonplasticized perchlorvinyl can serve as an example here. The adhesion – film depth relation obviously reflects the fact that internal shrinkage stresses are set up during film formation. These stresses tend to weaken the adhesion, especially in thicker films [8]. Plasticizing leads to partial, or even total, elimination of these stresses. Thus the adhesional strength of the plasticized system remains constant as the film depth is increased. Cases are even met in which the adhesion tends to rise with the film depth, one such instance being plasticized nitroenamel:

Film depth, δ, μ	15	25	60
Adhesion, kg/cm^2	0.23	0.29	0.63

Modern concepts of adhesional bonding suggest that the bond strength should be markedly affected by the presence and concentration of functional groups in the polymer. Figure 8 shows a curve developed with the centrifuge to cover the relation between the adhesion of films of copolymers of methylmethacrylate and methacrylic acid, and the content of acid in the polymer. Increase in the concentration of the active functional groups (—COOH—) furnished by the acid leads to a marked increase in the adhesional strength, just as was to be anticipated.

Although the ultracentrifuge method suffers from various defects (the apparatus is complex and the measurements can be made only with difficulty, the test film is heated in the course of the experiment, the rotor must be vacuum-housed, so that volatile components tend to vaporize from the film, etc.), it does permit a direct determination of the force of adhesion, the separating force increasing in moving from the outer surface of the coating to the base and reaching its maximum value at the interface itself. This reduces to a minimum the possibility of film rupture prior to breakdown of the adhesional bond. Breakdown does occur sometimes, however, especially if the film has been weakened by the introduction of large quantities of plasticizer.

Conclusions

Note has been made of the prime significance of the velocity factor in mechanical tests on the strength of adhesional bonding. This is a two-fold effect. First, the velocity determines whether disruption will be within one of the bound materials (cohesional breakdown) or along the interface (adhesional breakdown). Thus the velocity determines the value of the reported mechanical strength of the adhesional bond. Second, adhesional breakdown proves to take place in such a way that the strength of bonding is velocity dependent over a wide range of velocities, presumably because of electrostatic effects. Thus the rational selection of adhesional materials requires that account be taken of velocity effects on bulk mechanical processes and on surface electrical phenomena which develop at the adhesional bond.

AZS and APO adhesiometers have been constructed for detailed studies on the relation between the velocity and the strength of adhesional bonding. With these it has been possible to study the adhesion and mechanical properties of polymers over a rather wide range of velocities. The AZS adhesiometer permitted variation of the rate of mechanical interaction and was designed for study of adhesiograms developed in adhesional breakdown of the bond.

The APO adhesiometer was designed for studies on the relation between the type of breakdown (cohesional, adhesional) and the velocity program, and made it possible to predict the type of breakdown. Here study was made of the mechanical processes taking place in the body of the bonded material.

The determination of the strength of adhesional bonding of protective coatings based on polymerized resins is a problem of considerable experimental difficulty. Here measurement was made with systems which permitted the development of high accelerations at the interface between the polymer and the base. Experiments with the UTs-11-A ultracentrifuge showed the existence of a relation between the strength of adhesion and the depth of coating in perchlorvinyl based polymers; the adhesion was also shown to depend on the presence of the plasticizer in the film. This same method of measurement also disclosed that active polar groups, and especially the concentration of these groups in the chain, could have considerable influence on the adhesion of copolymers of methylmethacrylate and methacrylic acid to steel.

In concluding, the authors consider it their pleasant duty to express their thanks to B. V. Deryagin, Corresponding Member, Academy of Sciences, USSR, for his valuable advice and interest in this work. They would also like to thank S. V. Yakubovich and V. A. Zubchuk for having furnished the resin mixtures and helping in the work, and A. Ya. Korolev for having made available the samples of copolymers synthesized in his laboratory. The shop work was carried out by B. N. Parfanovich, A. I. Bessonov, and I. A. Shtykov.

Literature

1. B. V Deryagin and N. A. Krotova, Adhesion, Moscow, Izd. Akad. Nauk SSSR (1949).

2. Adhesion (Glues, Cements, and Solders), Collection [Russian translation under the editorship of Debroin], Moscow, IL (1954).

3. Adesion, D. D. Eley (ed.), Oxford University Press (1961).

4. N. A. Krotova and Yu. M. Kirillova, Proceedings, Third All-Union Conference on Colloidal Chemistry, Moscow, Izd. Akad. Nauk SSSR (1956), p. 329.
   N. A. Krotova, Concerning Gluing and Adhesion, Moscow, Izd. Akad. Nauk SSSR (1956).
   B. V. Deryagin, N. A. Krotova, V. V. Karasev, I. N. Aleinikova, and I. M. Kirillova, Proceedings of the International Congress of Surface Activity, Vol. 3, London (1957).

5. V. V. Karasev, N. A. Krotova, and B. V. Deryagin, Dokl. Akad. Nauk SSSR 89:109 (1953).

6. M. O. Kornfel'd and M. M. Ryvkin, Zh. Éksperim. i Teor. Fiz. 3:525 (1939).

7. H. Heidebrock, Angew. Chem., No. 5/6:77 (1941).

8. S. A. Shreiner and P. I. Zubov, Dokl. Akad. Nauk SSSR 124:1102 (1959).

9. W. D. May, N. D. Smith, and C. I. Snow, Trans. Inst. Metal Finishing 34:369 (1957); Khim. i Tekhnol. Polimerov, No. 6:122 (1959).

10. S. Moses and R. K. Witt, Ind. Eng. Chem., Vol. 41 (1949).

11. H. Alter and W. Soller, Ind. Eng. Chem. 50:922 (1958); Khim. i Tekhnol. Polimerov, No. 6:133 (1959).

THE INTERRELATION OF FRICTION AND ADHESION IN SOLID BODIES

Yu. P. Toporov

Institute of Physical Chemistry, Acad. Sci., USSR
Laboratory for Surface Phenomena

It is generally admitted at the present time that the force of external friction is of molecular origin. This does not, of course, imply that mechanical interaction has no influence on frictional processes. The term mechanical interaction is usually applied to those frictional deformation processes which involve masses in considerable excess of the mass of the monomolecular surface layer [1]. Here principal interest attaches to the bulk mechanical properties of the material. The term molecular interaction is usually reserved to designate the surface interaction resulting from unsaturated force fields on the surfaces. This distinction is, however, quite formal and arbitrary, since such bulk properties as shear strength are certainly determined by molecular interaction [2].

We will consider the interaction between the atoms and molecules of two contacted bodies in more detail, since this is a determining factor for frictional processes. Macrobodies separated by very small distances experience a mutual attraction because of the existence of attractive forces between the individual atoms and molecules. The first direct measurement by B. V. Deryagin and I. I. Abrikosova [3] showed the forces of attraction between solid surfaces to extend outward to distances of 0.04 μ. The energy of interaction proved to be inversely proportional to the third power of the distance of separation of the surfaces. These results were subsequently confirmed by the experiments of Kitchener and Prosser [4], Overbeek and Sparnaay [5], and De Jhong [6]. Thus it can be concluded that attractive forces of extended range exist between condensed phases. A general theory of this type of molecular attraction has been developed by E. M. Lifshits [7]. This theory considers the interaction between bodies to result from fluctuations of the electromagnetic fields which exist within the absorbing bodies themselves and extend beyond their confines. It follows that the force of attraction, F, is related to the distance of separation, H, of the surfaces by an equation of the form: $F = -k/H^3$ when $H \ll \lambda_i$ and by an equation of the form $F = -c/H^4$ when $H \gg \lambda_i$, the λ_i being the fundamental wavelengths in the absorption spectrum (or emission spectrum) of the microbody material.

It is impossible to determine the force of interaction over the entire range of separation distances at the present time because of the inevitable surface roughness and lack of physical chemical uniformity. For the same reason, it is not possible to extrapolate data applying to the forces acting between separated bodies to zero surface separation. The situation regarding the forces of interaction becomes quite complex as the distance between the surfaces is reduced. Forces of repulsion come into play, or chemical bonds are established as the result of an affinity, when external forces are applied and the separation of the bodies reduced to the point where it is comparable with the diameter of the unexcited electronic orbitals. Various types of chemical bonding (polar, hydrogen, homopolar) can then be set up, depending on the chemical properties of the materials involved. Metallic atoms undergo spontaneous collectivization of the valence electrons at a certain distance of approach to form bonds of the same strength as the bonds in the metal itself. The chemical bond energy is 1-2 orders higher than the van der Waals energy, falling in the 10-100 kcal/mole range. Two bodies will therefore unite to a single unit if chemical bonds are set up over any considerable fraction of their surface areas. On the other hand, it is rarely true that chemical bonding is established when solid bodies are brought into contact. Repulsive forces arise when two atoms of parallel total spin approach one another. The same is also true when the approaching atoms are already chemically bound to one another. The quantum mechanical interaction of the overlapping electron clouds of the atoms is the factor chiefly responsible for this repulsive force. For this reason, the range of action of the repulsive forces is limited and comparable to the atomic diameter. The forces of repulsion increase exponentially with diminishing distance of separation, however, so that the electron clouds can be considered as incompressible and a fixed radius assigned to each atom to mark out the closest distance of approach under ordinary conditions. The action of the repulsive forces sets up a type of molecular roughness in the solid surface.

All of these forces come into play in the interaction of contacted solid surfaces. The molecular interaction gives rise to surface forces with one component directed at right angles to the interface (force of adhesion) and the other component tangential to it (frictional force). Since the radius of action of the attractive forces is many times greater than that of the repulsive forces, the interaction between contacted bodies is manifest as a mutual attraction. Depending on the nature of the bodies and the character of their contact, the universal van der Waals forces will be accompanied by forces of homopolar chemical bonding and by forces arising from the double electric layer which is established when contact is made. These forces are completely adequate to assure strong adhesional bonding between the contacting surfaces. The fact that such bonding is not invariably observed in practice is due, first, to the screening of the attractive forces by surface adsorption layers, and, second, to the fact that the area of true body contact is only a vanishingly small fraction of the apparent contact area.

Most molecular theories of external friction give principal weight to the molecular attraction between the members of the frictional pair. This tendency is supported by the experimentally confirmed fact that the force of friction rises as the degree of surface roughness is reduced and the true contact area increased. It should be noted that many investigators still consider the force of attraction to be the only factor involved in external friction. Thus the Terzagni—Bowden theory (which is generally accepted abroad [8, 9]) equates the force of friction to the shear strength of the adhesional bonds formed at true contact points, determining this force in the absence of furrowing through the equation

$$F = \tau S, \tag{1}$$

in which S is the true contact area and τ the bond shear resistance.

On the other hand, the true contact area of plastic materials can be determined through the expression

$$S = N/p^{\neq}, \tag{2}$$

in which N is the normal load, and $p^{\neq}$ is the flow limit of the softer of the two substances. From this it follows that the coefficient of external friction can be represented in the form

$$\mu = \frac{F}{N} = \frac{\tau}{p^{\neq}} = \frac{\text{shear resistance}}{\text{flow limit}}. \tag{3}$$

Thus a theory of external friction based exclusively on forces of molecular attraction is capable of explaining the widely used Amonton Law ($F = \mu N$) of the proportionality of the force of friction to normal load for plastic contact. It is not, however, invariably true that plastic contact is set up between a pair of solids. Moreover, deviations from the Amonton Law are frequently observed in practice. In fact, deviations appear exactly in those cases where the attractive forces are so clearly expressed as to lead to low pressure grabbing of the two surfaces. Departure is most obvious when the force of friction fails to vanish even though N = 0. Deviations from the Amonton Law are observed in the friction of powders on a plane surface, in the friction between plane guage blocks with carefully worked surfaces, and so on. A friction theory based on attractive forces alone can give no adequate explanation of these departures from the $F = \mu N$ law.

Since high magnitude adhesion forces and departures from the Amonton Law are observed simultaneously, it is only natural to assume that the law itself cannot be adequately interpreted in terms of attractive forces.

Any friction theory must take account of all of the forces acting between the contacted surfaces, the repulsive forces as well as the forces of attraction.

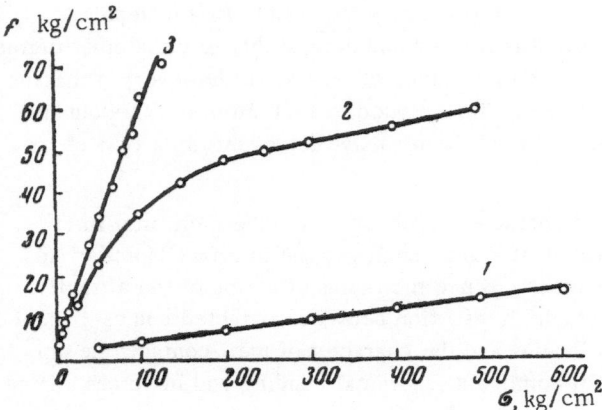

Fig. 1. The relation between the specific force of friction, f, and the specific pressure, σ, at the point of contact, for various pairs. 1) Steel—polyfluoroethylene-4; 2) steel—polyethylene; 3) steel—polyamide.

Tomlinson [10] was the first to attempt the introduction of repulsive molecular forces into the theory of the frictional force. The inadequacy of this work has been pointed out repeatedly [11, 12].

As early as 1934, B. V. Deryagin developed a molecular theory of friction in which account was taken of both repulsive and attractive molecular forces [13]. The basic idea of this theory was that the friction between smooth surfaces results from molecular roughness, which is to say, from Born repulsive forces between the electron clouds of the atoms of the contacted bodies, or of those portions of the bodies included in the slip plane. The adhesional force was introduced as a correction term to account for departures from the Amonton Law. This theory leads to a two-term equation relating the force of external friction, F, and the normal load, N:

$$F = \mu N + \mu AS, \tag{4}$$

μ being the true coefficient of friction, S the true contact area, and A the specific adhesion acting over the area S. This equation is referred to as the two-term Deryagin friction equation. From it the conclusion follows that the force of external friction is the sum of two component terms, μN representing the contribution arising from repulsive forces and μAS the contribution from forces of attraction. This two-term equation passes over to the Amonton Law when μAS is vanishingly small, or S is proportional to N. It has already been pointed out that the second term in the right-hand member of (4) is usually small in practice, the area of actual contact of the bodies being small and amounting to a vanishingly small fraction of the apparent contact area at lower pressures. Moreover, the value of the constant A is markedly reduced by the presence of adsorption layers which screen the surface force field of the condensed phase. At the same time, the two-term equation gives a quite simple explanation of the departures from the Amonton Law in all cases in which there is an appreciable effect from attractive forces and S is not proportional to N. The applicability of (4) to the friction and adhesion of plane gauge plates [14, 15] and highly dispersed powders [16] has been repeatedly demonstrated.

We will now describe the results obtained in a study of the frictional properties of polymers and thus prove the doubts concerning the applicability of Eq. (4) to polymers [17] to be without ground. Since the true contact surface increases with the normal load and the elasticity of the contacting bodies, it is to be expected that high areas of true contact under friction could be realized with certain polymers. Thus it is likely that the true contact area for frictional movement of a rather soft polymer over a very smooth metallic surface would become equal to the apparent contact area at a certain load and from then on remain constant as the load was increased. This would permit a test of Eq. (4) with fixed true contact area. It should be noted that study of the frictional properties of polymers under high compressing forces has practical as well as theoretical interest, these materials being frequently used in friction joints.

Tests on the external friction of polymers were carried out on a device similar to that previously described in [18]. This consisted of two screw presses fixed on a single base. The first of these was mounted vertically and served to apply the normal pressure to the frictional contact, while the second was mounted horizontally and served to produce a relative displacement in the contacting surfaces. The tests involved determination of the static force of friction, F, between the upper and lower surfaces of polymer films affixed to two gauge plates held in the horizontal plane but pressed against a moving gauge plate with force N by the vertical press. These gauge plates were made from ShKh-15 steel and worked to give Class 12-13 surfaces; they were cleaned prior to experiment by rubbing with degreased cotton moistened in distilled ether. The surfaces of the polymer films were cleaned in the same manner. Graphs showing the relation between the specific force of friction, $f = F/2S$,

314

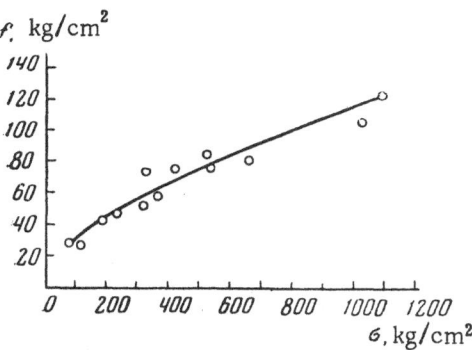

Fig. 2. The relation between the specific force of friction, f, and the specific pressure, σ, for steel-tread rubber.

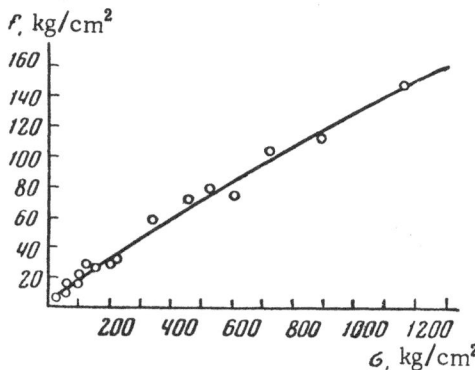

Fig. 3. The relation between the specific force of friction, f, and the specific pressure, σ, for steel-pure rubber.

and the specific pressure in the contact, $\sigma \approx N/S$ (S is the apparent contact area) for polyfluoroethylene-4 (1), polyethylene (2), and polyamide (3) films are presented in Fig. 1. It is seen that the force of friction continues to increase with the normal pressure, even after the true contact area has become constant. In actuality, the limit of pseudo-flow of polyfluoroethylene is approximately 150 kg/cm² at room temperature, recrystallization with cold flow setting in at higher pressures [19]. Thus it is to be anticipated that there would be no further increase in the contact area once a limiting specific pressure of some 200 kg/cm² had been reached (this calculation was based on the apparent contact area). The curve covering the relation between the force of friction and the normal pressure for polyfluoroethylene-4 (Fig. 1, 1) follows a two-term relation since it does not pass through the origin, and the force of friction continues to increase linearly with the normal load, even after the contact area has become constant.

Here, it seemed interesting to study the friction of a smooth steel surface moving over rubber, a substance which is known to make almost perfect elastic contact with a solid. Here the area of contact of metal and rubber (assumed bulk modulus, 12 kg/cm²) should cease to increase at a compressing load of 250 kg/cm² [20]. The applicability of the two-term equation to the friction of rubber on metal has been shown by S. B. Ratner [21]. Figures 2 and 3 are experimentally developed graphs showing the relation of the specific force of friction to the specific pressure for steel-tread rubber and steel-pure rubber. It is clear that the specific force of friction continues to increase, even after there is no longer any measurable increase in the true contact area.

The theoretical work of Terzagni, Bowden, and their followers [22] assumes the force of friction to depend only indirectly on the normal pressure, the dependence being through the true contact area which is a function of the pressure. If alteration in the normal load leads to no alteration in the true contact area, it would then follow that the force of friction should remain constant.

Conclusions

The results obtained here justify the conclusion that the normal load affects the force of friction not only through the true contact area, but indirectly, in the manner indicated by the two-term Deryagin equation. This, in our opinion, confirms the validity of the basic ideas of the theory of friction and, at the same time, demonstrates the possibility and necessity of resorting to the two-term friction equation in treating the frictional properties of polymers.

The author would like to express his sincere thanks to B. V. Deryagin, Corresponding Member, Academy of Science, USSR, for having directed this work.

Literature

1. I. V. Kragel'skii. Friction and Wear, Moscow, Mashgiz (1962).
2. B. V. Deryagin and V. P. Lazarev. Proceedings, Second All-Union Conference on Friction and Wear in Machines, Vol. 3, Moscow, Izd. Akad. Nauk SSSR (1949), p. 106.

3. B. V. Deryagin and I. I. Abrikosova. Zh. Éksperim. i Teor. Fiz. 30:993 (1956); 31:3 (1956); Dokl. Akad. Nauk SSSR 108:214 (1956).

4. I. A. Kitchener and A. P. Prosser. Proc. Roy. Soc., A 242:403 (1957).

5. I. T. G. Overbeek and M. I. Sparnaay. Discussion Farady Soc., No. 18:12 (1954).

6. De Jhong. Dissertation, Utrecht (1958).

7. E. M. Lifshits. Zh. Éksperim. i Teor. Fiz. 29:94 (1955); Dokl. Akad. Nauk SSSR 97:643 (1954).

8. K. Terzagni. Erdbaumechanik, Vienna (1925).

9. F. P. Bowden and D. Tabor. Friction and Lubrication of Solids, Oxford (1954).

10. G. A. Tomlinson. Phil. Mag. 7:205 (1929).

11. D. V. Konvisarov. Friction and Wear of Metals, Sverdlovsk, Mashgiz (1948).

12. I. V. Kragel'skii and V. S. Shchedrov. Development of the Science of Friction, Moscow, Izd. Akad. Nauk SSSR (1956).

13. B. V. Deryagin. Zh. Fiz. Khim. 5:1165 (1934); Z. Phys. 88:661 (1934).

14. P. V. Denisov. Dissertation, Moscow, Institute for Machine Instrumentation (1956).

15. A. S. Akhmatov. In collection: Research in Surface Forces, Moscow, Izd. Akad. Nauk (1961), p. 93. [English translation: Consultants Bureau, New York (1963).]

16. B. V. Deryagin and V. P. Lazarev. Kolloidn. Zh. 1:295 (1935).

17. G. V. Bartenev. Kolloidn. Zh. 18(2):249 (1956); Dokl. Akad. Nauk SSSR 103:1017 (1955).

18. B. V. Deryagin and Yu. P. Toporov. Kolloidn. Zh. 23:118 (1961).

19. D. D. Chegodaev, Z. K. Naumova, and Ts. S. Dunaevskaya. Polyfluoroethylene of Plastics, Moscow (1960).

20. G. M. Bartenev and V. V. Lavrent'ev. Dokl. Akad. Nauk SSSR 141:334 (1961).

21. S. B. Ratner. Kolloidn. Zh. 19(3):394 (1957); Proceedings, Third All-Union Conference on Friction and Wear, Moscow, Izd. Akad. Nauk SSSR 2:87 (1960).

22. P. Thirion. Rev. Gen. Caoutchoue 23:101 (1946).

THE FRICTION OF DUSTY SURFACES

Yu. M. Luzhnov

Institute of Physical Chemistry, Acad. Sci., USSR
Laboratory for Surface Phenomena

The effect of atmospheric moisture on the friction of solid surfaces has been studied in the Laboratory for Surface Phenomena of the Institute of Physical Chemistry of the Academy of Sciences of the USSR, the aim being to determine the factors responsible for the considerable alteration in the coefficient of static friction, which work carried out by S. I. Kosikov [1] in conjunction with Yu. P. Toporov and the author of the present article has shown to occur when the dusty metallic surface is exposed to the atmosphere.

These studies involved strict control of the coefficient of static friction between the dusty sheet and a slide, with variation of the relative humidity of the surrounding atmosphere from 5 to 100% [2].

The apparatus consisted of a 5 liter hermetically sealed chamber with a center mounted table to which the sample could be affixed, a drier and a humidifier, an arrangement for controlling the temperature of the sample and the air in the chamber, and elements for recording the coefficient of static friction and temperature (Fig. 1).

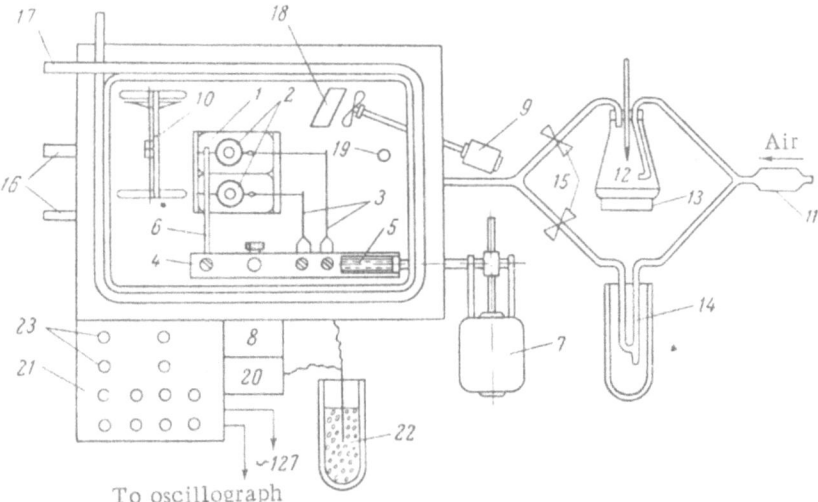

Fig. 1. Schematic representation of the system used in studying the coefficient of static friction in a chamber containing air of controlled humidity. 1) Sample table, with attached samples; 2) slides; 3) spring dynamometers; 4) console of the rotation-translation mechanism; 5) translation-rotation mechanism; 6) deflecting rod; 7) motor; 8) thermoregulator; 9) ventilator; 10) film hydrometer; 11) air filter; 12) humidifier; 13) humidifier heater; 14) nitrogen trap; 15) flow meter and dosing valves; 16) coils for sample table; 17) coils for chamber walls; 18) heater; 19) contact thermometer; 20) thermocouple switch; 21) control panel; 22) thermocouple reference junctions; 23) balancing screw for dynamometer.

317

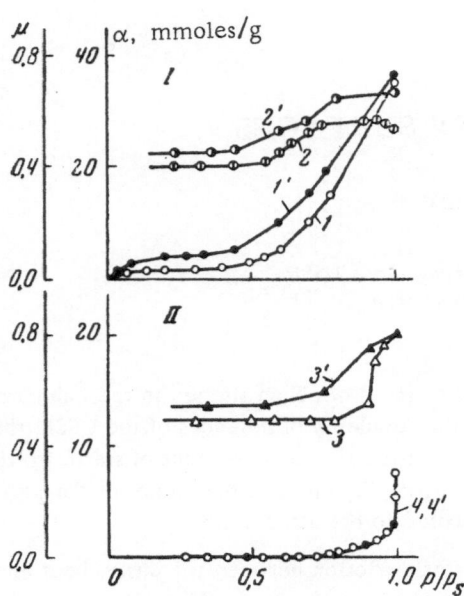

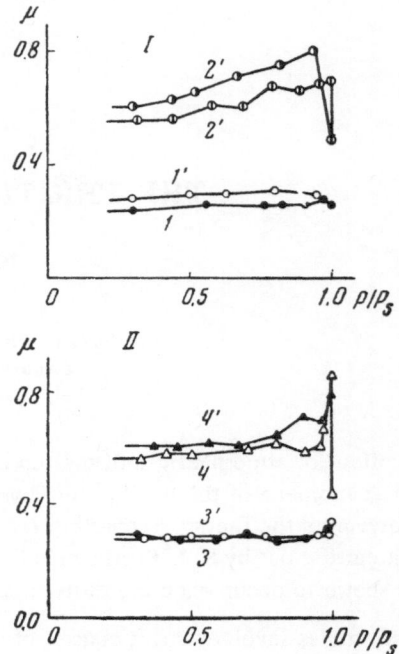

Fig. 2. The relation between the coefficient of static friction, μ, on sheets covered with films of hydrophilic and hydrophobic powders, the relative humidity of the atmosphere, and the water vapor adsorption isotherms for the same powders. I) Experiments with hydrophilic powders; II) experiments with hydrophobic powders. 1, 1', 4, 4') Water vapor adsorption isotherms; 2, 2', 3, 3') coefficients of static friction plotted against relative humidity.

Fig. 3. The relation between the coefficient of static friction and the relative humidity of the atmosphere. I) Experiments on sheets covered with monolayers of stearic acid (1, 1'), and with a hydrophilic powder (2, 2'); II) experiments on steel sheets covered with monolayers of stearic acid (3, 3'), and with a hydrophobic powder (4, 4').

Tests were generally performed on pairs of samples, 40 × 25 × 10 mm in dimensions, which were mounted on the table. Slides, each with three firmly attached spherical bearings, 0.5 mm in diameter, were set on each sample. These slides were attached by elastic wires to spring dynamometers which were, in turn, firmly affixed to the console of a rotation-translation mechanism.

Setting the motor of the spring dynamometers into operation caused these slides to bend and move over the sample surfaces under the action of the force of friction. The mechanical deformation of the dynamometer springs was translated into electrical impulses by four tensiometric resistance sensors attached to their surface and connected in a measuring bridge circuit, and these impulses were then recorded on an oscillograph tape. The force of static friction between slide and sample surface was reach from the maximum spring deflection recorded on the tape. Some 20-30 measurements were carried out in the course of each experiment. The coefficient of static friction was calculated from the measured value of the force of friction and the known weight of the slides.

The moisture in the chamber was varied by introducing portions of dried air and air saturated with water vapor, these being carefully mixed with the air in the chamber itself. Determination of the humidity of the chamber air was made through a hydrometer incorporating the film moisture sensor developed in the Institute of Hydrometric Equipment [3]; this has a 2-5 second inertia. The error in the measurement of positive temperatures was less than ±5%, under the working conditions.

All of the friction joints in this system were constructed of teflon to avoid contaminating the air with oil vapors and eliminate the possibility of oil molecule migration along the chamber walls.

By taking precautions, the coefficient of static friction of samples introduced into the chamber at fixed relative humidity could be held constant for more than 5-6 hours.

It was decided to first carry out experiments with powders deposited on clean metallic surfaces so as to obtain a basis for comparison with the results of later work on powders capable of affecting the coefficient of static friction under variation of the relative humidity. For this reason, samples and slides were subjected to a standard glow discharge cleaning before each experiment, the technique used being the same as that developed by V. V. Karasev [4]. The degree of cleaning of the surface was estimated from the coefficient of static friction, samples in which this coefficient was less than 0.7 being discarded.

Study was first made of the effect of hydrophobic and hydrophilic powders on the coefficient of static friction, and an understanding obtained of the powder processes occurring at the contact between two bodies working against one another in an atmosphere of high moisture content.

The materials used here were two specially treated fine grained powders of Ukhtinskii channel black, with specific surface areas of 80 and 130 m^2/g, respectively. The water vapor adsorption isotherms showed the one sample to be a hydrophilic powder and the other a hydrophobic powder* (Fig. 2).

In Fig. 2, a represents the amount of sorbate, μ, the coefficient of static friction, p/p_s, the vapor pressure of water vapor in the chamber; 1, 1', 4, and 4' are water vapor adsorption isotherms, while 2, 2', 3, and 3' represent the coefficient of static friction as a function of the vapor pressure of water. (The primed numbers are of forward movement and the imprimed, for reverse.)

The metallic surfaces were covered with a nickel film 18 μ deep in order to avoid corrosion effects in the course of the experiments [5].

Colloidal films of these powders were deposited on the clean sample surface by placing the sample under a hood which had been previously sprayed with a definite portion of the powder. The cleaned samples were put into the hood after the coarser particles had precipitated out. After remaining under the hood for a definite period of time, the samples became covered with a layer of powder and were removed as ready for experiment. The technique of depositing the colloidal powder films on the samples was such that the depth of the layer covering would not exceed 0.1-0.15 mm.

The experiments were begun with the lowest relative humidity in the chamber of the system. The humidity of the air was then gradually increased to saturation. Having reached saturation, the chamber air was again dried. Measurements of the force of friction on the dusty surfaces were made during this entire cycle.

These experiments indicated that an increase in the relative humidity of the air in the chamber led to an increase in the coefficient of static friction. This increase began with the initiation of capillary condensation of water vapor in the contaminating powders. The existence of this relation was confirmed for both rising and falling relative humidity of the chamber air.

In order to develop an explanation for this effect, control experiments were carried out in which the sample surface was covered with a monolayer of stearic acid before depositing the powders (Fig. 3).

The results showed the relation between the coefficient of static friction and the relative humidity of the air to be essentially the same for the clean metallic surface and for the surface covered with a monolayer of stearic acid prior to powder deposition. The coefficient of static friction increased in both cases.

Conclusions

The experiments carried out here justify the conclusion that the force of interaction between individual solid powder particles begins to increase at the instant of initiation of capillary condensation in the powder layer. The mechanical properties of the colloidal film itself then begin to alter and the true contact area between the sliding bodies and the separating powder film increases. This is the principal factor accounting for the increase of the static friction of the dusty surface with increasing humidity in the surrounding air.

---

*The samples of channel black and their water vapor adsorption isotherms were furnished by N. V. Kovaleva, to whom the author expresses his sincere thanks.

The author considers it his duty to express his thanks to B. V. Deryagin, Corresponding Member, Academy of Sciences, USSR, for having directed this work.

Literature

1. S. I. Kosikov, Izv. Akad. Nauk SSSR, Otd. Tekhn. Nauk, No. 4 (1958).
2. Yu. M. Luzhnov, Apparatus for studying the effect of the moisture of the air and the coefficient of static friction, TsITÉIN, No. P-62-72/10 (1962).
3. N. S. Varzhenevskii, "Film moisture sensors," Tr. NII GMP, No. 5 (1957).
4. V. V. Karasev and G. I. Izmailova, Zh. Tekhn. Fiz. 24:871 (1954).
5. K. M. Gorbunova and A. A. Nikiforova, Chemical Nickel Plating, Moscow, Izd. Akad. Nauk SSSR (1960); Physicochemical Principles of Nickel Plating, Washington, D. C. (1963).